Computer-Aided Techniques in Food Technology

FOOD SCIENCE

A Series of Monographs

Editors

STEVEN R. TANNENBAUM
Department of Nutrition and Food Science
Massachusetts Institute of Technology
Cambridge, Massachusetts

PIETER WALSTRA
Department of Food Science
Wageningen Agricultural University
Wageningen, The Netherlands

Other Volumes in Preparation

Computer-Aided Techniques in Food Technology

edited by

ISRAEL SAGUY

Agricultural Research Organization
The Volcani Center
Bet Dagan, Israel

MARCEL DEKKER, INC.   New York and Basel

Library of Congress Cataloging in Publication Data

Main entry under title:

Computer-aided techniques in food technology.

   (Food science ; v. 8)
   Includes indexes.
   1. Food industry and trade--Data processing.
I. Saguy, Israel, [date]. II. Series: Food science
(Marcel Dekker, Inc.) ; v. 8.
TP370.5.C65 1983   664   83-5129
ISBN 0-8247-1383-4

MARCEL DEKKER, INC.
270 Madison Avenue, New York, New York 10016

Current printing (last digit):
10 9 8 7 6 5 4 3 2 1

PRINTED IN THE UNITED STATES OF AMERICA

Preface

For many years, an acute need has existed for a computer book that is suitable for food practitioners. The subject of computers and food occupies a position of major importance both in science and industry, and this trend is very likely to expand in the near future. To meet those needs, this book is primarily designed as a textbook for graduate students and upper-level undergraduates who are starting to explore the enormous potential of computer applications and utilizations. Also, it is designed to be a reference book for professional food engineers, food technologists, food scientists, and industrial personnel who wish to expand and update their knowledge on computer-aided methods and techniques.

This book emphasizes the relationships between food, science, industry, and computers and includes some detailed case studies concerning operational research, design problems, data processing, and other related scientific and industrial topics solved by computer-aided programs and subroutines.

Careful thought was given to the number of topics and the contributors selected for this project. Special emphasis was focused on providing up-to-date knowledge and innovations encountered by computer-aided methodologies. Thus, a multiauthor contribution was chosen. Although this approach may result in different philosophies, uneven coverage, unwarranted duplication, and inadvertant omission of important material, it seemed necessary to envision a few subjects from an endless list dealing with computers and food. Furthermore, it enabled us ultimately to have an input from outstanding participants.

Having these constraints in mind, I believe that the outcome is quite satisfactory, especially if one considers the list of outstanding contributors.

The organization of the book is quite simple. Each chapter provides a separate unit which may be read independently. Covered in sequence are fundamentals of computers, programming numerical techniques, and model building followed by various applications and implementation in kinetics, simulation, heat transfer, linear programming, production control, rheology studies, optimization process control, and finally new approaches in computer system developments, interface and super-routines, and professional responsibility. The intent being to progress from basic fundamental topics to more specific systems. Of course, complete coverage of all computer applications has not been attempted due to the formidable task involved. Yet, it is hoped that most important topics have been covered.

During the publication process of this book, Dr. Amihud Kramer passed away. Ami, as he was known to his many friends, made many outstanding contributions to his profession. Among these was his pioneering work in the use of mathematics, statistics, and computers to solve problems in food science.

As a vigorous theoretical presentation of the computer-aided techniques can be somewhat overpowering to food practitioners who are merely seeking an efficient and effective technique, thus the material was presented in as simple a way as possible. Yet, every effort has been made to indicate the theoretical background and assumptions made.

Several illustrative examples and case studies are given to depict the applications of the principles to practical situations. Figures and tables have been used liberally with the aim of representing the material clearly and of facilitating understanding. Only a few computer programs were included. Nevertheless a list of numerous general-purpose "codes," programs, and subroutines are cited.

To all readers, I extend an invitation to report errors that no doubt have skipped our attention and to offer suggestions for improvements that can be incorporated. Furthermore, I would like to believe that this book will trigger the collection of programs, codes, and subroutines specially tailored and designed for the food science industry.

It is hoped that this book will enlighten readers with respect to the vast possibilities and the potential of computer applications, and thereby encourage and enhance the entreé into the world of computer-aided methods and techniques that are most needed and suitable to our ever-changing modern society.

                                                                          Israel Saguy

Acknowledgments

The number of individuals to whom we are indebted for helping in preparing this book is too large to mention by name. Yet, I would like to express my very special appreciation to the following:

My parents Clara and Eliezer Siminovitch, for everything they have done

My wife Irit and children Ami and Dan for their support and encouragement

My coauthors for their enthusiasm and dedication

My series editor Professor Steven R. Tannenbaum who started me on this project, for his faith and vision

The staff at Marcel Dekker for their skill and endless efforts

And most of all to Professor Marcus Karel for his encouragement, trust, collaboration, and assistance that made this project possible

Israel Saguy

Contributors

MALCOLM C. BEAVERSTOCK, Ph.D. Systems Project Manager, The Foxboro Co., Foxboro, Massachusetts

FILMORE E. BENDER, M.S., Ph.D. Associate Director, Maryland Agricultural Experiment Station, University of Maryland, College Park, Maryland

EDWARD W. BURNSIDE, M.Sc. Commercial Planning Manager, UK Household and Toiletry Division, Reckitt and Colman, Hull, England

NICHOLAS CATSIMPOOLAS, Ph.D. Professor, Department of Biochemistry, Boston University School of Medicine, Boston, Massachusetts

RONALD I. FRANK Staff Member, National Accounts Division, Cambridge Engineering/Scientific Support Center Staff, Cambridge; Visiting Lecturer, Department of Computer Science, Framingham State College, Framingham, Massachusetts

JAMES F. KAMMAN, M.Ag., Ph.D. Development Leader, James Ford Bell Technical Center, General Mills, Inc., Minneapolis, Minnesota

MARCUS KAREL, Ph.D. Professor, Department of Nutrition and Food Science, Massachusetts Institute of Technology, Cambridge, Massachusetts

AMIHUD KRAMER, Ph.D.* Professor Emeritus, Food Science Program, Department of Horticulture, University of Maryland, College Park, Maryland

---

*Deceased

THEODORE P. LABUZA, Ph.D.   Professor, Department of Food Science and Nutrition, University of Minnesota, St. Paul, Minnesota

DARYL B. LUND, M.S., Ph.D.   Professor, Department of Food Science, University of Wisconsin, Madison, Wisconsin

STEPHEN C. NICHOLSON, Ph.D.   Planning Manager, UK Pharmaceutical Division, Reckitt and Colman, Hull, England

JOHN P. NORBACK, Ph.D.   Assistant Professor, Department of Food Science, University of Wisconsin, Madison, Wisconsin

MICHA PELEG, D.Sc.   Associate Professor, Department of Food Engineering, University of Massachusetts, Amherst, Massachusetts

WILLIAM M. RAND, Ph.D.   Lecturer, Department of Nutrition and Food Science, Massachusetts Institute of Technology, Cambridge, Massachusetts

ISRAEL SAGUY, D.Sc.*   Senior Scientist, Division of Food Technology, The Volcani Center, Agricultural Research Organization, Bet Dagan, Israel

R. PAUL SINGH, Ph.D.   Associate Professor, Department of Agricultural Engineering, University of California, Davis, California

*Present affiliation:
Visiting Lecturer, Department of Nutrition and Food Science, Massachusetts Institute of Technology, Cambridge, Massachusetts

Contents

Computer-Aided Techniques in Food Technology

1

Introduction to Computers and Programming

ISRAEL SAGUY* The Volcani Center, Agricultural Research
Organization, Bet Dagan, Israel

I. INTRODUCTION

Computers have become a part of our everyday lives. We can see their
operation in banks, department stores, fast-food restaurants, modern food-
processing plants, warehouse management, airline terminals, and endless
other locations. In most of these sites computers have been installed
and utilized to take advantage of their enormous computational power
in processing data. During the last decade, based on figures released
by Business Week (June 8, 1981), the data processing industry grew

*Present affiliation:
Massachusetts Institute of Technology, Cambridge, Massachusetts

at breakneck speed from $11.7 billion in 1970 to $53.5 billion in 1980 (nearly fivefold).

In the engineering profession, the computer's impact is even more
significant, changing our way of thinking and becoming an integral part of
almost any problem-solving approach.

The computer not only enables us to make many computations, fast
and reliable, but allows us to create new approaches which were not con-
ceivable before. It is generally accepted that computers have moved into
the forefront of technology.

As expected, the sharp decrease in the price of computers and micro-
processors has sparked their use in most industries. However, the food
industry had been lagging behind in computer application and exploitation.
Recently, major efforts have been focused on providing the necessary tools
(e.g., hardware, improved "easy-to-use" specially designed software,
computer-aided techniques and modeling, optimization theories, control
theories) to enhance computer utilization in the food science and food engineering fields in the coming years.

This chapter addresses the following aims:

1. It introduces the reader to the terminology often used in the utilization of computers (a concise glossary is furnished in the Appendix).

2. It reviews several software libraries and packages which are designed to meet special needs and applications.

We would like to point out that this chapter is not a replacement for
any standard primer dealing with either computer programming or computing fundamentals and applications. For this purpose several references
are included.

II. THE COMPUTER SYSTEM

A computer is defined as any machine which processes data in a prescribed
form, and furnishes results. The latter could be either information (numeric or nonnumeric data) or signals used to control other processes.
More specifically, a computer performs a sequence of arithmetic and logical operations, following a set of internally stored instructions.

Three general main categories of computers are known: digital computers, analog computers, and hybrid computers. A digital computer performs operations on data represented in digital form. Analog computers,
on the other hand, accept data as a quantity varying over a length of time.
Hybrid computers are yet another type of machine combining some of the
properties of both digital and analog computers.

In most electronic digital computers the method of number presentation is based on the system of binary notation. The binary system is the

simplest possible and is based on representing numbers by only two digits 0 and 1. In other words, in the binary system, the base or the radix is 2. A most common abbreviation of a binary digit is the bit.

In the binary system each decimal digit from 0 to 9 is represented by four binary digits known as binary-coded decimal (BCD), and is described as follows:

Decimal digit	Binary code
0	0000
1	0001
2	0010
3	0011
4	0100
5	0101
6	0110
7	0111
8	1000
9	1001

The representation of information is somewhat more complicated. This complication arises from the fact that information represents a variety of types (i.e., numeric and nonnumeric). To overcome this problem, non-numeric (character) data is represented through a coding scheme whereby certain patterns of bits are by agreement taken to stand for certain characters. Two coding schemes are most popular, namely, the 8-bit Extended Binary-Coded Decimal Interchange Code (EBCDIC) and the 7-bit American Standard Code for Information Interchange (ASCII).

Each data element is stored in the computer memory as a word. The latter consists of a predetermined number of bits and is processed by the computer as an entity. Many computers use a fixed word length, while others allow a variable length by grouping several words according to the requirements. The size of a fixed-length word depends on the computer and may range from 16 bits in a small-scale computer to 60 bits in a large computer. Each word may be subdivided into several bytes. The word length is an important and typical design parameter of the computer. It determines, for instance, the largest and the smallest numbers that can be stored or the precision by which calculations are performed.

A digital computer system is a collection of components working together to process data. The computer system is a combination of hardware
and software. By hardware we mean all the electronic and mechanical gad-
getry, or in other words, the physical computing equipment. This includes
the computer itself, card reader, tape drive, plotter, line printer, etc. By
software we mean the collection of support programs written to provide
various services to the users of the system. These ready-made programs
may be put to use at once, thus minimizing time and effort required.

The software covers a great range of programs and areas. During
the last 10 years, major efforts have been focused on developing software
packages of mathematics, statistics, operation research, data-base management, etc. A special report from Business Week (Sept. 1, 1980) estimated that users of main-frame computers alone spent $8.4 billion for soft-
ware, a 50% increase over the $5.6 billion spent only 2 years earlier. They
reported also that most computer manufacturers already are spending 50%
of their research and development budget on software, and this percentage
is expected to climb. This trend is manifested especially in the minicomputer industry where vendors find it almost impossible to sell computers
without appropriate software. It is envisioned that software costs will approach 70% of total systems costs.

It is almost useless to talk about the computer without including the
computing environment. Hence, "computer system" was used to describe
what the user interacts with, rather than the term "computer," which has a
definite hardware overtones.

Four types of computer system operations are worth mentioning: batch
processing systems, interactive timesharing systems, on-line transaction
systems, and computer network. The differences between these systems
are based on the method of access and the user requirements.

Based on their size, computers may be divided into three main groups:
large-scale computers, minicomputers, and microcomputers. In the 1970s
a minicomputer was commonly recognized as a desk-top-size computer,
which typically weighed less than 50 lb and could be purchased at prices
ranging from $2000 to $20,000. Their internal memory was relatively
small (up to 32K words). Obviously the breakthroughs in technology and
minimization knowledge overcame the aforementioned limitations. To date,
the difference between a minicomputer and large-scale computer is not so
well defined. A 256K words internal memory is a common feature of a
minicomputer. While a large-scale computer is quite small in its physical
dimensions, it is supplied with thousands of K words internal memory. A
microcomputer, on the other hand, is mounted on a single semiconductor
chip or printed circuit board.

Two typical minicomputers are shown in Figs. 1 and 2. As many different configurations exist, we shall use Fig. 2 to illustrate several main
functional elements, namely, input, storage, arithmetic and logic, control,
output, and peripheral units.

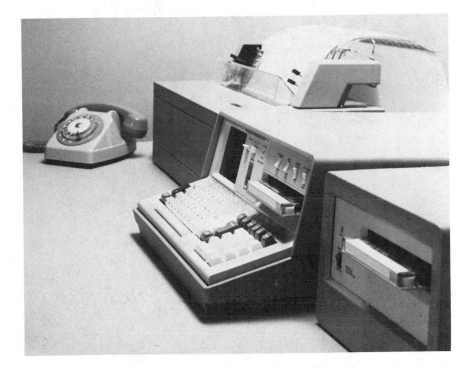

FIGURE 1   IBM 5100 Computer. (Courtesy of IBM, Israel.)

Input: The input consists of all the devices that provide the means by
which data is transmitted to the computer. A variety of devices are well
known. Among them we may include the punched card reader, tape reader,
computer console, diskette reader, and a cathode-ray tube (CRT) with a
keyboard as illustrated in Fig. 2. Input, therefore, reads or senses the
data that is recorded in a prescribed form.

Storage (memory): The memory of a digital computer, also known as
the storage section is used for storing the user's programs and the data to
be processed. Also, it stores the system library programs of instructions
coordinating the processing of the data in a predetermined and organized
fashion.

In every computer, from the smallest to the largest, the amount of
internal storage capacity (also known as the main memory or core memory,
expressed in K words or bytes) is expensive and a scarce resource. Thus,
information not immediately needed is normally stored on external storage.
The flexibility and expandability of modern automatic data processing systems, however, make it increasingly more difficult to distinguish an internal
from an external storage device. Moreover, while a certain type of storage

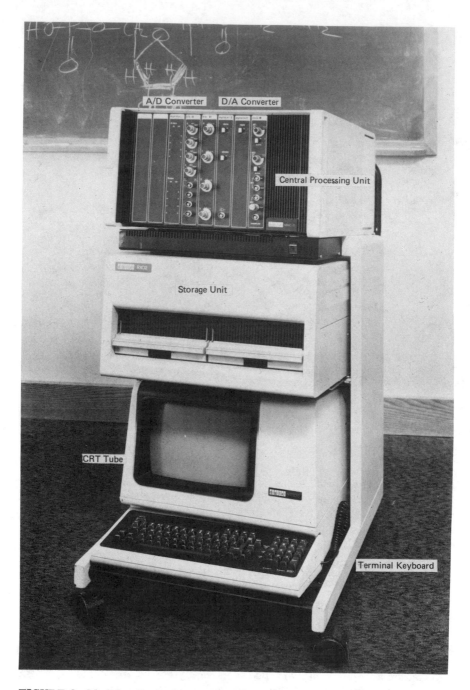

FIGURE 2   Modular Instrument Computer (MINC) RT-11/03 System.
(Courtesy of Digital Equipment Corp., Massachusetts.)

device may be internal in one system, the same type of device will be external in another configuration. A common method of classifying and evaluating storage devices involves capacity, access time (defined as the time required to transfer information from a storage location to the main memory, where it is available for processing), and cost in relation to capacity and speed. The internal memory was often constructed of a matrix of ferrite or magnetic cores. More recently the memory resides on semiconductor chips. New and advanced technology and developments promise much larger storage capacity on smaller and cheaper devices. The configuration of the storage devices are designed so that access time is minimized. The access time and computation time are expected to decrease, yielding, ultimately, circuits operating at speeds approaching the speed of light.

Figure 2 shows an external storage medium known as a floppy diskette. The information is stored on the diskette using a random access method (i.e., access time for data is independent of the location of the data on the storage medium).

Many other storage units are available in other systems, namely, magnetic disks, magnetic drums, tapes, etc.

Arithmetic and logic unit: The arithmetic and logic unit is in charge of performing the computations and to manipulate the data. Furthermore the arithmetic and logical unit has the ability to make simple logical decisions.

Control unit: The control unit is the coordinator or director of all operations within the computer. This is done through a detailed list of instructions, called a program, residing in the main memory. The control unit is in charge of the following operations:

Read data from the input devices.

Transfer the appropriate values from the storage units to the CPU (central processing unit) (i.e., a combination of the arithmetic and logic unit, the control unit, and the internal memory).

Control the execution of the required calculations.

Retrieve data and store intermediate results from the main memory.

Pass final results to the output devices.

Furthermore, the control unit ties together all the units necessary to perform any operation carried out by the computer.

Output: The output furnishes the finished product, thus working in the direction opposite to the input. The output can be provided on the same media as the input. One big difference is that readable hard copy can be provided by the output. The most common output devices are line printer plotter, card puncher, tape, and diskette. Once again the CRT may be used, indicating that it may serve as both input and output device.

Peripheral units: The final element in this supersimplified visualization of a digital computer is the input/output (I/O) interface. In order for the computer to detect and respond to external events, peripheral units are required. The interface in most computers consists of a set of bidirectional data lines and control lines, generally referred to as "busses." Figure 2 shows several important peripheral devices. These include a clock and two converters: analog to digital (A/D) and digital to analog (D/A). These two important features allow this computer configuration to process data on a "real-time" basis and also to control the process.

III. COMPUTER OPERATION AND LANGUAGES

A. Algorithm

For a computer to follow a specific order of operations from given data a strategy or a plan must be furnished. The exact sequence of operations is known as the algorithm. Hence, the computer will follow the exact sequence of operations according to the algorithm supplied. An algorithm has to provide all the necessary alternatives based upon the conditions. Once an algorithm is executed and is free from either syntax errors or logical errors, it guarantees the desired results.

The process of preparing algorithms is one of the most intellectually demanding tasks associated with computer usage.

B. Flow Charts

A flow chart is a pictorial representation of the algorithm. A flow chart is needed to describe in a graphic and simple way the sequence of the operations that are performed by an algorithm. Also, it expresses the relations existing among the various sections. Hence, they are used widely to aid in problem solving and also for documentation purposes. It is generally accepted that, once a problem becomes complex, a flow chart is needed to keep track of all the algorithm details. Moreover, as flow charts are independent of the programming language used for coding the algorithm, they allow direct communications among users from different backgrounds.

Five basic flow-chart symbols are normally applied to distinguish between different computation tasks. These symbols represent the following tasks:

1. Input/output symbols

2. Computation symbols

3. Decision symbols

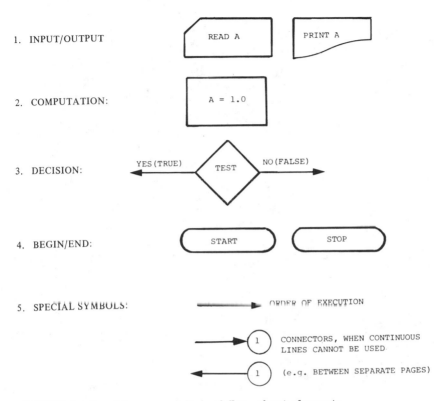

1. INPUT/OUTPUT

2. COMPUTATION:

3. DECISION:

4. BEGIN/END:

5. SPECIAL SYMBOLS:

FIGURE 3   Graphic presentation of flow-chart elements.

    4. Begin/end symbols

    5. Special sequence symbols

Their graphic representation is depicted in Fig. 3.

    Figure 4 is a typical flow chart illustrating the solution of a quadratic equation expressed as $Ax^2 + Bx + C = 0$ (A is assumed to be nonzero). Naturally, this flow chart applies to different values of A, B, and C (as long as $A \neq 0$).

    Even this simple example illustrates how several routes of action may be adopted, based on directives from the logical decisions.

    During their initial study period, many programmers regard flow charts as a waste of time. Eager to deal with the "real" program, they forget the flow chart. Nonetheless, once they begin to code algorithms for solving real-situation problems (unfortunately, they normally tend to be quite complex), the above attitude fades quickly. We

10

Saguy

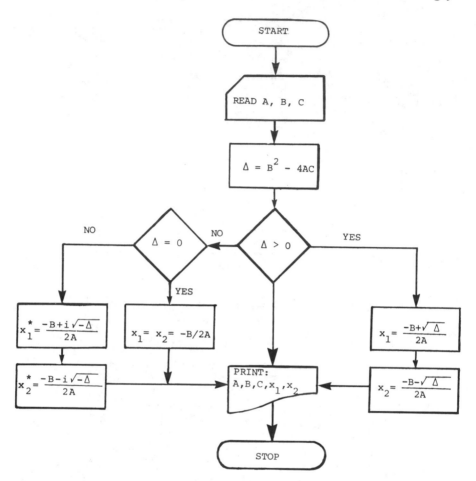

FIGURE 4 Flow chart for solving a quadratic equation ($Ax^2 + Bx + C = 0$, where A is assumed to be nonzero; *-complex solution).

strongly recommend starting with the flow chart before any "codes" are written.

C. The Computing Process

The method of presenting an algorithm to be executed by a computer is to write a program in a "high-level programming language" (e.g., FORTRAN, COBOL, BASIC, etc., to be discussed shortly). These programs are then translated by a "translator program" (known as the "compiler") into a

machine language, also known as "assembly language." Once in machine
language the program is ready for execution by the computer.

In the machine language, every operation (i.e., addition, subtraction,
etc.) is specified in a code understood by the CPU. The codes may vary
from one computer to another; hence, a high-level language enables the
programmer to communicate on common ground and limits the dependence
on the manufacturer's CPU design.

The compiler accepts as input the "source module" (also known as
"source program" or "source code") which is a meaningful collection of
statements in a given high-level language. The compiler output is the
translated machine-language form, known as the "object module" (or
"object code"). The compiler output will also indicate all syntax errors
found in the source module.

The syntax of a language defines what statements or combinations of
statements a computer will accept. This is the easiest part of a language
to specify. Therefore, most reference manuals define very clearly the
syntax of a language.

Several functions that are frequently used (such as calculation of
square roots, exponentials, and other arithmetic functions; handling of input and output operations, detection of several kinds of error conditions)
have been gathered into a special default "system library" (known as
"SYSLIB"). During the translation of the source module, the compiler
examines each statement to see whether an operation which is to be performed by the SYSLIB was included, and the compiler makes the appropriate reference to the SYSLIB. Obviously, if any other library is used, the
same action is followed. The result of the compilation is therefore an
object module which needs parts of the system library (SYSLIB) and frequently other object modules and libraries to form a complete functioning
program. All the required object modules must be linked together, before
the program can actually be executed.

The linkage is carried out by the "linkage editor." The linkage function also adjusts the object code to place all of the program units one after
the other. The derived load module is actually an image of the computer
core prior to execution. The final step is obviously to transfer this core
image to the central processing unit, and to execute it. Assuming that
neither syntax errors were detected nor logical mistakes were made, the
executed program will produce the output with the desired results. The
process outlined previously is depicted in Fig. 5.

Unless otherwise required, all steps needed to execute a program
are automatically performed by the computer or are controlled by the user,
that is, to follow his or her requests.

D. High-Level Programming Languages

As already pointed out, a high-level language is a notation with which
people can communicate algorithms to the computer and to one another.

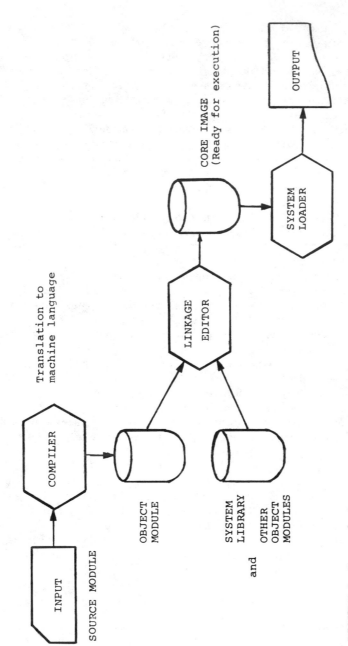

FIGURE 5   Stages during proprogram processing.

Hundreds of programming languages exist. They differ in their degree of closeness to natural or mathematical languages on the one hand and to machine language on the other.

Some of the aspects of high-level languages which make them preferable to machine or assembly language are the following (Aho and Ullman, 1979):

1. Ease of understanding

2. Naturalness (ease of expressing an algorithm)

3. Portability (the potential to run a program on a variety of computers)

4. Efficiency of use

No matter what language is being used, several drawbacks could be pointed out. Therefore, the main goal of each new language is to improve the efficiency of describing a given algorithm. Of all the languages that have been developed, only a few are of interest to us. A limited number of the most widespread languages are described here (each description includes at least one reference to a textbook if further details are required):

1. ALGOL (International Algebraic Language) was started in the late 1950s. The language is mainly suitable for describing computational processes. The syntax of ALGOL has a high level of internal consistency, hence permits rapid compilers and short elegant programs. Very popular outside the United States (Pagan, 1976).

2. APL (A Programming Language) is one of the most concise, consistent, and powerful programming languages ever designed. The normal operations of single scalars utilized by most languages were extended in a simple and natural way, to carry arrays of operations of any size and shape. Hence, handling of arrays is its central strength (Gilman and Rose, 1976). See also Chap. 14 by Burnside and Nicholson for further details and applications.

3. BASIC (Beginner's All-Purpose Symbolic Instruction Code), despite its name, is a high-level language. It achieves simplicity and ease in use through restrictions in the syntax. BASIC is useful for a wide range of jobs that can be approached through numeric computation. Its input and output provisions make it inconvenient for users with large amounts of input/output data. However, it is generally accepted as an introductory programming language (Kurtz and Kemeny, 1980).

4. COBOL (Common Business Oriented Language) is designed specially for handling commercial problems. The language uses a syntax closely resembling English, avoiding the use of special symbols, thus offering an easy and straightforward language capable of handling business-oriented problems (Stern and Stern, 1980).

5. FORTRAN (FORmula TRANslation) is an algebraically oriented
language, originally designed for solving mathematical equations and formulas. This was one of the first high-level languages initiated during the
late 1950s. Today it is by far the most widely used language for programming of scientific and engineering computations. In 1966 the American
National Standards Institute (ANSI) published what was regarded as the
definition of the language. Recently a new standard has been produced.
The new 1978 standard called FORTRAN 77 (ANSI, 1977; Meissner and
Organick, 1977) is an attempt to codify existing developments in FORTRAN
rather than to design a new language. Several compilers are not yet capable of complying with the new elements of FORTRAN 77. As this language is by far the most utilized by scientists and engineers, we will refer
to software written in FORTRAN only.

6. PASCAL is derived from ALGOL, but is more powerful and much
easier to use. PASCAL is now widely accepted as a useful and excellent
teaching tool (Grogono, 1980).

7. PL/I (Programming Language I) is a powerful language that can
be used to solve both business and scientific problems. One of its early
names was NPL (New Programming Language). It is primarily applied to
IBM computers; however, many other computers are equipped with an appropriate compiler (Hughes, 1980).

Finally, computer experts predict that in the 1980s a new language
Ada will come into frequent use (for details see Mayoh, 1981, and Computer J., June, 1981).

IV. LIBRARIES AND CODES

A. Available Software

One of the most exciting opportunities presented by the growth of computers is the transfer of research results from one field to another via
general-purpose software (Chan et al., 1978). Programs and subroutines
are becoming a central and leading channel between software, numerical
analysis, and application. The principal criterion of choosing a subroutine
is based generally on ease of use, machine efficiency and availability.

Numerical analysts and computer software experts often remark
(and argue) on the difficulty of comparing "codes," and only rarely provide
the potential users with clear-cut selection guidelines. Our aim will focus
on providing information on the most sophisticated, state-of-the-art subroutines collected in special libraries and packages available from commercial sources. Such sources are

1. General Applications

a. IMSL: International Mathematical and Statistical Libraries, Inc.,
Houston, Texas (IMSL, 1982), supplies a wide collection of routines that

include most of the problem areas of numerical analysis, including statistics. The library is tailored for use on IBM computers; but parallel versions are available for a variety of other machines.

b. NAG: Numerical Algorithm Group, Oxford, England (NAG, 1977), supplies a library similar in scope to IMSL, however, documentation is somewhat more emphasized. The NAG library also includes a part of the National Physical Laboratory (NPL) (NPL, 1978) library which provides superior software specially designed for optimization problems.

c. EISPACK: National Activity to Test Software (NATS), Argonne National Laboratory, Argonne, Illinois, is a comprehensive set of routines devoted to computation of eigenvectors and eigenvalues of many different types of matrices. It is considered to be an archetype of mathematical software packages but is much more specialized than IMSL or NAG.

d. FUMPACK: NATS, Argonne National Laboratory is a collection of routines to compute certain special functions. Like EISPACK, it is well designed (particularly in error handling). However, the number of special functions included in the package is limited.

e. LINPAK: National Energy Software Center, Argonne National Laboratory, Argonne, Illinois (Dongarra et al., 1980), is a collection of FORTRAN subroutines which analyze and solve various systems of simultaneous linear algebraic equations. The subroutines are designed to be completely machine independent and to run at near optimum efficiency in most operating environments.

f. MINPACK: National Energy Software Center, Argonne National Laboratory, Argonne, Illinois, is a package of FORTRAN subroutines providing numerical solutions for systems of nonlinear equations and nonlinear least-squares algorithms. The User's Guide (More, Garbow, and Hillstson, 1980) describes the algorithms and software of the package and includes documentation and program listings.

g. NPL: National Physical Laboratory, Division of Numerical Analysis and Computing, Teddington, England (NPL, 1978 and 1980), includes highly sophisticated software to handle most optimization problems.

h. HARWELL (AERE): Computer Science and System Division, Atomic Energy Research Establishment, Harwell, Oxfordshire, England (Hopper, 1981), includes many subroutines devoted to optimization.

i. SHARE: Share, Inc., Chicago, Illinois (SHARE, 1981), is a nonprofit organization, its principal purpose being to foster the development of free exchange and public dissemination of research data. The library includes a collection of generally useful programs created and administered to promote the exchange of technical information. Programs and their documentation are made available at distribution cost.

2. Statistical Packages

   a. BMDP: Biomedical Computer Programs, P-Series. This package resembles a library of main programs and provides an extensive number of advanced statistical analyses. It may be accessed through the SAS (SAS, 1981) or SPSS (Nie et al., 1975) packages. Application notes and guidelines are given in a single manual functioning as both user guide and reference (Dixon and Brown, 1979).

   b. MINITAB: Minitab Project Statistical Department, Pennsylvania State University (Rayan et al., 1981). MINITAB is a very easy to use, flexible, and powerful statistical computing system. It is machine compatible, and runs interactively or in batch. It is particularly suited for small data sets. The relatively small scope of the package leads to a simple English-like syntax.

   c. SAS: Statistical Analysis System (SAS, 1981). This library provides a unified and extensive environment for statistical data analysis, including data-management capabilities. It also permits data to be sorted, merged, updated and copied. It can read, also, system BMDP files as well as interface to BMDP programs directly. Several manuals and user guides are available (Reinhardt, 1980; Kirk, 1980).

   d. SPSS: Statistical Package for Social Sciences (Nie et al., 1975) provides almost the same capabilities as the SAS, but does not support data-management and report features. A special application guide is available (Hull and Nie, 1979).

3. Special Packages

Two examples illustrating some of the capabilities of these packages are given below.

   a. RS/1: The Research System, Bolt, Beranek and Newman, Inc., Cambridge, Massachusetts. This package is tailored for PDP-11 or VAX and is used as a notebook. Numeric and textual data can be permanently entered and stored. The data management enables support of large files. Also, it provides an easy method of data entry and retrieval, statistical analysis, curve fitting, and plotting. Also, an access to BMDP is possible.

   b. IDMS: Integrated Data Base Management System (IDMS, 1981). This package centralizes the data used by an organization, but does not prevent the data from being arranged in different ways to suit different applications. All the data records are treated as members of one cohesive storage facility, namely, the data base. Records are connected in different logical groups called "sets," expressing several data arrangements as required. When data is to be shared by many applications, conventional data-management techniques may result in one or more problems, namely,

Detar, D. F. (ed.) (1968). Computer Programs for Chemistry. Vols. I,
     II, and III. Benjamin, New York.
Dixon, W. J. and Brown, M. B. (1979). BMDP-79 Biomedical Computer
     Programs P-Series. University of California Press, Berkeley,
     Calif.
Dongarra, J. J., Bunch, J. R., Moler, C. B., and Stewart, G. W. (1980).
     LINPACK User's Guide. SIAM, Philadelphia, Pa.
Druck, S. (1978). Weizmann Institute Computer Center User's Guide.
     The Weizmann Institute of Science, Rehovot, Israel, September.
Gilman, L. and Rose, A. J. (1976). APL An Interactive Approach, 2d ed.
     Wiley, New York.
Grogono, P. (1980). Programming in PASCAL, 2d ed. Addison-Wesley,
     Reading, Mass.
Hetzel, W. C. (ed.) (1972). Program Test Methods. Prentice-Hall,
     Englewood Cliffs, N.J.
Hopper, M. J. (1981). Harwell Subroutine Library—A Catalogue of Sub-
     routines. Computer Science and Systems Division, AERE, Harwell,
     Oxfordshire, England, November.
Hughes, J. K. (1980). PL/I Structured Programming, 2d ed. Wiley, New
     York.
Hull, C. H. and Nie, N. H. (1979). SPSS Update: New Procedures and
     Facilities for Release 7 and 8. McGraw-Hill, New York.
IDMS (1981). System Overview (Release 5.5). Cullinane Database Sys-
     tems, Inc., Westwood, Mass., March.
IMSL (1982). International Mathematicals Statistical Library, 9th ed.
     IMSL, Inc., Houston, Tex.
Ingber, D. (1981). Computer addicts, Science Digest (July):88.
Jarosch, H. S. and Druck, S. J. (1982). Numerical Analysis User's Guide.
     The Weizmann Institute of Science, Rehovot, Israel, January.
Kirk, H. J. (ed.) (1980). SAS Views. SAS Inst. Inc., Cary, N.C.
Kurtz, T. E. and Kemeny, J. C. (1980). BASIC Programming, 3d ed.
     Wiley, New York.
Mayoh, B. (1981). ADA. Wiley, New York.
Meissner, L. P. and Organick, E. I. (1980). Fortran 77. Addison-
     Wesley, Reading, Mass.
MIT (1981). Numerical Analysis User's Guide. Application Program
     Series AP-3, IPS, Cambridge, Mass.
More, J. J., Garbow, B. S., and Hillstron, K. E. (1980). User Guide
     for Minpack-1. Argonne National Laboratory, Argonne, Ill.
NAG (Numerical Algorithms Group) (1977). NAG Fortran Library Manual,
     Mark 6. Oxford, U.K.
Nie, N. H., Hull, C. H., Jenkins, J. G., Steinbrenner, K., and Bent,
     D. H. (1975). Statistical Package for Social Sciences. McGraw-Hill,
     New York.
NPL (National Physical Laboratory) (1978). Introduction to the NPL Nu-
     merical Optimization Software Library, Vols. I and II. Teddington,
     Middlesex, England.

NPL (National Physical Laboratory) (1980). A Brief Guide to the NPL
  Numerical Optimization Software Library. Teddington, Middlesex,
  England. December.
Pagan, F. G. (1976). A Practical Guide to Algol 68. Wiley, New York.
Ray, D. H. (1976). A methodology for computer software selection, in
  Managing Data Processing (R. A. Bassler and E. O. Joslin, eds.).
  Colledge Reading, Alexandria, Va., p. 178.
Reinhardt, P. S. (ed.) (1980). The SAS Supplemental Library User's
  Guide. SAS Inst. Inc., Cary, N.C.
Ryan, T. A., Jr., Joiner, B. L., and Ryan, B. F. (1981). Minitab Ref-
  erence Manual. Pennsylvania State University, University Park, Pa.
SAS (1981). SAS Programmer's Guide. SAS Inst. Inc., Cary, N.C.
SHARE, Program Library Agency (1981). User's Guide and Catalog of
  Programs. Triangle University Computer Center, Research Triangle
  Park, N.C.
Stern, N. B. and Stern, R. A. (1980). Structured COBOL Programming,
  3d ed. Wiley, New York.
U.S. Department of Energy (1982). Compilation of Program Abstract
  (ANL-7411). National Energy Software Center, Argonne National Lab-
  oratory, Argonne, Ill.

reduction in the time required for the implementation rather than the "fastest" software. It is therefore recommended that whenever the decision is between an available and well-documented but slow subroutine, and a faster code which, however, has to be purchased and implemented, the first choice should be taken, with the option of using the faster program at a later date if desired.

Obviously, each case should be weighed separately and optimization of time, money, labor, and effort should be made. The reader should consult Chap. 15 by Frank, which further discusses the subject.

ACKNOWLEDGMENTS

We are indebted to many people who have helped in the development of this chapter. Special thanks are due to Marty Mishkin and Steve Haralampu, who carefully read this manuscript at various stages of its development and offered a number of very helpful suggestions and comments.

We would like to express our gratitude to Digital Equipment Corporation for granting permission to reproduce portions of their copyright material.

Special thanks are due to Ms. Janet Copley for her dedication in editing and typing this manuscript.

REFERENCES

ACM (1978). Collected Algorithms from ACM. Vol. 1, 2. ACM.
Aho, A. V. and Ullman, J. D. (1979). Principle of Compiler Design.
    Addison-Wesley, Reading, Mass.
ANSI (1966). American National Standard Fortran. Pub. X3.9, American
    National Standard Institute, New York.
ANSI (1978). American National Standard Programming Language Fortran.
    Pub. X3.9, American National Standards Institute, New York.
Barnes, J. E. and Waring, A. J. (1980). Pocket Programmable Calcula-
    tors in Biochemistry. Wiley, New York.
Bolstad, J. H., Chan, T. F., Coughran, W. M., Jr., Grosse, E. H.,
    Heath, M. T., Luk, F. T., Nash, S. G., and Trefethen, L. N. (1978).
    Numerical Analysis Program Library User's Guide NAPLUG User
    note 82, Computer Science Department and Stanford Linear Accelera-
    tor Center, Stanford University, Menlo Park, Calif.
Chan, T. F., Coughran, W. M., Jr., Grosse, E. H., and Heath, M. T.
    (1978). A Numerical Library and Its Support. STAN-CS-78-673,
    Computer Science Department, Stanford University, Menlo Park,
    Calif., November.
Coughran, W. M., Jr. (1977). A note concerning the construction of a
    numerical analysis program library, Tech. Memo 107. SCS-SCIP,
    Stanford University, Stanford, Calif., September.

SAS are excellent choices. Several computer centers publish a user guide
which in many cases covers frequently encountered topics and lists recommended software. Several user's guides include Stanford University (Bolstad et al., 1978), MIT (1981), and Weizmann Institute (Druck 1978;
Jarosch and Druck, 1982). Similar guides are available in most computer
centers.

The guides just quoted identify numerical areas and discuss the best
algorithms to be used for a specific problem. They also give an elaborate
list of references which contains an in-depth discussion of most numerical
aspects.

When your system is lacking a specific program, there is always the
temptation to use the next-door neighbor's software even if he or she is
using a different computer. Based on experience, it should be emphasized
that only in special cases is it worth starting with software written for
another system, thus going through the "hassle" of implementing it on your
machine, as computers from different manufacturers don't have the same
exponent range, and/or significant digits, and/or media used (e.g., tape,
floppy diskettes). Hence, this possibility of utilizing outside software may
lead to complete disaster, and all efforts will be wasted.

If you plan to use a commercial source to supply your software needs,
the following checklist is recommended before any software acquisition is
made (Ray, 1976):

1. The package and its specification

2. The terms

3. Documentation

4. Installation

5. Modification, maintenance, and improvements

6. Copyrights and patents

7. Nondisclosure clause

8. Penalties

9. Termination

In all cases when new software is implemented, several test problems
should be run, in order to verify that an error-free version was implemented.
This precaution is always recommended (Hetzel, 1972).

The most crucial decision for a food engineer or a food practitioner
is how to formulate the problem to represent and simulate the process, and
which algorithm to use. Hence, in most cases one is more interested in
solving the numerical problem at hand rather than choosing the "most"
efficient subroutine reported in the literature. As a rule of thumb, the
decision on which software to use has to be judged, in this case, by the

Most existing FORTRAN libraries include the following information:

1. Purpose. Provides a perspective on the specific program and its potential usage.

2. Description. The algorithm and method used are explained.

3. Specification. Summarizes the information about the order, type, and dimension of the parameters (arguments).

4. Parameters. A brief description of each parameter (in the order of its appearance in the program) is given. Special attention should be focused on entry/exit values, array dimensions, etc.

5. Auxiliary routines. Specifies other routines called directly. In most practical cases, this description will not include other subroutines called by auxiliary routines. User-supplied subroutines are also described.

6. Error indicator. Provides all the information on the following: errors observed during calculation, parameters outside of their expected range, conditions not met, overflow, and many others.

7. Timing. In most cases a rough estimation of the CPU time is given. The latter may vary widely between different machines and applications. However, it may be used as an indicator for evaluating the effect of changing some parameter values (e.g., step size, matrix dimension).

8. Storage. FORTRAN subroutines avoid the need of declaring large internal arrays. The required core storage is thus mainly affected by the user.

9. Accuracy. Specifies the precision of the parameters. Single or double precision may be used. Also it may specify the decimal accuracy of the solution.

10. This section presents an example problem, shows the input, and explains the output results. In most cases it indicates the number of iterations and the CPU required. For most practical cases, users are urged to use this example before running their own program.

C. Choosing the "Right" Code or Subroutine

In our review of available software, no attempt was made to choose the most appropriate library. The decision should be based on previous experience, special needs and requirements, availability, and implementation suitability. The selection of the "right" subroutine is a complicated task and requires skillful experience. When this expertise is missing, the IMSL, NAG, and

redundancy, reprogramming, and dispersion. IDMS, on the other hand, maintains the data integrity and is readily accessible to applications.

4. Other Sources

Many other sources beside those already mentioned are now available. These commercial sources are distributing software for the solution of a variety of problems on a wide range of computers. Another plausible avenue is to check with government bureaus (e.g., U.S. Department of Energy, 1982) or computer centers at most universities. Also, several computer journals provide a wide range of possibilities (e.g., ACM, 1978) and vendors.

Several years ago a three-volume book entitled Computer Programs for Chemistry (Detar, 1968) was published. The purpose of this publication was to make available in convenient form a wide range of programs of interest to chemists, thus providing a medium for publication to encourage the discovery and application of quantitative mathematical models to problems in chemistry. Tapes with these programs were available from the publishers.

Many computer books and journals include numerous algorithms which may be of interest. Finally, for those who are keeping their pocket programmable calculators, the book of Barnes and Waring (1980) provides useful programming for biochemistry applications.

B. The Structure of a Program Library

The program library is a set of well-designed, reliable, robust, and easy-to-use routines (e.g., subroutines and functions) that are conceived and written within a unified framework to be available to the general community of potential users. Furthermore they are supported by the computer center and its software experts.

Program libraries are designed to solve general problems and to provide the "best" available up-to-date numerical techniques, thus avoiding the difficulties and efforts otherwise required. Chan et al. (1978) specify the reasons for using libraries as follows: duplication of efforts is reduced; well-tested, well-tuned routines are used; state-of-the-art algorithms are available; storage and compilation costs are reduced; implementation details are done correctly; and elapsed time to get a working program is reduced. To this list we should add robustness, which describes the character of a routine that allows it to recognize a wide variety of pathologies and to react to them (Coughran, 1977).

Therefore we may conclude that it is strongly recommended whenever possible to use only a library software. The latter is true for all potential users including computer programmers and numerical analysis experts.

NPL (National Physical Laboratory) (1980). A Brief Guide to the NPL
    Numerical Optimization Software Library. Teddington, Middlesex,
    England. December.
Pagan, F. G. (1976). A Practical Guide to Algol 68. Wiley, New York.
Ray, D. H. (1976). A methodology for computer software selection, in
    Managing Data Processing (R. A. Bassler and E. O. Joslin, eds.).
    Colledge Reading, Alexandria, Va., p. 178.
Reinhardt, P. S. (ed.) (1980). The SAS Supplemental Library User's
    Guide. SAS Inst. Inc., Cary, N.C.
Ryan, T. A., Jr., Joiner, B. L., and Ryan, B. F. (1981). Minitab Ref-
    erence Manual. Pennsylvania State University, University Park, Pa.
SAS (1981). SAS Programmer's Guide. SAS Inst. Inc., Cary, N.C.
SHARE, Program Library Agency (1981). User's Guide and Catalog of
    Programs. Triangle University Computer Center, Research Triangle
    Park, N.C.
Stern, N. B. and Stern, R. A. (1980). Structured COBOL Programming,
    3d ed. Wiley, New York.
U.S. Department of Energy (1982). Compilation of Program Abstract
    (ANL-7411). National Energy Software Center, Argonne National Lab-
    oratory, Argonne, Ill.

2

Numerical Techniques

R. PAUL SINGH   University of California, Davis, California

I. INTRODUCTION

Numerical methods have become extremely useful tools for engineers in analyzing a variety of physical problems. With the advances made in the field of digital computers, problems that were virtually impossible to solve a few decades back are being solved by students in undergraduate courses.

The intent of this chapter is to acquaint the reader with a variety of topics important in numerical techniques. Due to space limitations, it is difficult to cover these topics in extensive detail. A list of reference books is given at the end of the chapter for those who want to know more about background theory and analysis of errors associated with approximations. Each method, whenever possible, is presented with procedural steps to make the material practical and useful. It is assumed that the reader has a working knowledge of introductory calculus.

The concepts related to Taylor's series and difference equations are reviewed in the following section. Both of these appear often in various numerical methods to be discussed later.

A. Taylor's Series Expansion

The Taylor's series forms the backbone of numerical methods. Among various applications, the series expansion is useful in approximating functions and analyzing errors.

Consider a function $f(z)$ which is analytic in the neighborhood of a point $z = a$; then the value of the function can be expressed by an infinite power series

$$f(z) = f(a) + (z - a) f'(a) + \frac{(z - a)^2}{2!} f''(a) + \cdots + \frac{(z - a)^n}{n!} f^{(n)}(a) + \cdots \tag{1}$$

If the series is truncated after the first three terms, then Eq. (1) can be represented as

$$f(z) = f(a) + (z - a) f'(a) + \frac{(z - a)^2}{2!} f''(a) + O(z - a)^3 \tag{2}$$

where $O(z - a)^3$ means that the error due to truncation of the series is of the order $(z - a)^3$

B. Forward, Backward, and Central Differences

The primary applications of forward, backward, and central differences include interpolation of polynomials and transformation of differential equations into simple algebraic equations. The following steps are useful to obtain these differences.

1. Consider a function f(z) which can be expanded by a Taylor's series. In the neighborhood of point z expand f(z) to find f(z + h) as

$$f(z + h) = f(z) + hf'(z) + \frac{h^2}{2!} f''(z) + \cdots \tag{3}$$

2. Solve for f'(z) when Eq. (3) is truncated to two terms.

$$f'(z) = \frac{f(z + h) - f(z)}{h} + O(h) \tag{4}$$

The first term in Eq. (4) is the first forward difference approximation of f'(z) with the order of error being h.

3. The first forward difference can also be written as

$$\frac{\Delta f_k}{h} + O(h)$$

where

$$\Delta f_k = f_{k+1} - f_k \tag{5}$$

4. Similarly, the first backward difference approximation of f'(z) with the order of error h is given by

$$\frac{f(z) - f(z - h)}{h} + O(h) \tag{6}$$

5. The first backward difference can also be written as

$$\frac{\nabla f_k}{h} + O(h)$$

where

$$\nabla f_k = f_k - f_{k-1} \tag{7}$$

6. The second forward and backward difference approximations for f''(z) can be computed, respectively, as follows:

$$f''(z) = \frac{f_{k+2} - 2f_{k+1} + f_k}{h^2} + O(h) \tag{8}$$

and

$$f''(z) = \frac{f_k - 2f_{k-1} + f_{k-2}}{h^2} + O(h) \tag{9}$$

7. The central differences can be determined by subtracting the forward and backward Taylor's series expansions about z. Thus

$$f(z + h) - f(z - h) = 2hf'(z) + \frac{h^3}{3!} f'''(z) + \cdots \tag{10}$$

8. Solving for $f'(z)$,

$$f'(z) = \frac{f(z + h) - f(z - h)}{2h} + O(h^2) \tag{11}$$

The first term on the right-hand side is the central difference representation of $f'(z)$. This term can also be written as

$$\frac{f_{k+1} - f_{k-1}}{2h} + O(h^2) \tag{12}$$

9. Similarly the central difference representation for second order is

$$f''(z) = \frac{f_{k+1} - 2f_k + f_{k-1}}{h^2} + O(h^2) \tag{13}$$

The application of the difference representations developed in this section will be discussed later in Secs. V and VI in this chapter.

II. APPROXIMATION OF FUNCTIONS

A. Interpolation

If a function is known only at discrete points, Gregory-Newton interpolation formulas are useful in interpolations. Assuming that the function $f(z)$ can be expanded by Taylor's series, then the value of $f(z)$ can be found at any point between the tabulated discrete values.

1. Assume the point is at $z = 0$; the Taylor's series expansion about $z = 0$ is

$$f(z) = f(0) + zf'(0) + \frac{z^2}{2!} f''(0) + \cdots \tag{14}$$

2. Using forward difference expressions for the derivatives in Eq. (14) such as

$$f'(0) = \frac{\Delta f_0}{h} - \frac{h}{2} f''(0) + O(h^2) \tag{15}$$

we get

$$f(z) = f(0) + \frac{z}{h} \Delta f_0 + \frac{z(z - h)}{2!h^2} \Delta^2 f_0 + \cdots \tag{16}$$

Equation (16) is called the Gregory-Newton forward interpolation formula.

3. Similarly, the Gregory-Newton backward interpolation formula is

$$f(z) = f(0) + \frac{z}{h} \Delta f_0 + \frac{z(z + h)}{2!h^2} \Delta^2 f_0 + \cdots \tag{17}$$

Central differences are also used in interpolations. Bessel's formula and Stirling's formula (Hornbeck, 1975) use central differences. These formulas are recommended when the interpolation is to be done near the center of evenly spaced tabulated values. Lagrange polynomials are useful for unequally spaced data (Carnahan, Luther, and Wilkes, 1969; Hildebrand, 1956).

III. NUMERICAL INTEGRATION

Numerical integration (or quadrature) is valuable in evaluating integrals when analytical evaluation is either difficult or impossible. The following two methods are commonly used.

A. The Trapezoidal Rule

Consider a function $f(z)$ integrable over an interval $a \leq z \leq b$. The integral

$$N = \int_a^b f(z) \, dz \tag{18}$$

is to be evaluated numerically. The following steps are used.

1. Divide the interval $a \leq z \leq b$ into n subintervals of equal width

$$\Delta z = \frac{b - a}{n} \tag{19}$$

2. Assume that function is linear over each subinterval.

3. The area for subinterval z_{k-1} to z_k, assuming a straight line approximates the function, is then

$$\int_{z_{k-1}}^{z_k} f(z)\, dz \approx \frac{f_{k-1} + f_k}{2}\, \Delta z$$

4. The area for the next subinterval z_k to z_{k+1} is

$$\int_{z_k}^{z_{k+1}} f(z)\, dz \approx \frac{f_k + f_{k+1}}{2}\, \Delta z \tag{21}$$

5. Since the integral over the two subintervals is

$$\int_{z_{k-1}}^{z_{k+1}} f(z)\, dz = \int_{z_{k-1}}^{z_k} f(z)\, dz + \int_{z_k}^{z_{k+1}} f(z)\, dz \tag{22}$$

The approximation gives

$$\int_{z_{k-1}}^{z_{k+1}} f(z)\, dz \approx \frac{\Delta z}{2}\, (f_{k-1} + 2f_k + f_{k+1}) \tag{23}$$

6. Equation (23) can be generalized over the entire domain as

$$\int_a^b f(z)\, dz \approx \frac{\Delta z}{2} \left(f_0 + f_n + 2 \sum_{k=1}^{n-1} f_k \right) + O(\Delta z)^2 \tag{24}$$

B. Simpson's Rule

Simpson's rule uses a parabolic arc instead of straight line for approximating the function in the subintervals.

Simpson's rule for even subintervals is expressed by the following:

$$\int_a^b f(z)\, dz = \frac{\Delta z}{3} \left(f_0 + f_n + 4 \sum_{\substack{k=1 \\ odd}}^{n-1} f_k + 2 \sum_{\substack{k=2 \\ even}}^{n-2} f_k \right) + O(\Delta z)^5 \tag{25}$$

where

$$a \le z \le b$$

Other numerical integration techniques used often are Romberg integration and Gauss quadrature [see Henrici (1964) and Hornbeck (1975)].

IV. SOLUTION OF EQUATIONS

Most numerical procedures allow reducing differential equations into simple algebraic equations. The solution of equations, particularly simultaneous equations, by efficient means is often necessary. A working knowledge of matrix algebra is necessary to write efficient computer programs useful in solving simultaneous equations. In this section, various methods commonly used in solving simultaneous equations will be presented.

A. Gaussian Elimination

Consider the following set of linear equations with unknowns x_1, x_2, and x_3 and forcing functions u_1, u_2, and u_3:

$$a_{11}x_1 + a_{12}x_2 + a_{13}x_3 = u_1 \tag{26}$$

$$a_{21}x_1 + a_{22}x_2 + a_{23}x_3 = u_2 \tag{27}$$

$$a_{31}x_1 + a_{32}x_2 + a_{33}x_3 = u_3 \tag{28}$$

The following steps are used in the gaussian elimination procedure to solve the above functions.

1. Multiply Eq. (26) by $-a_{21}/a_{11}$ and add to Eq. (27).

2. Multiply Eq. (26) by $-a_{31}/a_{11}$ and add to Eq. (28).

3. Steps 1 and 2 yield the following:

$$a_{11}x_1 + a_{12}x_2 + a_{13}x_3 = u_1 \tag{29}$$

$$a'_{22}x_2 + a'_{23}x_3 = u'_2 \tag{30}$$

$$a'_{32}x_2 + a'_{33}x_3 = u'_3 \tag{31}$$

where a' and u' are new coefficients and forcing functions, respectively.

4. Multiply Eq. (30) by $-a'_{32}/a'_{22}$ and add result to Eq. (31). The following triangular system of equations is obtained:

$$a_{11}x_1 + a_{12}x_2 + a_{13}x_3 = u_1 \tag{32}$$

$$a'_{22}x_2 + a'_{23}x_3 = u'_2 \tag{33}$$

$$a''_{33}x_3 = u''_3 \tag{34}$$

where a''_{33} and u''_3 are new coefficients and forcing functions, respectively.

5. Solve Eqs. (32) to (34) by backward substitution; that is, $x_3 = u''_3/a''_{33}$, substitute x_3 in Eq. (33) and calculate x_2, substitute x_3 and x_2 in Eq. (32) and calculate x_1.

The above steps can be easily programmed on a digital computer in a matrix formulation.

B. Gauss-Jordan Elimination

A variation of gaussian elimination is the Gauss-Jordan elimination procedure.

Starting with the triangular set of Eqs. (32) to (34), the following steps are used.

1. Divide Eq. (32) by a_{11}, Eq. (33) by a'_{22}, and Eq. (34) by a''_{33}.

$$x_1 + a'_{12}x_2 + a'_{13}x_3 = u'_1 \tag{35}$$

$$x_2 + a''_{23}x_3 = u''_2 \tag{36}$$

$$x_3 = u'''_3 \tag{37}$$

where a'_{12}, a'_{13}, a''_{23}, u''_2, u'_1, and u'''_3 are new coefficients and forcing functions, respectively.

2. Multiply Eq. (36) by a'_{12} and subtract from Eq. (35).

$$x_1 + a''_{13}x_3 = u''_1 \tag{38}$$

$$x_2 + a''_{23}x_3 = u''_2 \tag{39}$$

$$x_3 = u'''_3 \tag{40}$$

where a''_{13} and u''_1 are new coefficients and forcing functions, respectively.

3. Multiply Eq. (40) by a_{23}'' and subtract from Eq. (39) and multiply Eq. (40) by a_{13}'' and subtract from Eq. (38)

$$x_1 = u_1''' \tag{41}$$

$$x_2 = u_2''' \tag{42}$$

$$x_3 = u_3''' \tag{43}$$

Often the triangular set of equations as expressed by Eqs. (32) to (34) are not necessary as the intermediate step in Gauss-Jordan procedure. An efficient computer algorithm is given by Hornbeck (1975).

C. Gauss-Siedel Partial Iteration Scheme

Consider the set of equations earlier presented as Eqs. (26) to (28). The following steps are used.

1. Solve Eq. (26) for x_1, Eq. (27) for x_2, and Eq. (28) for x_3. Thus

$$x_1 = \frac{u_1 - a_{12}x_2 - a_{13}x_3}{a_{11}} \tag{44}$$

$$x_2 = \frac{u_2 - a_{21}x_1 - a_{23}x_3}{a_{22}} \tag{45}$$

$$x_3 = \frac{u_3 - a_{31}x_1 - a_{32}x_2}{a_{33}} \tag{46}$$

2. The computational procedure begins by using initial guesses for x_1, x_2, and x_3, expressed by $x_1^{(0)}$, $x_2^{(0)}$, and $x_3^{(0)}$, respectively.

3. Using step 2 in Eq. (44),

$$x_1^{(1)} = \frac{u_1 - a_{12}x_2^{(0)} - a_{13}x_3^{(0)}}{a_{11}}$$

4. Using the new value $x_1^{(1)}$ in Eq. (45),

$$x_2^{(1)} = \frac{u_2 - a_{21}x_1^{(1)} - a_{23}x_3^{(0)}}{a_{22}} \tag{47}$$

5. Using the new value $x_1^{(1)}$ and $x_2^{(1)}$ in Eq. (46),

$$x_3^{(1)} = \frac{u_3 - a_{31} x_1^{(1)} - a_{32} x_2^{(1)}}{a_{33}} \qquad (48)$$

6. The iteration procedure presented in steps 3 to 5 is continued until convergence is obtained. A convergence criterion is often used to terminate the iteration process.

V. SOLUTION OF ORDINARY DIFFERENTIAL EQUATIONS

The discussion in this section will emphasize the numerical solution of initial-value problems. The methods to be discussed can be easily adapted to boundary-value problems.

1. Consider an initial-value problem given by

$$\frac{du}{dt} = f(u, \ t) \qquad (49)$$

$$u(0) = u_0 \qquad (50)$$

2. Assume that the solution is known over a range $0 \leq t \leq t_k$ as shown in Fig. 1.

3. A numerical procedure is required to estimate the solution u_{k+1} at $t_{k+1} = t_k + \Delta t$ as shown in Fig. 1. The following methods allow the development of such numerical procedure.

A. Euler Method

In Eqs. (49) and (50), du/dt may be replaced by a simple forward difference. Thus,

$$\frac{u_{k+1} - u_k}{\Delta t} = f(u_k, \ t_k) \qquad (51)$$

Solving for the unknown u_{k+1}, gives

$$u_{k+1} = u_k + \Delta t \ f(u_k, \ t_k) \qquad (52)$$

Equation (52) is used to calculate the new value of u_{k+1} from the previous value of u_k. The solution can be started from $t = 0$. Euler method is not recommended due to inherent high errors.

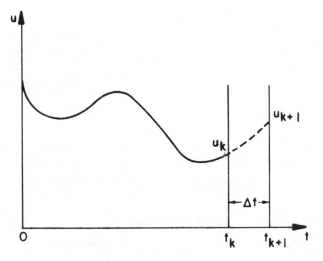

FIGURE 1   Approximation of a function.

B.   Runge-Kutta Methods

Runge-Kutta formulas are widely used in numerical solutions of ordinary differential equations. A second-order Runge-Kutta formula is given by

$$u_{k+1} = u_k + \Delta t \; f(u^*_{k+1/2}, \; t_{k+1/2}) \tag{53}$$

where

$$u^*_{k+1/2} = u_k + \frac{\Delta t}{2} f(u_k, \; t_k) \tag{54}$$

$$t_{k+1/2} = t_k + \frac{\Delta t}{2} \tag{55}$$

The method is self-starting. The procedure is as follows:

1. From Eq. (54), for known $f(u_k, t_k)$, calculate $u^*_{k+1/2}$ (note the similarity with Euler's method).

2. Evaluate $f(u_{k+1/2}, \; t_{k+1/2})$ and substitute in Eq. (53) to calculate u_{k+1}.

A widely used Runge-Kutta method is of fourth order.

$$u_{k+1} = u_k + \Delta t \left[\frac{1}{6} f(u_k, \; t_k) + \frac{1}{3} f(u^*_{k+1/2}, \; t_{k+1/2}) \right.$$
$$\left. + \frac{1}{3} f(u^{**}_{k+1/2}, \; t_{k+1/2}) + \frac{1}{6} f(u^*_{k+1}, \; t_{k+1}) \right] \tag{56}$$

where

$$u^*_{k+1/2} = u_k + \frac{\Delta t}{2} f(u_k, t_k) \tag{57}$$

$$u^{**}_{k+1/2} = u_k + \frac{\Delta t}{2} f(u^*_{k+1/2}, t_{k+1/2}) \tag{58}$$

$$u^*_{k+1} = u_k + \Delta t\, f(u^{**}_{k+1/2}, t_{k+1/2}) \tag{59}$$

The procedure involves calculation of $u^*_{k+1/2}$ from Eq. (57), followed by $u^{**}_{k+1/2}$ from Eq. (58), followed by u^*_{k+1} from Eq. (59). These values are then substituted in Eq. (56) to calculate u_{k+1}.

C. Multistep Formulas and Predictor-Corrector Methods

Multistep formulas are used widely in packaged computer subroutines. The derivation of these formulas is simple and will be presented in the following.
    Consider the initial-value problem given by Eqs. (49) and (50). The following steps are used to obtain the multistep formulas known as "Adams' formulas."

1. Perform a forward Taylor's series expansion about t.

$$u(t + \Delta t) = u(t) + \Delta t\, \frac{du}{dt} + \frac{(\Delta t)^2}{2!} \frac{d^2 u}{dt^2} + \cdots \tag{60}$$

2. Using u_k notation,

$$u_{k+1} = u_k + \left.\frac{du}{dt}\right|_k \Delta t + \left.\frac{d^2 u}{dt^2}\right|_k \frac{(\Delta t)^2}{2!} + \cdots \tag{61}$$

3. The equation can be written as

$$u_{k+1} = u_k + \Delta t \left(\left.\frac{du}{dt}\right|_k + \frac{\Delta t}{2} \left.\frac{d^2 u}{dt^2}\right|_k + \frac{(\Delta t)^2}{3!} \left.\frac{d^3 u}{dt^3}\right|_k + \cdots \right) \tag{62}$$

4. Similarly, using the backward-difference approach, one gets the open Adams' formula of second order.

$$u_{k+1} = u_k + \Delta t \left(\frac{3}{2} \left.\frac{du}{dt}\right|_k - \frac{1}{2} \left.\frac{du}{dt}\right|_{k-1} \right) + O(\Delta t)^3 \tag{63}$$

5. In general the Adams' open formulas can be written as

$$u_{k+1} = u_k + \Delta t \sum_{n=1}^{m} \alpha_{mn} \left.\frac{du}{dt}\right|_{k-n} + O(\Delta t)^{m+2} \tag{64}$$

The values of α_{mn} are given in Table 1.

6. Similarly, the Adams' closed formulas can be written in general form as

$$u_{k+1} = u_k + \Delta t \sum_{n=1}^{m} \alpha^*_{mn} \left.\frac{du}{dt}\right|_{k+1-n} + O(\Delta t)^{m+2} \tag{65}$$

The values of α^*_{mn} are also given in Table 1. Although the calculation time required for closed formulas is more than for open formulas, the actual error of closed formulas is considerably smaller than that of open formulas of the same order.

A predictor-corrector formulation is developed from these formulas as follows. Consider as a "predictor" the third-order open Adams' formula

$$u^{(p)}_{k+1} = u_k + \Delta t \left(\frac{23}{12} \left.\frac{du}{dt}\right|_k - \frac{16}{12} \left.\frac{du}{dt}\right|_{k-1} + \frac{5}{12} \left.\frac{du}{dt}\right|_{k-2} \right) \tag{66}$$

The "corrector" may be chosen as third-order closed Adams' formula

$$u_{k+1} = u_k + \Delta t \left(\frac{5}{12} \left.\frac{du}{dt}\right|^p_{k+1} + \frac{8}{12} \left.\frac{du}{dt}\right|_k - \frac{1}{12} \left.\frac{du}{dt}\right|_{k-1} \right) \tag{67}$$

TABLE 1 Coefficients of α_{mn} and α^*_{mn}

α_{mn}						α^*_{mn}			
m \ n	0	1	2	3		m \ n	0	1	2	3
0	1					0	1			
1	3/2	-1/2				1	1/2	1/2		
2	23/12	-16/12	5/12			2	5/12	8/12	-1/12	
3	55/24	-59/24	37/24	-9/24		3	9/24	19/24	-5/24	1/24

The procedure can be summarized in the following steps:

1. Use a third-order (or higher) Runge-Kutta-type formula to obtain values of u and du/dt for the first two steps of Δt beyond the initial condition.

2. With the values from step 1 use Eq. (66) to estimate next value of $u_{k+1}^{(p)}$.

3. Using the estimated of $u_{k+1}^{(p)}$, use "corrector" Eq. (67) and iterate until the desired degree of convergence is obtained. A modifier may be used to reduce the number of iterations (Hornbeck, 1975).

VI. SOLUTION OF PARTIAL DIFFERENTIAL EQUATIONS

The common classification of linear partial differential equations (PDEs) are the following:

$$\frac{\partial^2 u}{\partial x^2} = \Psi \qquad \text{(parabolic)} \tag{68}$$

$$\frac{\partial^2 u}{\partial x^2} + \frac{\partial^2 u}{\partial y^2} = \Psi \qquad \text{(elliptic)} \tag{69}$$

$$\frac{\partial^2 u}{\partial x^2} - \frac{\partial^2 u}{\partial y^2} = \Psi \qquad \text{(hyperbolic)} \tag{70}$$

where

$$\Psi = \, . (x, \, y, \, u, \, \frac{\partial u}{\partial x}, \, \frac{\partial u}{\partial y}) \tag{71}$$

The commonly occurring PDEs are

$$\frac{\partial u}{\partial t} = \alpha \frac{\partial^2 u}{\partial x^2} \qquad \text{(diffusion equation)} \tag{72}$$

$$\frac{\partial^2 u}{\partial x^2} + \frac{\partial^2 u}{\partial y^2} = \text{constant} \qquad \text{(Poisson's equation)} \tag{73}$$

$$\frac{\partial^2 u}{\partial x^2} + \frac{\partial^2 u}{\partial y^2} = 0 \qquad \text{(Laplace's equation)} \tag{74}$$

The following general procedure using finite-difference technique is used in solving PDEs.

1. Establish a network of grid points throughout region occupied by the independent variables. For example, for two space coordinates x and y, and time t as independent variables, grid spacings are x, y, and t. In this section i, j, and k will be used as subscripts denoting coordinates i Δx, j Δy, and k Δt.

2. Let the exact solution of the PDE be u = u(x, y, t). The approximation, yet to be determined, is $v_{i,j,k}$.

3. Approximate the partial derivatives of original PDE by finite-difference expressions.

4. Step 3 will result in a set of algebraic equations in $v_{i,j,k}$, the value of which can be determined by solving the simultaneous equations.

5. If the grid spacings are sufficiently small, the approximation $v_{i,j,k}$ will be sufficiently close to $u_{i,j,k}$ at any grid point (i, j, k).

From Taylor's series expansion for $u_{i-1,j}$ and $u_{i+1,j}$ about the central value $u_{i,j}$ for the grid shown in Fig. 2,

$$u_{i-1,j,k} = u_{i,j,k} - \Delta x \frac{\partial u}{\partial x} + \frac{(\Delta x)^2}{2!} \frac{\partial^2 u}{\partial x} - \frac{(\Delta x)^3}{3!} \frac{\partial^3 u}{\partial x^3} + \frac{(\Delta x)^4}{4!} \frac{\partial^4 u}{\partial x^4}$$

(75)

$$u_{i+1,j,k} = u_{i,j,k} + \Delta x \frac{\partial u}{\partial x} + \frac{(\Delta x)^2}{2!} \frac{\partial^2 u}{\partial x^2} + \frac{(\Delta x)^3}{3!} \frac{\partial^3 u}{\partial x^3} + \frac{(\Delta x)^4}{4!} \frac{\partial^4 u}{\partial x^4}$$

(76)

By rearranging Eqs. (75) and (76),

$$\frac{\partial u}{\partial x} = \frac{u_{i+1,j,k} - u_{i,j,k}}{\Delta x} + O(\Delta x)$$

(77)

$$\frac{\partial u}{\partial x} = \frac{u_{i,j,k} - u_{i-1,j,k}}{\Delta x} + O(\Delta x)$$

(78)

$$\frac{\partial u}{\partial x} = \frac{u_{i+1,j,k} - u_{i-1,j,k}}{2\Delta x} + O[(\Delta x)^2]$$

(79)

$$\frac{\partial^2 u}{\partial x^2} = \frac{u_{i-1,j,k} - 2u_{i,j,k} + u_{i+1,j,k}}{(\Delta x)^2} + O[(\Delta x)^2]$$

(80)

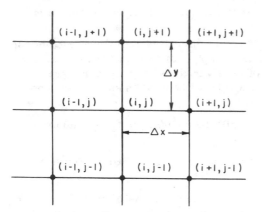

FIGURE 2   Grid arrangement for numerical solutions of partial differential equations.

$$\frac{\partial^2 u}{\partial x \partial y} = \frac{u_{i+1,j+1,k} - u_{i-1,j+1,k} - u_{i+1,j-1,k} + u_{i-1,j-1,k}}{4\Delta x \Delta y} + O\,(\Delta x + \Delta y)^2]$$

(81)

where Eqs. (77) to (79) are forward, backward, and central differences, respectively.

A. Explicit Method

To illustrate the explicit method, the following physical problem will be examined:

$$\alpha\,\frac{\partial^2 u}{\partial x^2} = \frac{\partial u}{\partial t}$$

(82)

with boundary conditions

$$u(a,\ t) = u_a$$

(83)

$$u(b,\ t) = u_b$$

(84)

and initial condition

$$u(x,\ 0) = u_0$$

(85)

1. Establish a grid as shown in Fig. 3.

2. The grid spacing $\Delta x = 1/M$, and $\Delta t = 1/N$, where M and N are arbitrary integers.

3. Using the forward difference Eq. (77) in Eq. (82),

$$\frac{v_{i,k+1} - v_{i,k}}{\alpha \, \Delta t} = \frac{v_{i-1,k} - 2v_{i,k} + v_{i+1,k}}{(\Delta x)^2} \tag{86}$$

4. Define

$$\lambda = \frac{\alpha \, \Delta t}{(\Delta x)^2} \tag{87}$$

$$v_{i,k+1} = \lambda v_{i-1,k} + (1 - 2\lambda)v_{i,k} + \lambda v_{i+1,k} \tag{88}$$

5. Equation (87) is an explicit formulation since if all $v_{i,k}$ are known at any time t_k, then $v_{i,k+1}$ can be calculated explicitly at time t_{k+1}.

FIGURE 3   A finite-difference grid for the example problem. (Adapted from Carnahan, Luther, and Wilkes, 1969.)

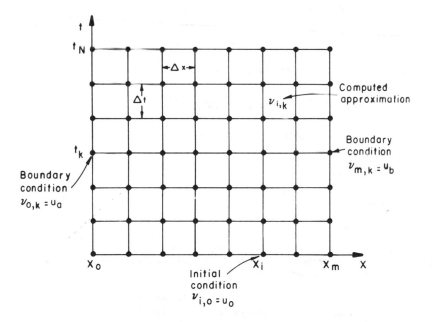

6. The boundary conditions are

$$v_{0,k+1} = u_a \tag{89}$$

$$v_{M,k+1} = u_b \tag{90}$$

and the initial condition is

$$v_{i,0} = u_0 \tag{91}$$

7. A condition necessary for convergence of this solution is $0 < \lambda \leq 1/2$ (Carnahan, Luther, and Wilkes, 1969).

8. Repeated application of Eq. (88) and Eqs. (89) and (90), gives the values of v.

B. Implicit Method

Referring to the preceding example problem discussed to illustrate explicit method, the procedure for implicit method is as follows:

1. Evaluate $\partial^2 u/\partial x^2$ at the next time increment t_{k+1} instead of t_k as used in the explicit form. Thus,

$$\frac{v_{i,k+1} - v_{i,k}}{\alpha \, \Delta t} = \frac{v_{i-1,k+1} - 2v_{i,k+1} + v_{i+1,k+1}}{(\Delta x)^2} \tag{92}$$

2. Equation (92) can be rearranged as

$$-\lambda v_{i-1,k+1} + (1 + 2\lambda)v_{i,k+1} - \lambda v_{i+1,k+1} = v_{i,k} \tag{93}$$

3. The boundary and initial conditions are

$$v_{0,k+1} = u_a \tag{94}$$

$$v_{M,k+1} = u_b \tag{95}$$

$$v_{i,0} = u_0 \tag{96}$$

4. The implicit method converges to the solution of PDE as $\Delta x \rightarrow 0$ and $\Delta t \rightarrow 0$, regardless of the value of λ.

5. At any one time level, Eq. (93) is written once for each $1 < i \leq M - 1$. This procedure yields a system of $M - 1$ simultaneous equations in the $M - 1$ unknowns $v_{i,k+1}$.

6. The set of simultaneous equations can be solved using the gaussian elimination method.

C. Crank-Nicolson Method

The explicit and implicit methods described in this section result in discretization error of order $\Delta t + (\Delta x)^2$. The Crank-Nicolson method decreases the error order on time increment to $(\Delta t)^2$. The following steps are suggested for the example problem described in Sec. VI.A on the explicit method.

1. Write the derivative $\partial u/\partial t$ with respect to halfway point $(i, k + 1/2)$

$$\frac{\partial u}{\partial t} = \frac{u_{i,k+1} - u_{i,k}}{\Delta t} + O[(\Delta t)^2] \qquad (97)$$

2. From Taylor's expansion

$$\frac{\partial^2 u}{\partial x^2} = \beta \, \delta_x^2 \, u_{i,k+1} + (1 - \beta) \, \delta_x^2 \, u_{i,k} \qquad (98)$$

where $0 \leq \beta \leq 1$. In the Crank-Nicolson method, $\beta = 1/2$ and δ_x is the central difference operator.

$$\delta_x u_{i,k} = \frac{u_{i+1/2,k} - u_{i-1/2,k}}{\Delta x} \qquad (99)$$

and

$$\delta_x^2 u_{i,k} = \frac{u_{i-1,k} - 2u_{i,j} + u_{i+1,k}}{(\Delta x)^2} \qquad (100)$$

3. Thus, the governing equation in Crank-Nicolson formulation is

$$\frac{v_{i,k+1} - v_{i,k}}{\alpha \, \Delta t} = \frac{1}{2} \delta_x^2 \, v_{i,k+1} + \frac{1}{2} \delta_x^2 \, v_{i,k} \qquad (101)$$

The above equation can be solved as shown previously for the implicit method.

In addition to the methods discussed above, there are other unconditionally stable explicit procedures such as the DuFort-Frankel method, the Brakat and Clark method, and implicit alternating-direction methods, useful in numerical approximations of partial differential equations.

VII. THE FINITE ELEMENT METHOD

The finite element method, since its first application in aerospace industry
in the 1950s, has been gaining considerable attention in the areas of structural and solid mechanics, fluid mechanics, and heat transfer. The advances
in finite element methods would not have been possible without the development of high-speed digital computers in the last two decades.

A. Fundamental Concept

The fundamental concept of the finite element is that approximation of any
continuous function can be done by a discrete model which is composed of
a set of piecewise continuous functions defined over a finite number of sub-
domains (or elements). The following steps are used to construct a discrete
model:

1. By using a finite number of nodes, a domain is identified.

2. The values of a continuous function are specified at the nodes.

3. The domain is divided into finite number of elements. The elements
   are connected by common nodes.

4. Then the continuous function is approximated over each element by
   a function which is defined in terms of the values of the continuous
   function at the nodal points. It should be noted that the element
   function is continuous only over the domain of the element.

B. Properties of a Single Element

Unknown parameters, such as temperature, pressure, displacement, etc.,
are approximated over the domain of the element using a mathematical
function. The most common function used is a polynomial. The order of
the polynomial will depend on the number of nodes and the number of unknowns at each node.

Thus, temperature may be approximated by the following expression
derived for a triangular element:

$$T = \alpha_1 + \alpha_2 x + \alpha_3 y \tag{102}$$

The coefficients α_1, α_2, and α_3 must be determined in order to find
the physical value for T. In the following, a one-dimensional simplex element is discussed.

As shown in Fig. 4, a one-dimensional simplex element is a line seg-
ment and two nodes. Let the nodal values be scalar quantities Φ_i and Φ_j.
The origin of the coordinate system is assumed to be outside the element.
The polynomial function for the scalar quantity Φ is

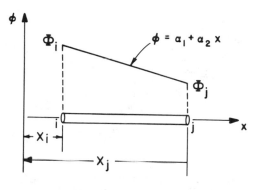

FIGURE 4 A simplex one-
dimensional element. (Adapted
from Segerlind, 1976.)

$$\phi = \alpha_1 + \alpha_2 x \tag{103}$$

Since

$$\phi = \Phi_i \qquad \text{at } x = X_i \tag{104}$$

$$\phi = \Phi_j \qquad \text{at } x = X_j \tag{105}$$

$$\alpha_1 = \frac{\Phi_i X_j - \Phi_j X_i}{L} \tag{106}$$

$$\alpha_2 = \frac{\Phi_j - \Phi_i}{L} \tag{107}$$

where

$$L = X_j - X_i$$

Thus

$$\phi = \left(\frac{X_j - x}{L}\right)\Phi_i + \left(\frac{x - X_i}{L}\right)\Phi_j \tag{108}$$

The linear functions enclosed in the parentheses in Eq. (108) are called
"shape functions" and are defined as follows:

$$N_i = \frac{X_j - x}{L} \tag{109}$$

$$N_j = \frac{x - X_i}{L} \tag{110}$$

In matrix notation,

$$\phi = [N]\{\Phi\} \tag{111}$$

where

$$[N] = [N_i \ N_j] \tag{112}$$

$$\{\Phi\} = \begin{Bmatrix} \Phi_i \\ \Phi_j \end{Bmatrix} \tag{113}$$

For a two-dimensional simplex element such as a triangular element, the equation comparable to Eq. (108) can be obtained as follows (Segerlind, 1976):

$$\phi = N_i\Phi_i + N_j\Phi_j + N_k\Phi_k \tag{114}$$

where

$$N_i = \frac{1}{2A} [a_i + b_i x + c_i y], \qquad a_i = X_j Y_k - X_k Y_j$$

$$b_i = Y_j - Y_k$$

$$c_i = X_k - X_j$$

$$N_j = \frac{1}{2A} [a_j + b_j x + c_j y] \qquad a_j = X_k Y_i - Y_k X_i$$

$$b_j = Y_k - Y_i$$

$$c_j = X_i - X_k$$

$$N_k = \frac{1}{2A} [a_k + b_k x + c_k y] \qquad a_k = X_i Y_j - X_j Y_i$$

$$b_k = Y_i - Y_j$$

$$c_k = X_j - X_i$$

$$2A = \begin{vmatrix} 1 & X_i & Y_i \\ 1 & X_j & Y_j \\ 1 & X_k & Y_k \end{vmatrix}$$

The shape functions have the following properties: N_i has a value of one at node i and zero at node j and k. Similarly, N_j has a value of one at node j and zero at node i and k. The following characteristics are important to note.

1. The scalar quantity ϕ varies linearly between the two nodes.

2. Since the gradient of ϕ is constant over an element, many small elements have to be used to approximate whenever high gradients are involved.

3. There is continuity between ϕ along the boundary between elements.

C. Interpolation Equations

The interpolation polynomials discussed above are arbitrary since the numerical values of the coordinates were not specified. The fact that element size and orientation can be changed to match the problem needs is a key advantage of the finite element method as irregular boundaries can be easily analyzed with the same general computer subroutines.

The polynomial functions developed for individual elements can be transformed into the entire region by the use of global coordinates and global degrees of freedom (unknown quantities at each node).

A general form for Eq. (111) is

$$\phi^{(e)} = [N]\{\Phi\} = \left[N_i^{(e)} \, N_j^{(e)} \, N_k^{(e)} \, \cdots \, N_n^{(e)} \right] \begin{Bmatrix} \Phi_i \\ \Phi_j \\ \cdot \\ \cdot \\ \cdot \\ \Phi_n \end{Bmatrix} \qquad (115)$$

for an element with n nodes. The superscript e refers to the element number.

Consider a simple configuration consisting of three triangles as shown in Fig. 5. The element indices i, j, k can be related to the node numbers as soon as the location of the starting node in each element has been identified. Node i in Fig. 5 is identified by an asterisk. Thus, using counterclockwise convention,

Element 1: i = 1, j = 2, k = 3.

Element 2: i = 2, j = 4, k = 3.

Element 3: i = 4, j = 5, k = 3.

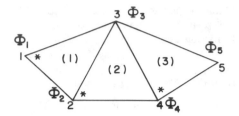

FIGURE 5   A three-element domain.

The interpolation functions are

$$\phi^{(1)} = N_1^{(1)}\Phi_1 + N_2^{(1)}\Phi_2 + N_3^{(1)}\Phi_3 \tag{116}$$

$$\phi^{(2)} = N_2^{(2)}\Phi_2 + N_4^{(2)}\Phi_4 + N_3^{(2)}\Phi_3 \tag{117}$$

$$\phi^{(3)} = N_4^{(3)}\Phi_4 + N_5^{(3)}\Phi_5 + N_3^{(3)}\Phi_3 \tag{118}$$

The shape functions can be evaluated using the expressions given in Eq. (114).

Equations (116), (117), and (118) serve the objective of embedding the individual element equations into the complete domain. These equations in matrix form are programmed into a digital computer.

The above discussion has shown how to approximate a continuous function over a single element and has given a procedure to combine the results for several elements to obtain a set of piecewise continuous equations that approximate the continuous function over the body. In order to obtain numerical values for the nodal quantities that define these piecewise continuous functions, the minimization of a functional is conducted. Calculus of variations is used to obtain the functional that, when minimized, satisfies the differential equation under consideration. More discussion on this aspect of finite element procedures can be found in Segerlind (1976), Zienkiewicz (1971), and Hubner (1975).

In summary, the various numerical methods important in analyzing physical problems have been presented in this chapter. It should be emphasized that although the analysis of errors associated with these methods has not been discussed, it is equally important. It is left to the reader to review the various reference books listed for more discussion on such topics. The book by Carnahan, Luther, and Wilkes (1969) includes several computer programs, written in FORTRAN, that the readers will find useful. It is anticipated that these numerical techniques will continue to play a major role in the design and analysis of food processing systems.

REFERENCES

Carnahan, B. C., Luther, H. A., and Wilkes, J. O. (1969). Applied
    Numerical Methods. Wiley, New York.
Henrici, P. H. (1964). Elements of Numerical Analysis. Wiley, New
    York.
Hildebrand, F. B. (1956). Introduction to Numerical Analysis.
    McGraw-Hill, New York.
Hornbeck, R. W. (1975). Numerical Methods. Quantum, New York.
Hubner, K. H. (1975). The Finite Element Method for Engineers. Wiley,
    New York.
Segerlind, L. J. (1976). Applied Finite Element Analysis. Wiley, New
    York.
Zienkiewicz, O. C. (1971). The Finite Element Method in Engineering
    Science. McGraw-Hill, London.

3
Development and Analysis of Empirical Mathematical Kinetic Models Pertinent to Food Processing and Storage

WILLIAM M. RAND   Massachusetts Institute of Technology, Cambridge, Massachusetts

Mathematical modeling is an important approach to a diversity of problems encountered in the food industry. The procedure draws upon techniques developed by mathematicians, statisticians, and computer scientists, yet its successful application to real problems requires first and foremost a familiarity with those problems. Thus it is important that those most aware of the problems of the food industry be at least acquainted with the

procedures of mathematical modeling. There are a number of textbooks
and journal articles available which detail the various aspects of the modeling, and there also are a number of computer programs which will perform
the often laborious calculations involved. This chapter is intended as a
general introduction to mathematical modeling and a practical guide to the
techniques and resources available.

I. INTRODUCTION

How changes in one set of variables can or do effect changes in another set
of variables is a familiar problem; it is encountered by such different groups
as those doing research on product development to those investigating ways
of increasing profits. In these situations there are, in general, two approaches: those which emphasize the empirical, or "trial and error," and
those which emphasize the theoretical, focusing on the analysis of a formalization of the phenomena of interest.

Neither of these approaches works well in isolation. Experimentation
guided only by intuition is usually not efficient in its use of resources and
often not even effective. It can be too expensive and too time consuming
unless there is some sort of framework to guide the search and to organize
and specify the patterns which emerge. This, of course, can be provided
by a formalized model. Modeling, by itself, most frequently suffers from
the fault of simplifying or modifying the situation so that it no longer represents the problem of interest. Any modeling procedure needs to be tied to
reality through actual experience: it needs the results of experiment and
observation to guide the form of the model and to insure that the model does
mimic the situation.

Mathematical modeling, in its broad sense, includes both the empirical and theoretical aspects of solving those problems which can be viewed
as investigations of how specific variables are related. The term includes
using the investigator's experience to suggest both an initial mathematical
formalization of the process of interest and what experiments would provide insight into the process and measure how well it can be mimicked by
a model. The modeling procedure is an iterative procedure, going back
and forth between the real world of experimentation and the theoretical
world of formalization, each step giving direction to the next, until a model
is achieved which represents the phenomena to a desired degree of precision.

Mathematical modeling rests on the assumption that a process can be
described in terms of a fixed number of variables. The values of these
variables at any specific time form a vector which describes the system,
a static description. A dynamic description of a process requires that the
relationships between each variable be described. The modeling of any
system requires that we develop mathematical equations or functions which
formalize these relationships.

Since we are interested in modeling real-world phenomena, we must additionally formalize the variability (or lack of precise reproducibility) which is present in our observations. This explains the importance of the statistical techniques which form the bridge between the models (mathematical idealizations of the real world) and the data (measurements of the real world).

The mathematical model, the goal of the mathematical modeling procedure, serves several functions. It summarizes the process we are interested in, and often that alone can improve our understanding. It may suggest underlying processes that were not obvious before, or it may direct our attention toward critical points or regions or relationships. Once we have specified the mathematics and assured ourselves that the equations do indeed represent the phenomena over some region of interest, we can proceed to solve specific problems, such as prediction, projection, interpolation, extrapolation, or optimization, by solving the same problems in the model.

II. MATHEMATICAL MODELING: GENERAL

Formally, mathematical modeling is a procedure leading to the description of a process or phenomena of interest by one or more mathematical equations:

$$\underline{y} = f(\underline{x}; \underline{\theta}, \underline{e})$$

These equations relate certain variables of special interest, the y's (called outcome or dependent variables), to other variables, the x's, which are considered more fundamental, perhaps under our control (called the independent variables). This relationship is formalized by the functional form f of the mathematical equations, and the parameters of these equations, the θ's. In addition, we formalize the uncertainty with which we can observe the system by the e's, the errors, irrespective of whether we consider these inherent error or measurement error. We will discuss systems which have a single outcome variable (at least at a time) and have additive error:

$$y = f(\underline{x}; \underline{\theta}) + e$$

The process of modeling involves iterating between two tasks: choosing a model and gathering data. These tasks are linked by the fitting of the model to the data. This fitting has two purposes: the estimation of the parameters of the model in order to find the specific model most consistent with the data, and the estimation of how well the model fits the data in order to see if that model really is consistent with the data which are assumed to represent the phenomena of interest. The inherent circularity here will be discussed in Sec. VI.

Thus, given a process or phenomenon of interest, the modeling procedure involves the following steps:

1. Selecting the form of the model (such as linear, exponential, sigmoid) which might represent the process (Sec. III).

2. Collecting data from the process, over the range of interest, and at a density comparable to the precision required (Sec. IV).

Often these two steps are interchanged.

3. Fitting the model to the data (Sec. V) to determine

   a. Estimates of the parameters

   b. Estimates of how well the model fits

4. Deciding whether the fitted model describes the situation of interest adequately for the purpose at hand. If it does, one is finished with the mathematical modeling phase of the problem and moves into the interpretation phase. If the fitted model is judged inadequate, one returns to step 1, and a new model is investigated, usually a modification of the initial model; or one returns to step 2 where new data are gathered; or one does both, and the fitting procedure is repeated (Sec. VI).

This chapter closes (Sec. VIII) with a numeric example of the procedures described.

Good general references to the whole process of modeling are Draper and Smith (1981), Box, Hunter, and Hunter (1978), and Himmelblau (1970). These authors provide detailed exposition, numerous examples, and extensive references. See Dempster (1971) for a good survey of the practical aspects of analyzing data.

III. MATHEMATICAL MODELS

At the heart of the modeling procedure is the choice of model, and several factors guide this.

1. Theoretical considerations: The underlying chemical/physical/biological processes may suggest certain forms of models. Thus, constant rates in chemical reactions suggest linearity, while rates proportional to amounts suggest exponentials. Growth processes often start slowly, proceed rapidly until some limit is approached, and then slow again; this suggests a sigmoid model. Kittrell (1970) presents a good survey of various models and their chemical motivations.

2. Empirical considerations: Examination of the preliminary data frequently gives a first insight into the form of the model. Thus, the data may look linear or appear to approach an asymptote. Draper and Smith

(1981) and Daniel and Wood (1980) are both good references for preliminary inspection of data.

3. Practical considerations: Mathematical functions have different characteristics; certain equations are much easier to work with than others. For example, polynomials are especially nice, and for some applications they (or their ratios) can be used to approximate quite complex behavior. Table 1 shows some useful mathematical models.

This idea of mathematical simplicity so pervades the mathematical modeling that models are first classified on the basis of how easy they are to work with; they are either linear or nonlinear. "Linear" means linear in the parameters, for example,

$$y = A + Bx_1 + C \exp(x_1) + D \frac{x_1}{x_2}$$

where we can redefine $x_3 = \exp(x_1)$, $x_4 = x_1/x_2$, and we have $y = A + Bx_1 + Cx_3 + Dx_4$.

Nonlinear equations include a variety of different types, ranging from exponential, for example,

$$y = A - B \exp(-Ct)$$

to the periodic, for example,

$$y = A \sin Bt + C \cos Dt$$

One especially interesting subset of nonlinear models consists of those which can be transformed into linear models. For example, if the dynamics of the system suggest that the relation

$$y = A \exp(-Bt)$$

describes the system, then we might be tempted to take the logarithm of both sides to get

$$\log y = \log A - Bt$$

which is a linear equation with $A' = \log A$ and $y' = \log y$:

$$y' = A' - Bt$$

This procedure does linearize the deterministic equation; however, if error is considered, this can lead to difficulties. If the error is multiplicative, then there is no problem; after transformation, it becomes additive and we can proceed. However, if the error was originally additive (as it often is), then taking logarithms does not produce a linear equation.

TABLE 1 Examples of Mathematical Models

Form	Description
$y = A + Bx$	Linear
$y = A + Bx + Cx^2 + \cdots + Dx^p$	Polynomial
$y = Ax^B$	Power
$y = \dfrac{x}{Ax - B}$	Hyperbola: asymptotes at $y = b/a$ and $x = 1/A$
$y = A \exp(-Bx)$	Exponential: descends to an asymptote at $y = 0$
$y = A[1 - \exp(-Bx)]$	Exponential: rises from 0 to an asymptote at $y = A$
$y = \dfrac{A}{A - B}[\exp(-Bx)]$	Double exponential: rises from 0 to a maximum and then returns asymptotically to $y = 0$
$y = \dfrac{A}{1 + \exp(B - Cx)}$	Logistic: sigmoid between asymptotes at $y = 0$ and $y = A$

Other models which have been found useful in modeling systems with a single independent variable are piecewise polynomials. These models, called splines, consist of a number of polynomials joined together at "knots." It is usually required that at the knots the splines themselves be continuous and that some of their derivatives also be continuous. Thus with splines, some quite complex phenomena can be represented by quite simple functions over much of their range. For some purposes, especially interpolation, splines are quite adequate; their main deficiency is that they fail to give any insight into the underlying process. References to this literature can be found in Draper and Smith (1981).

Another special type of inherently nonlinear function is that which is periodic, that is, which presents the same pattern repeatedly over some fixed interval. Approaches for dealing with these types of models are those of time series analysis [see Box and Jenkins (1970)].

In addition to the shape of the initial model, an important consideration is that of the size of the region over which it is desired to model the phenomena. In general, the smaller the region, the better. If discontinuities, singularities, and even boundaries can be avoided, simpler models can usually be derived. It is often impossible to avoid these portions of the domain of the phenomena, since often it is just these neighborhoods that are of interest. However, the more complex the situation becomes, the more complex the mathematical models must be, and thus more data needed to gain any measure of confidence that the phenomena are adequately approximated by the model.

Thus the modeling procedure starts with the simplest mathematical equations that both satisfy the theoretical understanding of the process (or at least do it no violence) and fit the preliminary data. In addition, we should restrict ourselves to exploring the smallest possible region.

IV. DATA GATHERING

The field of experimental design considers the general question of what data to gather. Good introductions to this field are found in Box, Hunter, and Hunter (1978), Davies (1967), and Cochran and Cox (1957). Initially one gathers data to gain a feel for the pattern of the phenomena of interest in order to choose an initial mathematical approximation of the process. Care must be taken that the data cover the full range of interest for each of the independent variables and that enough data are gathered so that the investigator feels confident that the data do represent the phenomena of interest—that nothing significant is missed. Moreover, one must be careful that practical correlations between independent variables do not restrict the data collected. For example, in a study of the effects of heat and humidity on a process, special care must be taken to "uncouple" these two variables so that the full range of each is explored.

Once fully involved with the modeling procedure, the data serve two purposes: estimation of model parameters and validation of the correctness of the model. Each of these purposes places somewhat different requirements on the sorts of data to be collected, requirements which vary as the model does. General considerations for experimental design are:

1. The data should cover the region of interest and only that region. Given no other constraints, the data should cover the region of interest evenly. There should be strong theoretical reasons that the process is "smooth" between the data points. (A point to be remembered is that it is the data that are analyzed and modeled, not the phenomena.)

2. Enough distinct data points would be gathered so that the model of the system is overdetermined. Thus at the very least, three points are needed to fit a straight line, and at least four are needed for a plane. This permits the calculation of how well the model fits the data (the standard error of the fit) and of the estimated parameters.

3. There should be replicates of the data, preferably at several different places in the experiment space. From these replicates a measure of

the error inherent in the system can be estimated. The error from 2 above, representing how well the model fits the data, can be compared to the error in the data itself from this step to see if the model is adequate or sufficient to describe the data.

There are schemes for optimal design; for example, see Chernoff (1979). Most seek to minimize the number of data points gathered, and there is a definite tradeoff between the number of data points taken and the number of assumptions made. Thus if a straight line is known to fit a phenomena, and the error is negligible, only two points are needed at the edges of the range of the independent variable. If there is any question of the linearity of the phenomena, then a third, intermediate point can be chosen. The presence of error suggests that that midpoint be replicated; question as to the additivity of error suggests that the endpoints be replicated as well. Thus as each assumption is questioned, data must be added to the design to test that assumption. For more complex models the analysis becomes more complex, until one is often left with only the three general suggestions given above.

V. MODEL/DATA FITTING

Given data and a model, the investigator is faced with two tasks:

1. Determine which specific model of the family of models under consideration best fits the data. Here we are asking for the best estimates of the parameters of the model, and in addition, we need to know how sure we can be of these estimates. These procedures for this step are called, in general, "regression" [see, for example, Seber (1977)].

2. Determine whether one of the family of models under consideration does indeed fit the data. More formally, this can be stated: Could the data have come from a phenomena which can be described by the model? There is both a necessity and sufficiency aspect to this. First, it is appropriate to ask if a model of this complexity is needed for the data at hand. This involves testing the various individual parameters of the model and their combinations to see if they could be zero. Second, the question is whether the model under consideration, the best one of a certain family, is sufficiently complex to describe the phenomena. The general approach used is that of comparing the error of fit to the error inherent in the system as estimated from replication, and the general procedure is that of analysis of variance (ANOVA). The classic reference is Scheffe (1959); see also Winer (1971). A good general reference to the statistics is Afifi and Azen (1979).

A. Parameter Estimation

Given a set of data $\{(x_1, x_2, \ldots, x_p, y)\}$ and a function $y = f(\underline{x}, \underline{\theta}, e)$, we need to find the "best" estimates of the θ's and some indication of how good these estimates are. The problem is the definition of "best." While there are many definitions, ranging from the very theoretical to the very practical, the most useful is the "least-squares" approach. Here the best estimate of the coefficients are defined as those which minimize the function:

$$SS = \Sigma_i \, [y_i - f(\underline{x}_i, \, \underline{\theta})]^2 \tag{1}$$

This definition gives estimates with very nice mathematical properties when the errors are independent of the values of the x's and when the errors are normally distributed. In fact, when these restrictions can be assumed, the least-squares estimates are the same as those produced by the more mathematically acceptable "maximum likelihood," and thus for most modeling purposes the least-squares estimates are sufficient. The main concern of the investigator is that the size of the error is independent of the values of $\underline{x}$. Formally we would write that the error at x_i has variance independent of i. The check of this assumption is the motivation for replication of several distinct experimental situations. Discussion of this problem is found in Seber (1977).

In the application of the least squares method the distinction between linear and nonlinear models becomes important. The distinction between these two classes of models is that linear models have closed solutions for their least-squares estimators, while nonlinear models require iterative techniques for estimation, and thus the situation is more complicated.

B. Linear Models

Minimizing Eq. (1) leads to solving a system of linear equations in the case of linear models. This usually gives estimates of the parameters as linear combinations of the observations, and the calculations can be carried out by any regression program. In addition, such a program will estimate the overall error of fitting, the standard error of the fit, by $SS/(n - p - 1)$, where SS is from Eq. (1) and we have n data points and p variables. This estimates how well the best-fitting model fits the data in absolute terms. In addition, we can calculate a measure of relative fit. This is called the coefficient of determination, R^2; is calculated as the ratio of the variability of y as predicted by x, denoted as y, to that of $\hat{y}$ alone:

$$R^2 = \frac{\Sigma \, (\hat{y} - \bar{y})^2}{\Sigma \, (y - \bar{y})^2}$$

This is the square of the multiple correlation coefficient. See the example in Sec. VIII.

In addition to these two measures of overall fit, it is important to examine the fit between the data and the model locally throughout the region

of interest, by examining the residuals (y - $\hat{y}$, the difference between the
observed and the predicted value of y) plotted against each x in turn and
against the predicted y. This check of residual patterns is for major defi-
ciencies in the model. In addition, one must always watch for "bad" data.
Techniques exist which explore the data to identify points which have undue
influence in the determination of the parameter estimates. One must carefully examine these points, or areas of the sample space, to decide whether
the data are in error or the model is deficient. See Belsey, Kuh, and
Welsch (1980) for a good survey of these techniques.

C. Nonlinear Models

For nonlinear models analogous measures exist; however, the situation is
much more complex. Whereas estimating the parameters in a linear model
might be compared to finding the top of a single smooth hill in the middle
of a flat plane in broad daylight, parameter estimation of nonlinear equations might be likened to finding the highest point of a mountain range in a
snow storm. Searching a complex, perhaps ragged surface for its highest
point requires that we proceed in a stepwise fashion. The problems are
(1) where to start, (2) how to determine what direction to go, and even
(3) how to know when to stop. Afifi and Azen (1979) present a brief introduction to the use of nonlinear regression, Draper and Smith (1981) go into
it in more detail, and Bard (1974) discusses the situation in full detail.
    Having decided these problems for any particular situation, computer
programs exist which will produce estimates and asymptotic errors; for
example, see Dixon (1979) or Helwig and Council (1979). Then quality-of-
fit procedures can be followed which are the same as those used for linear
models.

D. Weighted Least Squares

One modification of the least-squares method which is important is that of
weighted least squares. This is a procedure dealing with situations in which
some of the data points or regions of the experimental space are better
known than others. Usual least squares will not take this into account, assuming that the variability is the same everywhere. The modification of
the procedures is straightforward; however, it often needs to be iteratively
applied. Draper and Smith (1981) discusses this technique.

VI. MATHEMATICAL MODELING: SPECIFIC

A. Model Building

Given these tools and concepts, the actual procedure of mathematically
modeling as a process is fairly straightforward. One starts by fitting a

first approximation of the model to some initially gathered data and looking to see how and where it fits or does not fit. The best model or models are decided upon given that data set, and then new data are gathered, to reduce the uncertainty of the estimates to an acceptable level, to extend the region of the model, or to discriminate between competing models [for this particular important point, see Hill (1978)].

For a given set of data, modifications to the model take the form of adding variables or modifying the function. For each case the models are fitted, and the decisions are based on the various statistics mentioned above. There are formalized stepwise regression programs which do this automatically for the investigator. These are quite useful, although the resultant equations need to be carefully examined by the investigator to see if they are reasonable. See Draper and Smith (1981) and Hocking (1976).

One point which must be stressed is that the inherent variability of the system can never be entirely eliminated. One can never identify a particular model as being the only possible one, and thus the investigator must always balance the cost and time involved in gathering data with the return from knowing the answers to another decimal place. Similarly, reality and simplicity must be balanced in the formulation of the model itself.

Ideally, once the best model has been found for a specific set of data, one should gather another set of data to test that fit. This is because we have calculated the model that best fits the data, not the phenomena. While the standard errors of the estimates of the parameters and of the model itself do give us some feel for the fit to the phenomena, still the fit may well be in some part determined by that specific set of data. Ideally then, on the basis of an initial set or data, a model should be hypothesized, and a new set of data should be gathered to test this hypothesis. Often this is not practical, and other approaches are required. These include jackknifing (sequentially refitting the data leaving out a single different data point each time), bootstrapping (an extension of jackknifing), and cross-validation (using part of the data to fit and frame the hypothesis and the rest to test that hypothesis) [see the paper by Snee (1977) for some discussion of this topic]. These considerations are especially important in the case of nonlinear models where estimation is often based on linear approximations.

B. Computer Assistance

No longer is modeling done with paper and pencil. Even the age of physically taking one's data to a large central computer facility is quickly passing. But while many calculations can be done by the investigator using a hand calculator or a desk-top minicomputer, some of the calculations necessary for modeling can only be carried out using the resources of a computer center. This is in part because of the complexity of the computer programs; further, since modeling is an iterative procedure, the data need be retained within the machine for examination of residuals and for recalculations [see Chambers (1977) and Nash (1979)].

The investigator must thus have access to a large computer (at the present time). Given this, the question is what to do about the programs. There are three options: (1) write one's own programs, (2) use already written subroutines, writing the drivers for them, and (3) use package programs. The complexity of the procedures and the costs of inefficient programs effectively rule out the first option, and one is left with choosing between using subroutines from libraries such as IMSL (International Mathematical and Statistical Libraries, Inc., Houston, Texas) or GLIM (Numerical Algorithms Group, Oxford, England) and the use of program packages such as BMD (Dixon and Brown, 1979), SAS (Helwig and Council, 1979) or MINITAB (Ryan, Joiner, and Ryan, 1981). The packaged programs are easy to use by investigators who have neither the time nor inclination to learn how to program. One essentially has only to provide the input and set the various options, and the program does the rest. The disadvantages of these packaged programs are that they are often inefficient since they must allow for different users wanting different options. Further, given the time it takes to put together a package, the coding does not include recent advances in either statistics or programming. Subroutines have the disadvantage that one must do some coding to use them, and this often necessitates a significant programming commitment. The subroutines however often embody current algorithms and thinking and can be tailored to specific applications to take advantage of special features of one's specific problems. For further details see Chapter 1.

VII. SUMMARY

Modeling is thus the mathematical formalization of a process in order to gain more information about that process. Unfortunately there is not a nice closed algorithm which says explicitly how to proceed; there is a fair amount of art in the procedure. Successful modeling has two resource requirements which parallel the two levels of iteration which make up the modeling.

First, modeling requires the resources for applying an array of statistical techniques to a data set. This is directed toward finding the model which best fits a given set of data, and often necessitates examining various relationships between the data and potential models before a best fit can be chosen.

Second, experiments can be run to either explore the process space or confirm the best model. The fitting of a model to data does just that; it does not fit to the phenomena.

Once a satisfactory model has been derived, the investigator can relax momentarily. As the model is used to answer questions, many more will arise. More accurate, more comprehensive models will be called for as people try to expand the situation originally modeled. Like all science and engineering, often the most visible result of solving one problem is the discovering of several more.

VIII. EXAMPLE

Consider a particular process about which we know little more than that there is a single outcome variable, y, and four independent variables which are probably relevant, x_1, x_2, x_3, and x_4. We have no insight into the mechanics of the situation and, in fact, hope that the model can provide this. As a start we gather a limited amount of data, 20 points, randomly over the region of interest. (We choose four sets of 20 uniformly random numbers, assign each set to a variable, and scale them to fit the range of interest.) The outcome variable was then measured at each combination of x's. These data are shown in Table 2.

We start with the consideration of the simplest model possible, a linear model:

$$y = \beta_0 + \beta_1 x_1 + \beta_2 x_2 + \beta_3 x_3 + \beta_4 x_4 + e$$

where the error is normally distributed.

In order to gain some feel as to what to expect, we first examine the simple linear correlations between y and

x_1	x_2	x_3	x_4
0.80	-0.02	0.33	0.44

[With our sample size of 20 a correlation coefficient must exceed 0.44 to be significantly different (at 0.05 level) from zero.] These correlations suggest that y is linearly related to x_1, not linearly related to x_2, and perhaps linearly related to x_3 and x_4.

A useful interpretation of the correlation coefficient is that it measures how well an independent variable "explains" the variability of a dependent variable in that its square is the coefficient of determination. This interpretation follows from considering that if we have no information about any of the x's, a natural prediction of y would be its average value, here 530.3. Moreover, our uncertainty of that estimate could be measured by the standard error of y, $s_y = 95.6$. Were we given a value of a particular x, say x_1, we could improve our estimate of y by basing it on the simple linear regression of y on x_1, here $166.7 + 5.0 x_1$. The standard error of this fit is the standard error of y given x_1, here $s_{y|x_1} = 58.8$. Thus we can say that knowledge of x_1 permits us to reduce our uncertainty of the situation from $s_y = 95.6$ to $s_{y|x_1} = 58.8$. This information can be

TABLE 2 Data

No.	x_1	x_2	x_3	x_4	y
1	53.7	1.5	6.52	4.8	520.1
2	54.2	0.2	9.56	4.7	546.2
3	54.6	1.4	4.88	3.5	422.2
4	55.9	0.5	4.96	1.5	402.3
5	57.8	0.4	0.77	3.5	421.8
6	58.0	1.8	9.59	2.0	463.5
7	61.0	1.1	8.58	3.2	480.9
8	63.7	0.1	1.70	4.0	474.6
9	64.0	1.4	6.93	1.2	458.1
10	69.4	0.5	0.06	2.6	460.1
11	72.8	0.7	9.79	2.5	544.2
12	75.9	0.8	5.80	1.9	510.9
13	76.4	1.2	8.33	0.8	532.8
14	77.0	1.1	4.00	3.7	541.7
15	81.3	1.9	5.36	4.9	657.8
16	82.0	0.3	0.08	0.2	511.3
17	93.0	0.0	0.10	2.6	578.3
18	96.1	1.0	6.65	2.5	622.9
19	96.2	0.6	9.17	5.0	786.2
20	97.4	0.5	7.85	3.7	670.0

summarized by the square of the correlation coefficient (the coefficient of determination, or multiple correlation coefficient), which can be calculated as

$$R^2 = \frac{(n-1)sy^2 - (n-2)sy^2 1x_1}{(n-1)sy^2}$$

This easily generalizes to multiple independent variables.

The information on how each variable relates independently and linearly to y can thus be summarized (where we follow our motivation and express R^2 as a percentage).

	r	R^2 (%)	s_{y1x}
x_1	0.80	64.2	58.8
x_2	-0.02	0.1	98.2
x_3	0.33	10.6	92.9
x_4	0.44	18.9	88.4

The fitting of the full model by multiple linear regression produces estimates of the coefficients of the model and estimates of their standard errors (indicated parenthetically).

$$y = 28.2(\pm 36.8) + 5.2(\pm 0.4)x_1 - 0.6(\pm 12.7)x_2$$

$$+ 8.3(\pm 2.1)x_3 + 27.9(\pm 4.5)x_4$$

Moreover, as part of this procedure we also estimate the standard error of the fit, $s_{y1x_1x_2x_3x_4} = 27.2$, and the corresponding $R^2 = 93.6\%$. This confirms what our examination of the simple linear correlations told us, namely, that x_1 is the most useful of the four in linearly predicting y and that x_2 appears useless (its coefficient has a standard error 20 times its value). In addition, we have the information that these variables explain more than 90% of the original variability.

An alternative approach is that of stepwise regression. One stepwise approach first identifies the best single independent variable in terms of reducing the variability of y, at each successive step is added the variable which reduces the variability of y the most. The results of such a program applied to our data are

Step	Variable added	SE of fit	R^2	Change in R^2
0	None	95.6	—	—
1	x_1	58.7	64.3	64.3
2	x_4	39.0	85.1	20.8
3	x_3	26.4	93.6	8.5

TABLE 3   Results of Regression of y on
x_1, x_3, and x_4

No.	y	Pred. y	Residual
1	520.1	495.01	25.09
2	546.2	519.98	26.22
3	422.2	449.84	-27.64
4	402.3	401.47	.83
5	421.8	432.47	-10.67
6	463.5	464.67	-1.17
7	480.9	505.41	-24.51
8	474.6	484.83	-10.23
9	458.1	451.56	6.54
10	460.1	461.86	-1.76
11	544.2	557.31	-13.11
12	510.9	523.67	-12.77
13	532.8	516.53	16.27
14	541.7	564.72	-23.02
15	657.8	631.84	25.96
16	511.3	460.64	50.66
17	578.3	585.02	-6.72
18	622.9	652.58	-29.68
19	786.2	743.72	42.48
20	670.0	702.77	-32.77

This again shows that x_1 explains 64% of the variability of y. Moreover, the addition of x_4 to x_1 results in an explanation of 85% of the variability (reducing the standard error of the fit to 39.0). At the next step the inclusion of x_3 in the model reduces the standard error of the fit to 26.4 and increases the R^2 to 94%. The program stops at this point, finding that the addition of x_2 does not significantly improve the predictability of y. The resultant equation was

$$y = 27.6(\pm33.5) + 5.2(\pm0.4)x_1 + 8.3(\pm1.8)x_3 + 27.9(\pm4.4)x_4$$

Given these calculations, we next turn to the examination of how well the model fits the individual data points. Table 3 lists the residuals (the observed minus the predicted y values) for each of the experimental points. This shows that the 16th point has a residual almost twice as large as the standard error of fit. While this point is not deviant enough to be statistically excluded as an outlier, it should be checked that a mistake was not made somewhere between reality and the computer.

The plots of the residuals give further information (Fig. 1). Looking at the residuals versus y itself, we see no obvious region of misfitting. The residuals plotted versus x_1, x_2, and x_3 show no indication of any pattern other than that already found—linearity between x_1 and y and x_3 and y, with no relationship between x_2 and y. The plot of the residuals versus x_4, however, show a definite pattern, with the residuals being positive at both low and high values of x_4 and negative at intermediate values. This patterned deviation from linearity suggests that we should add a quadratic term, x_4^2, to the variables available to the stepwise procedure to see if it can improve the fit.

This next stepwise regression produces a new equation

$$y = 105.6(\pm 16.2) + 5.1(\pm 0.2)x_1 + 7.8(\pm 0.7)x_3$$
$$- 34.4(\pm 7.1)x_4 + 11.1(\pm 1.2)x_4^2$$

Step	Variable added	SE of fit	R^2	Change in R^2
0	—	95.6	—	—
1	x_1	58.6	64.2	64.2
2	x_4^2	32.4	89.7	25.5
3	x_3	16.6	97.5	7.8
4	x_4	10.8	99.0	1.5

This model explains most of the original variability, 99%, reducing the uncertainty to a standard error of fit of 10.8. Examining the residuals (Table 4) and the residual plots from this model (Fig. 2) shows again no discernable pattern for y, x_1, or x_2. However, it now appears that there may be a more complex relationship between x_3 and y and that the inclusion of x_3^2 might be useful. Moreover, it looks as if the addition of x_4^2 was an oversimplification of the relationship between the fourth independent variable and y; we should investigate more complex transformations of x_4.

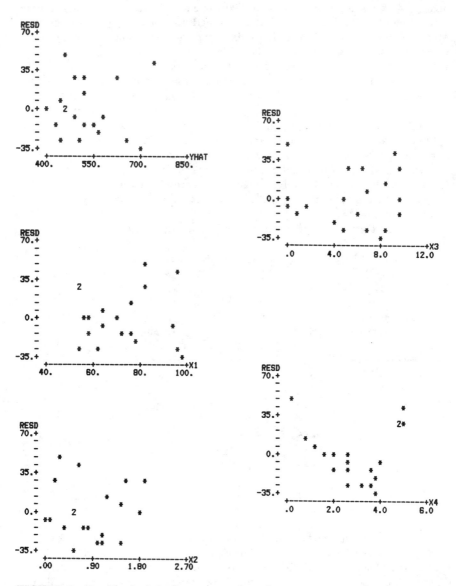

FIGURE 1   Residual plots for regression of y on x_1, x_3, and x_4.

TABLE 4 Results of Regression of y on
x_1, x_3, x_4, and x_4^2

Point no.	Y	Pred. Y	Residual
1	520.1	520.61	-0.51
2	546.2	539.85	6.35
3	422.2	436.88	-14.68
4	402.3	401.53	0.77
5	421.8	420.89	0.91
6	463.5	450.79	12.71
7	480.9	486.30	-5.40
8	474.6	482.67	-8.07
9	458.1	459.36	-1.26
10	460.1	443.98	16.12
11	544.2	535.29	8.91
12	510.9	510.97	-0.07
13	532.8	538.09	-5.29
14	541.7	552.77	-11.07
15	657.8	658.89	-1.09
16	511.3	515.75	-4.45
17	578.3	564.02	14.28
18	622.9	628.87	-5.97
19	786.2	771.93	14.27
20	670.0	686.45	-16.45

Further stepwise regressions should probably also include the interactions among the independent variables (the cross products) to see if they further reduce the mismatch between data and model.

At this stage, before embarking on a major investigation of second-order effects, it is essential that the fitting situation be reviewed. We have found that

1. x_1 is probably linearly related to y.

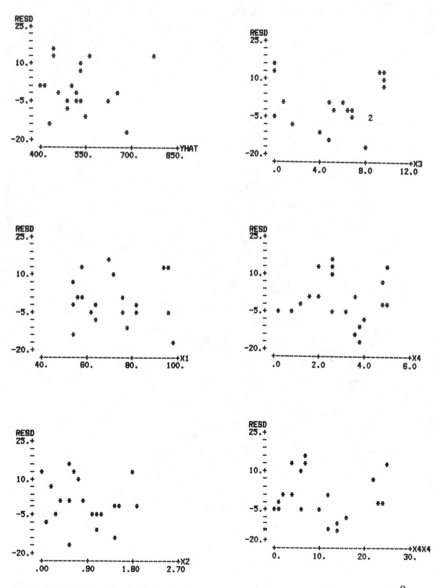

FIGURE 2   Residual plots for regression of y on x_1, x_3, x_4, and x_4^2.

2. x_2 is probably not related to y.

3. x_3 is related to y, and this relationship may be more complex than simple linearity.

4. x_4 is related to y in a more complex manner than quadratic.

Further, the variables x_1, x_3, x_4, and x_4^2, in a linear equation, account for 99% of the variability of y, leaving a standard error of fit of 10.8.

The major deficiency of the original experimental design now becomes obvious. We have explained almost all the variability of y; however, we do not know whether the error left, 10.8, represents inherent error or model misspecification. Before we proceed with further analyses, we need the results of two separate lines of investigation.

First, we need to run some replicate experiments, that is, to observe multiple y's at several different positions in the experimental space. This is necessary to see if the inherent variability (measurement error, etc.) for these data is significantly less than 10.8. If so, our model is definitely not sufficient to mimic the situation, and we need to investigate either whether a more complex function of the variables at hand can produce a smaller standard error of fit, or whether there are independent variables not measured which are important to the process of interest. Additionally, the running of replicates at several different x vectors gives a check on the independence of error, a check on the additivity of the model.

Second, we need to investigate the biological/physical/chemical processes involved to try and formulate theoretical relationships between the variables to give guidance to the search for more complex relationships.

The model building should probably pause at this stage until there is input from these two lines of investigation. With input from them it can proceed, but more cautiously as the situation becomes more complex and the alternatives multiply.

[Note: These data were simulated following the model

$$y = 100 + 5x_1 + 0.7x_3^2 + \exp(x_4) + e$$

where the error was normally distributed with mean equal to 0 and standard deviation equal to 1. They were analyzed using MINITAB on a Hewlett-Packard 3000 (Ryan, Joiner, and Ryan, 1981).]

REFERENCES

Afifi, A. A. and Azen, S. P. (1979). Statistical Analysis: A Computer Oriented Approach. Academic Press, New York.
Bard, Y. (1974). Nonlinear Parameter Estimation. Academic Press, New York.

Belsley, D. A., Kuh, E., and Welsch, R. E. (1980). Regression Diagnostics. Wiley, New York.

Box, G. E. P., Hunter, W. G., and Hunter, J. S. (1978). Statistics for Experimenters. Wiley, New York.

Box, G. E. P. and Jenkins, G. M. (1970). Time Series Analysis Forecasting and Control. Holden-Day, San Francisco.

Chambers, J. M. (1977). Computational Methods for Data Analysis. Wiley, New York.

Chernoff, H. (1979). Sequential Analysis and Optimal Design, SIAM Monograph 8, 2d ed. Society for Industrial and Applied Mathematics, Philadelphia.

Cochran, W. G. and Cox, G. M. (1957). Experimental Designs. Wiley, New York.

Daniel, C. and Wood, F. S. (1980). Fitting Equations to Data. Wiley, New York.

Davies, O. L., ed. (1967). The Design and Analysis of Industrial Experiments. Hafner, New York.

Dempster, A. P. (1971). An overview of multivariate data analysis, J. Multivariate Analysis 1:316-346.

Dixon, W. J. and Brown, M. B., eds. (1979). BMD P-79 Biomedical Computer Programs, P Series. University of California Press, Berkeley.

Draper, N. and Smith, H. (1981). Applied Regression Analysis, 2d ed. Wiley, New York.

Helwig, J. T. and Council, K. A., eds. (1979). SAS User's Guide, 1979 Edition. SAS Institute Inc., Cary, N. C.

Hill, P. D. H. (1978). A review of experimental design procedures for regression model discrimination, Technometrics 20(1):15-21.

Himmelblau, D. M. (1970). Process Analysis by Statistical Methods. Wiley, New York.

Hocking, R. F. (1976). The analysis and selection of variables in linear regression. Biometrics 32(1):1-49.

Kittrell, J. R. (1970). Mathematical modeling of chemical reactions, in Advances in Chemical Engineering, Vol. 8. Academic Press, New York.

Nash, J. C. (1979). Compact Numerical Methods for Computers. Wiley, New York.

Ryan, T. A., Joiner, B. L., and Ryan, B. F. (1981). MINITAB Reference Manual. Duxbury, North Scituate, Mass.

Scheffe, H. (1959). The Analysis of Variance. Wiley, New York.

Seber, G. A. F. (1977). Linear Regression Analysis. New York.

Snee, R. D. (1977). Validation of regression models: Methods and examples, Technometrics 19(4):415-428.

Winer, B. J. (1971). Statistical Principles in Experimental Design. McGraw-Hill, New York.

4

Reaction Kinetics and Accelerated Tests Simulation as a Function of Temperature

THEODORE P. LABUZA   University of Minnesota, St. Paul, Minnesota

JAMES F. KAMMAN   James Ford Bell Technical Center, General Mills, Inc., Minneapolis, Minnesota

I. INTRODUCTION

In order to predict the effect of temperature on the extent of deterioration of nutritive or quality factors during storage of foods, one can either make guesses based on general principles or apply kinetic analysis to the results of actual data collected under specific conditions. It is the analysis of these results that presents the researcher with many problems; however, the introduction of the computer has diminished this considerably. This chapter, rather than giving many different computer solutions for shelf-life deterioration, will detail some of the problems that need to be or can be solved, leaving it up to the computer programmers to write the necessary formats for solutions.

It should be noted that in testing of shelf life, the basic problem is to correlate measurable objective tests with sensory analyses. Not only is this a real problem, as recently reviewed by Sontag, Pangborn, and Little (1981), but since sensory testing is a costly procedure, some simple objective tests may be the only way in which a change in shelf life can be measured (e.g., peroxide value or Agtron reading). Whether these measurements are a measure of true shelf life may be up to the judgment of the researcher. It should also be noted that the actual food product in processing and distribution may experience a variable temperature/moisture/light/oxygen environment. Although it may be easy to duplicate processing conditions in the laboratory, this is not so for the food package which is in a carton/case/pallet/pallet load undergoing national distribution at any time in the United States. Thus, the scientist is faced with deciding how to select those conditions which will best test food shelf life in the laboratory.

This chapter will consider those conditions for which a simple chemical or sensory test can be used to measure a quality attribute of the food and for which the change in this attribute follows typical chemical kinetics as a function of temperature. Some cautions will be presented for limiting cases.

II. RATE LAW

A. General Rate Law

Chemical kinetics is the study of rates and rates of change of chemical reactions under various conditions. Since the chemical reaction in a food system can be very complex, it is usually easier to examine a reaction from a purely mathematical or semiempirical approach based on chemical laws rather than on a mechanistic approach in which each step must be known. The empirical approach permits the description of a molecular change by the deterioration of a single component A, in which

$$A \longrightarrow \text{end products} \longrightarrow B \tag{1}$$

while disregarding the actual mechanism or number of steps involved. The loss of A or production of end product B causes a decrease in quality. The basic principles of thermodynamics and chemical kinetics can be applied to the many modes of food deterioration, such as nutrient loss, color change, and flavor change. Several texts (Amdur and Hammes, 1966; Daniels and Alberty, 1975) exist which review these mathematical methods, but no examples of food systems are given.

The basic equation for rate of change of A with time is

$$\pm \frac{dA}{d\theta} = kA^n \tag{2}$$

where

> A = concentration or quality factor measured
>
> θ = time
>
> k = rate constant which is dependent upon temperature
>
> n = reaction order
>
> $\pm \dfrac{dA}{d\theta}$ = rate of change of A with time (+ denotes an increase; - denotes a decrease)

It must be noted that the equation implies that all things are held constant, that is, constant light, constant temperature, and constant concentration of other components such as constant moisture content. Much data in the literature on food shelf life imply this when, in fact, it may not be true, for example, dry foods stored in semipermeable packages instead of glass or cans, which can thus pick up water, and foods stored in glass in rooms where the lights go on and off, thereby causing photooxidation. This builds further errors into the analysis and computation.

For quality changes in biological systems such as foods or drugs, the reaction order n has generally been shown to be either 0 or 1, depending on the reaction involved (Labuza and Kreisman, 1978; Labuza, 1982a; Pope, 1980).

B. Zero-order Reactions

When n = 0, the reaction is said to be pseudo-zero-order with respect to A. Equation (3) for a loss in quality will then appear as

$$-\frac{dA}{d\theta} = k \tag{3}$$

This implies that the rate of loss of A is constant with time and independent of the concentration of A. Rearranging and integrating Eq. (3)

between A_0, the concentration of A at $\theta = 0$, and A, the concentration of A at time θ,

$$\int_{A_0}^{A} dA = -k \int_{0}^{\theta} d\theta \tag{4}$$

yields

$$A = A_0 - k\theta \tag{5}$$

or

$$A_0 - A = k\theta \tag{5a}$$

or

$$A_0 - A_s = k\theta_s \tag{5b}$$

where θ_s is the shelf life when the product reaches a quality level of A_s.

For a zero-order reaction, a plot of the amount of A left versus time yields a straight line (Fig. 1) with the slope equal to the rate constant k in units of concentration per unit time.

Typical pseudo-zero-order reactions include nonenzymatic browning (Labuza, Warren, and Warmbier, 1977; Labuza and Saltmarch, 1981), inhibited rancidity (Labuza, Tsuyuki, and Karel, 1969; Labuza, 1971), and many quality losses in frozen foods (Van Arsdel, 1957; Van Arsdel and Guadagni, 1959). It must be realized that a pseudo-zero-order reaction does not imply that the mechanism is a monomolecular breakdown independent of concentration of reacting species. It suggests only that A as a function of time will give a linear plot with a high correlation. The test of this correlation is one of the simple tasks the computer or the desk-top calculator can be used for. One must remember, however, that a good fit (high r^2) does not necessarily mean that the reaction is zero order or that the thing measured (A) is the cause of loss of quality; it only means that some statistical correlation exists.

C. First-order Reactions

While the pseudo-reaction order n can range over any fractional value from 0 to 2 for different reactions, foods that do not follow a pseudo-zero-order reaction generally deteriorate according to a pseudo-first-order ($n = 1$) pattern in which the rate of loss is dependent on the amount left. In this case, the solution is

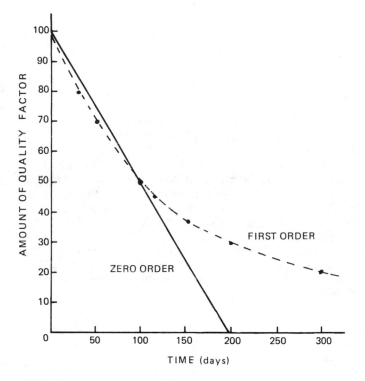

FIGURE 1 Extent of quality loss as a function of time for zero- and first-order reactions with both passing through 50% loss in 100 days.

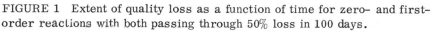

$$\pm \frac{dA}{d\theta} = kA \tag{6}$$

Rearrangement and integration of Eq. (6) between the same limits as Eq. (4) for a loss of A gives

$$\int_{A_0}^{A} \frac{dA}{A} = -k \int_{0}^{\theta} d\theta \tag{7}$$

and

$$\ln \frac{A}{A_0} = -k\theta \tag{8}$$

or

$$\ln A = \ln A_0 - k\theta \tag{9}$$

or

$$A = A_0 e^{-k\theta} \tag{10}$$

When first-order data are plotted as amount left or A versus time, a curved line is derived. However, the data will follow a straight line when plotted on a semilog scale as in Fig. 2. In this case, the slope of the straight line is equal to $-k/2.303$. The units of k on such a plot are $(time)^{-1}$.

As a simple example, if a hypothetical situation is assumed as shown in Figs. 1 and 2, in which the initial concentration of A is 100 (at $\theta = 0$) and after 100 days of storage A = 50, then the difference between the reaction orders is apparent. Table 1 compares this rate constant data when analyzed by either Eq. (5) or Eq. (8).

When plotted as in Fig. 1, the deviation between the orders increases as the reaction proceeds. For example, at $\theta = 200$ days, there is a 100% loss of A with the zero-order reaction, whereas the first-order reaction shows only a 75% loss.

FIGURE 2   First-order semilog plot of quality loss as a function of time with 50% loss in 100 days.

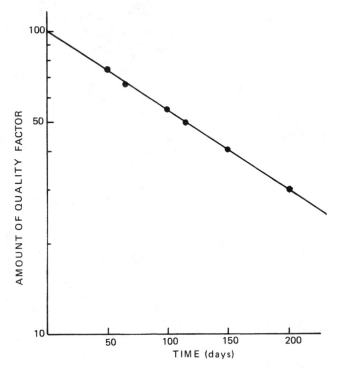

TABLE 1 Calculation of Rate Constants

Assumptions: $\theta = 0$ days, $A_0 = 100$; $\theta = 100$ days, $A = 50$	
Zero order	First order
$k_z = \dfrac{A_0 - A}{\theta}$	$k_f = -\dfrac{\ln A/A_0}{\theta}$
$= \dfrac{100 - 50}{100}$	$= -\dfrac{\ln 50/100}{100}$
$= 0.5$ units/day	$= 0.00693$ day^{-1}

One difficulty with much of the literature for food deterioration is that the reaction order is usually impossible to establish from the data. Many studies only present data determined by endpoint analysis, that is, time to some fractional loss or loss at some single time during storage. As a result, much of this work remains food and process specific. Labuza (1979), for foods, and Pope (1980), for drugs, have noted a further complication in establishing reaction order. Because of the mathematics of Eqs. (5) and (8) and the variability in many analytical techniques, for up to 20 to 30% loss, little or no difference may be noted between zero- and first-order plots of the data. This is further complicated by the correlation coefficient r^2 for log plots because the statistical process weighs toward the lower value more heavily. Optimally the determination of reaction order involves following the course of component change to greater than 50% loss. At this point, sufficient differences should exist between zero- and first-order reactions to enable order determination by the graphical methods just depicted unless some complication in reaction mechanism occurs. For example, Wolf, Thompson, and Reineccius (1977) showed that lysine loss in a soy protein system undergoing heat processing followed pseudo-first-order kinetics to about 40 to 50% loss and then underwent a no-loss period. Thus, some common sense must also be introduced into the order determination in terms of the practical loss percent that would be deemed unacceptable in normal processing or storage. For drugs, usually only a 10% loss in quality is allowed; thus, zero-order kinetics can usually be applied (Pope, 1980; Yang and Roy, 1980).

For first-order reactions, the rate constant k can also be determined from the half-life

$$k = \frac{0.693}{\theta_{1/2}} \tag{11}$$

where $\theta_{1/2}$ = half-life, or time when A = 0.5 A_0. Thus at a given temperature, the rate constant will be the same despite different values of A_0 (assuming everything else is equal). If the amount of degradation at any time
for a given temperature is known, and if the reaction is known to be first
order, a plot such as Fig. 2 can be easily constructed.

Equation (11) is also useful in proving that a reaction is indeed first
order. If the experimental time period for significant measurable loss
(> 50%) is too long, which could be the real situation for many foods under
normal storage conditions ($\theta_{1/2}$ from 6 months to 2 years), different initial
concentrations of A (e.g., spiking the system when possible to different
A_0's) can be utilized to determine if the half-lives are the same. This would
not be the case for a zero-order reaction, nor would it be possible for a
situation in which A_0 is a quality factor such as a hedonic value determined
by a sensory taste panel. Its use is best illustrated in the case of nutrient
fortification, but even here caution must be used if the nutrient causes other
changes at the higher concentration added.

III. GOODNESS OF FIT FOR RATE-LAW MODEL

A. General Least-squares Analysis

Once data are collected for change of quality with time, the first step is to
determine the order, by calculator or computer techniques, if shelf-life
predictions are going to eventually be made. Most universities have pre-
programmed packages for this (such as KINFIT at Michigan State or SAS
at University of Minnesota) and programmable calculators such as the
Hewlett-Packard or the Texas Instrument series which can do this with
either hard-wired or card programs. The basic equation is a regression of
A (the dependent variable y) against time (the independent variable x). Care
must be taken that x is really independent, that is, if the reaction is fairly
fast, half-day or even hourly divisions should be used, not just reports of
the value on day 2, 3, 4, etc. The regression equation is then

$$y = a + bx \tag{12}$$

which, when computed on the calculators or computer by least squares, gives
the following equations for n pairs of values:

$$\text{Intercept} = a = \frac{\Sigma y (\Sigma x^2) - (\Sigma x)(\Sigma xy)}{n(\Sigma x^2) - (\Sigma x)^2} \tag{13}$$

$$\text{Slope} = b = \frac{n \Sigma xy - (\Sigma x)(\Sigma y)}{n(\Sigma x^2) - (\Sigma x)^2} \tag{14}$$

Note that for a first-order reaction, y is ln A and a is ln A_0, while in zero order, y = A and a = A_0.

Several computational tests can then be made to determine the best order as well as the goodness of fit of the data (Freund, 1967). The standard method is to compute the coefficient of determination r^2, where

$$r^2 = \frac{[n \ (\Sigma \ xy) - (\Sigma \ x)(\Sigma \ y)]^2}{[n(\Sigma \ x^2) - (\Sigma \ x)^2][n \ \Sigma \ y^2 - (\Sigma \ y)^2]} \tag{15}$$

In pure chemical studies, values of r^2 are usually greater than 0.95, while in agronomy research, values of 0.6 to 0.7 are acceptable. Analysis of food quality data usually gives r^2 values which fall in between. It is here that the food scientist must use judgment as to whether the model fits the data, especially in terms of end use. Certainly polynomial expressions could be used for curvilinear data, but this precludes use of theoretical thermodynamic models. Knowledge of certain problems that can occur in foods makes personal judgment important. For example, Wolf, Thompson, and Reineccius (1977) showed that in thermal processing of soy protein systems after 50% loss which followed first order, the data show a no-loss period. Thus, it should be obvious that data beyond this time should not be included in the regression. Unfortunately, whether this phenomenon occurs in other foods is not known.

Of other interest in the data analysis for which there is no straightforward answer are the following points:

1. Should multiple data points taken at the same time be pooled as a single average or treated separately? This is especially a problem for the zero-time quality value A_0. Some statisticians, along with the author, feel that each collected value should be treated separately.

2. With the first-order model, the values at extended times are more heavily weighted in the standard least-squares method, which then may result in a poorer fit, especially for early values. One untested but practical way to overcome this is to analyze for more points earlier. However, this may not be acceptable if one is trying to find the end of shelf life based on the change in quality to some value of A which is unacceptable, that is, the time when A = A_s. Another suggestion to overcome the weighting is to do as many initial values as there are data points along the time scale. This would be nearly impossible if sensory testing were being used; however, it also could be used to determine the variability in the method. When few data points are available and the calculated r^2 values are close for both zero and first order, a question arises as to what is the proper order. Since use of the logarithmic first-order model will straighten out data, it invariably may be weighted toward a higher r^2 value unless there is a lot of data beyond the first half-life. Thus, if very little data is available, common sense should be used as to order choice. In many cases, use of zero order

is much easier for subsequent analysis. Finally several weight regression analysis techniques are available (Freund 1967).

B. Confidence of the Reaction-rate Value

An additional question is how much confidence should be put on the values of A_0 and k (the slope of the line)? In most kinetic studies, especially in physical chemistry, confidence limits are rarely presented, and the confidence limit on the initial quality A_0 from Eq. (13) is never presented. If the data are to be used in further computations, especially for quality loss under variable conditions, some measure of the goodness of the fit is needed. One method is to compute the standard error as

$$\text{Standard error} = S_e = \left(\frac{\Sigma y^2 - a\, \Sigma y - b\, \Sigma xy}{n-2} \right)^{1/2} \tag{16}$$

and then use it to determine the confidence limits of the slope and intercept given some degree of statistical significance. The true slope and intercepts are

$$\text{True slope} = \beta = b \pm \frac{t_{\alpha/2} S_e}{\{[n\, \Sigma x^2 - (\Sigma x)^2]/n\}^{1/2}} \tag{17}$$

$$\text{True intercept} = \alpha = a \pm t_{\alpha/2} S_e \left(\frac{1}{n} + \frac{(\Sigma x)^2}{n^2 \Sigma x^2 - n(\Sigma x)^2} \right)^{1/2} \tag{18}$$

where

$$y = \alpha + \beta x \tag{19}$$

A degree of confidence, $t_{\alpha/2}$, is selected from statistical tables such as in Table 2. If, for example, the 95% confidence limits are used, then P = 0.05. For least-squares linear regression, the degrees of freedom F are equal to the number of data points less 2.

Examining the middle column then, it can be seen that after 20 data points (F = 18), more sampling times will have very little effect on the statistical confidence, assuming they will not deviate anymore than the other points (i.e., S_e is constant). Going to 100 samplings gives a ±5% decrease in the magnitude of the confidence interval. This decrease usually may be less than the variation in the methodology. An examination of Table 2 suggests that at least 13 samplings should be made (F = 11), which gives only a ±10% larger confidence interval as compared to 60 to 100 samplings. Here it is obvious that the cost of doing more sampling would not be worth the increased confidence in the data unless the methodology variation is so large that increased sampling significantly reduces the standard error S_e.

TABLE 2 $t_{\alpha/2}$ Values

F degrees of freedom	P = 0.1	P = 0.05	P = 0.01
1	6.31	12.71	63.66
2	2.92	4.30	9.93
3	2.35	3.18	5.84
4	2.13	2.78	4.60
5	2.02	2.57	4.03
6	1.94	2.45	3.71
7	1.90	2.37	3.50
8	1.86	2.31	3.36
9	1.83	2.26	3.25
10	1.81	2.23	3.17
11	1.80	2.20	3.11
12	1.78	2.18	3.06
13	1.77	2.16	3.01
14	1.76	2.15	2.98
15	1.75	2.13	2.95
16	1.75	2.12	2.92
17	1.74	2.11	2.90
18	1.73	2.10	2.88
19	1.73	2.09	2.86
20	1.73	2.09	2.85
25	1.71	2.06	2.79
30	1.70	2.04	2.75
40	1.68	2.02	2.70
50	1.68	2.01	2.68
60	1.67	2.00	2.66
80	1.66	1.99	2.64
100	1.66	1.98	2.63

TABLE 3 Browning Data for Nonhygroscopic Whey (a_w 0.44) as a Function of Storage Temperature

Day	Sample weight (g)	Optical density OD $\times 10^2$		Browning value (OD/g solid) $\times 10^2$	
		25° C			
0	3.0412	5.5	5.8	1.8	1.9
	3.0426	5.5	5.5	1.8	1.8
	3.0418	5.5	5.8	1.8	1.9
30	3.0424	12.8	12.2	4.3	4.1
60	3.0431	18.6	18.0	6.3	6.1
90	3.0431	21.9	22.5	7.4	7.6
120	3.0446	28.3	28.9	9.6	9.8
150	3.0472	35.3	34.7	12.0	11.8
180	3.0427	36.8	37.4	12.5	12.7
210	3.0410	42.6	43.5	14.5	14.8
		35° C			
0	3.0424	5.2	4.9	1.7	1.6
	3.0412	5.5	4.9	1.8	1.6
	3.0463	4.9	4.7	1.6	1.5
10	3.0397	15.2	17.9	5.2	5.0
20	3.0402	23.1	23.1	7.9	7.9
30	3.0424	31.0	30.7	10.6	10.5
40	3.0436	40.2	40.5	13.7	13.8
50	3.0431	47.8	48.4	16.3	16.5
60	3.0412	59.3	59.0	20.2	20.1
70	3.0421	68.1	68.8	23.2	23.4
95	3.0428	81.5	81.2	27.8	27.7
		45° C			
0	3.0432	5.2	5.2	1.7	1.7
	3.0412	5.8	5.5	1.9	1.8
	3.0444	5.2	4.9	1.7	1.6
2	3.0496	15.2	15.2	5.2	5.2
4	3.0455	21.0	20.7	7.1	7.0
7	3.0413	65.7	65.8	22.4	22.4
11	3.0422	73.9	74.2	25.2	25.3
18	3.0463	93.2	92.9	31.7	31.7
28	3.0486	130.4	129.9	44.4	44.2
35	3.0473	148.7	149.3	50.7	50.9

In order to predict the range of values of y for any given value of x_0, then

$$y_0 = (a + bx_0) \pm t_{\alpha/2} S_e \sqrt{1 + \frac{1}{n} + \frac{n(x_0 - \bar{x})^2}{n \Sigma x^2 - (\Sigma x)^2}} \tag{20}$$

where $\bar{x}$ is the average of all x.

As an example in applying the above concept the data for browning of non-hygroscopic whey powder shown in Table 3 can be used. As noted at zero time, three samples were examined with duplicate readings, and at each subsequent time period, only single samples were taken with duplicate readings. Figure 3 shows a visual plot of the data for each temperature using zero-order kinetics, and Fig. 4 shows the data at 35°C plotted as first order. It is obvious that a pseudo-zero-order reaction is followed. To prove this quantitatively, the regression equations were calculated and compared to the "eyeball" or visual plotting of the data. The results are shown in Table 4.

As can be seen in Table 4, in this case the zero-order plot gives a higher r^2 than does the first-order analysis. Utilizing the 95% confidence limits and assuming zero-order equations for the amount of browning, the maximum rate (highest B_0 and highest k) and the minimum rate (lowest B_0 and lowest k) were compared to the amount of browning, assuming the average rate and the amount as found drawing the line visually. The time to reach a browning value of 10×10^{-2} ($\theta_{0.1}$ value) was also calculated. Table 4 shows that the eyeball technique gives an equation falling within the minimum/maximum limits. In addition, the eyeball technique predicts 27.9 days, while the zero-order kinetics predicts 23.8 to 32.3 days with an average value of 27.8 days to reach the $\theta_{0.1}$ value. In comparison, the first-order plot gives 18.5 to 85.1 days with the average value of 40.5 days for the same extent of browning. These data show that the zero-order equations give the best estimation of actual browning with an error of about ±4.1 - 27.8, or ±15.2%, if one bases the error on the extremes of the 95% confidence limits. This value seems high, but experience with food systems suggests it is realistic. Of course, if the confidence limit evaluation were eliminated, then the zero-order average rate equation gives 27.8 days versus the eyeball value of 27.9 days, a prediction which is quite good. The researcher should set up simple programs to test data in this way in order to maximize the evaluation of shelf-life results, especially for industrial situations. However, the limitations of statistical analysis should always be kept in mind.

A second method of analyzing for a rate constant, which some kineticists and statisticians use, is to assume each data point to be a separate

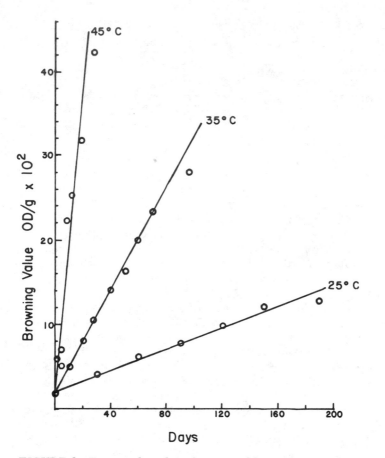

FIGURE 3   Zero-order plot of extent of browning of nonhygroscopic whey powder held at water activity of 0.44 and three temperatures (see data in Table 3).

experiment. The rate constant is then determined by application of the proper rate equation using this value and the averaged zero-time value $(B_0)_{avg}$. The method is called point-by-point analysis. For zero order, the point-method rate constant is

$$k = \frac{B - (B_0)_{avg}}{\theta} \tag{21}$$

where B is the value determined at time θ. Applying this to 35° C and calculating the confidence limits as $\pm 1.96\ \sigma/\sqrt{n}$ where σ is the standard deviation gives:

$$B_0 = (1.667 \pm 0.022) \times 10^{-2}$$

$$k = (0.3049 \pm 0.015) \times 10^{-2}$$

$$\theta_{0.1} = 27.3 \text{ days for average rate}$$

$$\theta_{0.1} = 25.9 \text{ days for maximum rate}$$

$$\theta_{0.1} = 28.8 \text{ days for minimum rate}$$

As seen, using only simple averages and the 95% confidence limits this method gives a much narrower range for the time to a given degree of quality change. This standard error gives an estimate that the data

FIGURE 4 First-order plot of extent of browning of nonhygroscopic whey powder held at 35°C and a water activity of 0.44 (see Table 3).

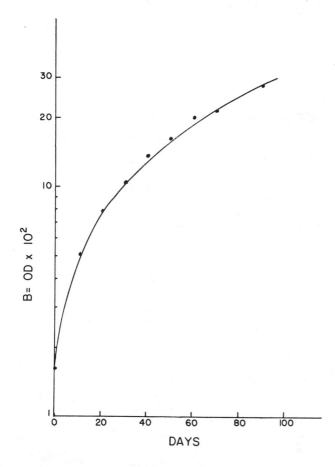

TABLE 4   Statistical Analysis of Shelf-Life Data of Table 3 at 35°C

Visual method

Rate constant = $k = 0.3 \times 10^{-2}$ OD/g day

Initial value = $B_0 = 1.63 \times 10^{-2}$ OD/g

Rate equation = $B = (1.63 \times 10^{-2}) + 0.3 \times 10^{-2}\,\theta$,

where B = browning value at time θ in days

Time for B = 0.1 = 27.9 days = $\theta_{0.1}$

Zero-order regression method

$r^2 = 0.994$

$S_e = 0.76$

$t_{\alpha/2}$ = 2.2 at 95% for F = 9, where n = 11

$k = (0.288 \pm 0.0165) \times 10^{-2}$ range = $(0.271$ to $0.304) \times 10^{-2}$

$B_0 = (1.99 \pm 0.76) \times 10^{-2}$ range $B_0 = (1.24$ to $2.75) \times 10^{-2}$

Max rate $B = (2.75 \times 10^{-2}) + 0.304 \times 10^{-2}$ $\theta_{0.1}$ = 23.8 days

Min rate $B = (1.24 \times 10^{-2}) + 0.271 \times 10^{-2}$ $\theta_{0.1}$ = 32.1 days

Avg rate $B = (1.99 \times 10^{-2}) + 0.288 \times 10^{-2}$ $\theta_{0.1}$ = 27.8 days

First-order regression method

$r^2 = 0.813$

$S_e = 0.501$

$t_{\alpha/2}$ = 2.2 at 95% for F = 9, where n = 11

$k = 0.032 \pm 0.011$ range = $(0.021$ to $0.043)$

$B_0 = 2.73 \times 10^{-2}$ range of $B_0 = (1.67$ to $4.47) \times 10^{-2}$

Max rate ln B = -3.10 + 0.043 $\theta_{0.1}$ = 18.5 days

Min rate ln B = -4.09 + 0.021 $\theta_{0.1}$ = 85.1 days

Avg rate ln B = -3.60 + 0.032 $\theta_{0.1}$ = 40.5 days

fall within some confidence range, in this case we chose the 95% confidence limits. One advantage of this latter method is that it gives more rate constant values which can then be used in evaluating temperature effects. However, the method gives more weight to values that have a large deviation from the average. Thus in analyzing the data of Table 3 at 45° C, the following is found:

Least squares k = $(1.439 \pm 0.224) \times 10^{-2}$ ($\pm 95\%$ confidence)

Point analysis k = $(1.82 \pm 0.43) \times 10^{-2}$ ($\pm 1.96/\sigma/\sqrt{n}$)

Thus, the point-by-point averaging method in this case gives a larger rate constant with a seemingly unacceptable 95% confidence range. However, the lowest rate constant at 45°C would be 1.39 OD/(g day) $\times 10^{-2}$, which is still larger than the values at 35°C. Again judgment must be combined with the statistics to make a decision as to how to treat the data.

IV. TEMPERATURE DEPENDENCE OF REACTION RATE

A. Arrhenius Relation

For a chemical reaction, the rate constant k can be related to temperature dependence according to the Arrhenius relation:

$$k = k_0 e^{-E_A/RT} \tag{22}$$

where

k_0 = pre-exponential (absolute) rate constant

E_A = activation energy in cal/mol

R = gas constant, 1.986 cal/(mol K)

T = temperature in Kelvin (°C + 273)

The equation basically states that the rate constant is an exponentiatial function of absolute temperature. Other expressions can be used, but at least this one has a thermodynamic basis. Thus, a plot of the log of the rate constant versus the reciprocal absolute temperature should give a straight line for any particular reaction (Eyring, 1935). Figure 5 shows plots for two reactions, A and B, which give straight lines. The Arrhenius equation assumes the following:

1. Only one reaction is responsible for loss of shelf life or quality. This may not be true since as one goes to higher temperatures, other reactions causing quality change may be more important, as is seen in Fig. 5 for two reactions, A and B. In this figure the

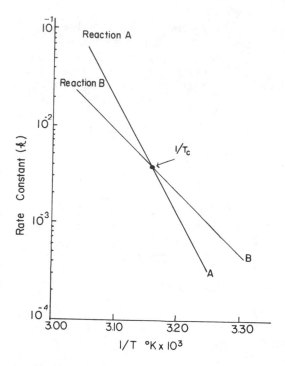

FIGURE 5   Arrhenius plot of log of rate constant versus reciprocal absolute temperature for two reactions showing a crossover at temperature T_c.

rate for B is faster as one crosses over and goes to higher temperatures. Thus, if both reactions occurred in a food, it would predominate above T_c, and A would predominate below T_c (Ragnarsson and Labuza, 1977).

2. No phase change takes place which could change reactant concentrations.

3. The partitioning of a reactant or product into an oil or fat phase does not change with temperature (Silvius and McElhaney, 1981).

4. There is no history effect on the reaction, that is, no matter how the food is prepared or held, when put at a certain temperature it will always show the same reaction rate.

5. Other unknown factors can also occur, especially when sensory testing is used which will make Eq. (22) invalid.

B. Goodness of Fit of Arrhenius Relation

Unfortunately, there is no a priori way for one to know whether these assumptions are valid, and thus, the only way to determine how good a fit occurs for Eq. (22) is to run experiments at several temperatures and then extrapolate the results to some other temperature to determine a presumed k, as denoted in Fig. 6. Then another test is made at that new temperature to see how good an estimate the extrapolated value was. In shelf life of foods, this is the desired method since one would want to use high-temperature data to predict the rate at some lower temperature, presumably the

FIGURE 6   Arrhenius plot for the rate constant of a reaction causing food quality loss showing hypothetical confidence limits of the highest rate (k_{upper}), lowest rate (k_{lower}) and average rate (k_{avg}) at some projected lower temperature.

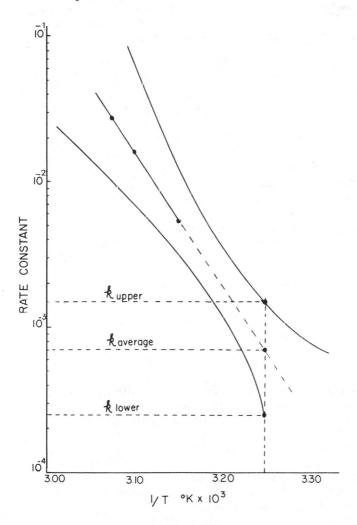

average storage condition for the food. The problem, of course, then goes back to how good is the estimate of the rate constants or what is the confidence interval on the Arrhenius plot? As shown in Fig. 6 for a hypothetical case, since it is a logarithmic function, the error in the value of k could be quite large. Thus, enough data points must be collected to ensure a small confidence interval. It is obvious that if only two temperatures are used, a straight line will always be found, and thus, no confidence limits can be calculated.

Several approaches can be used to determine the confidence interval of the Arrhenius plot. One would be to determine the rate constant at each temperature studied and then apply the standard linear regression analyses coupled with the standard error determination. Unfortunately, if the usual three temperatures are used in a kinetic study, $n = 3$ and $F = 1$. From Table 2, then $t_{\alpha/2} = 12.71$, and any calculated E_A could have a confidence range almost as large as the actual E_A value, depending on S_e. This suggests using more temperatures to reduce $t_{\alpha/2}$. Although this is logical, at least six temperatures would be required to get a significant reduction, which would be impossible for many laboratories to do because of equipment availability and cost. This would also require the same number of samplings at each temperature so that the amount of analyses would double, thereby doubling the cost of the study.

A second approach is to use only three temperatures and do a regression on the upper and lower confidence limits, giving an n of 6 and an F of 4. This reduces the confidence range considerably and may be acceptable for shelf-life testing. Another approach is to analyze each data point as a separate rate constant value, as mentioned before. This significantly increases n and F, usually with a corresponding reduction in the 95% confidence limits. It would be logical for anyone involved in reaction kinetics and shelf-life studies as a function of temperature to obtain or write computer or calculator programs for these methods. The former would be more acceptable since the storage space is limited on most calculators and the data usually have to be reentered for each analysis, which could lead to human error.

C. Example of Test for Goodness of Fit

Using the data in Table 3, the activation energy and $\ln k_0$ values were calculated by the above approaches including just drawing the line visually. The results are shown in Table 5. If confidence limits are not considered, all the methods including the eyeballed line, as drawn in Fig. 7, give activation energies and $\ln k_0$ values which are not very different from each other. As seen, when only three values are used to calculate E_A and $\ln k_0$, the 95% confidence limits are exceedingly large (methods 1 and 3) even though the r^2 is high. This is the result of the logarithmic transformations. The lowest confidence limits are from method 2, in which the 95% confidence limit extremes were used, and method 5, in which each data point was treated as a separate experiment.

TABLE 5 Statistical Evaluation of Activation Energies for Browning of Norhygroscopic Whey Powder Stored at $a_w = 0.44$

	Rate constants $\times 10^2$	
	k ± 95% confidence limit from regression analysis	k ± 95% confidence limits from point analysis
25°C	k = 0.0613 ± 0.00393 (r² = 0.994)	k = 0.0669 ± 0.0052
35°C	k = 0.283 ± 0.0165 (r² = 0.994)	k = 0.3049 ± 0.0072
45°C	k = 1.439 ± 0.224 (r² = 0.957)	k = 1.82 ± 0.43

Arrhenius kinetic values

Method	E_A (kcal/mol)[a]	ln k_0[a]	r^2	S_e
1. From average value of rate constants from regression	29.68 ± 11.48[a]	47.38 ± 18.8	0.999	0.068
2. From end points of 95% confidence limits	29.59 ± 3.51	47.20 ± 5.76	0.993	0.134
3. From mean values determined by point-by-point analysis	31.05 ± 26.2	49.72 ± 42.8	0.996	0.155
4. From ± 2σ limits of point-by-point analysis	28.99 ± 14.05	46.26 ± 22.98	0.892	0.537
5. Using each k from point-by-point analysis	30.79 ± 1.93	49.24 ± 3.15	0.982	0.183
6. Eyeball (visual) through average value from regression	30.45	—	—	—

[a] Confidence limits, ±95%.

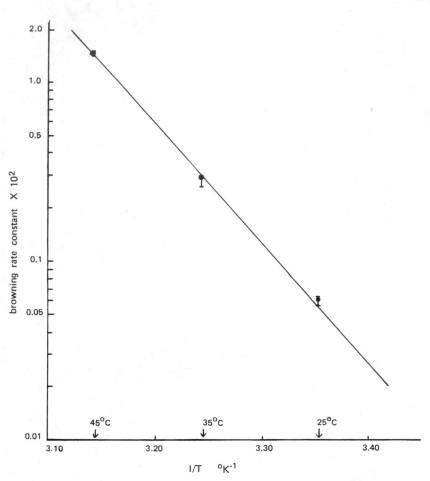

FIGURE 7   Arrhenius plot for rate of browning of nonhygroscopic whey powder (data from Table 5).

Table 6 compares the predicted rate constant at 20°C using the above techniques for the Arrhenius best straight line. As can be seen, each of the mathematical procedures gives about the same value with less than a 3% difference, except for method 4. When compared to the visual plot in Fig. 7, they were all about 12% higher, which is still within most analytical method sampling errors. The interesting factor is the extreme range of the maximum upper and lower limits as well as the 95% confidence limit, based on Eq. (20) and shown in Fig. 6. What this suggests is that statistics cannot be meaningfully used, at least with a 95% confidence limit, and thus,

TABLE 6 Predicted Rate Constants[a] at 20°C from Browning of Whey Based on Methods of Table 5

Methods[b]	k_{avg}[c]	Maximum limits (OD/g day) $\times 10^2$		95% confidence range[f]	
		k_{upper}[d]	k_{lower}[e]	k_{upper}	k_{lower}
1	0.0255	1.3×10^{15}	5×10^{-19}	0.104	0.006
2	0.0257	1.27×10^6	1.96×10^{-7}	0.043	0.016
3	0.0260	3.6×10^{36}	1.8×10^{-40}	0.613	0.0011
4	0.02799	8.13×10^{18}	9.6×10^{-23}	0.206	0.0039
5	0.0264	2.47×10^6	2.75×10^{-10}	0.040	0.0173
6	0.0231	—	—		

[a] Times 10^2.
[b] See Table 5 for explanation of methods.
[c] From regression line of Table 5.
[d] Using upper 95% confidence limit based on highest k_0, lowest E_A.
[e] Using lower 95% confidence limit based on lowest k_0, highest E_A.
[f] From Eq. (20).

in fact, one might just as well use visual plotting of the rate constants. This is not to say statistics are not applicable. In this type of analysis, however, the mathematics belie the usefulness of the data. One conclusion of this analysis is that one could then justify using only two temperatures in a shelf-life study as long as there is personal confidence (not statistical) and that there are no breaks in the Arrhenius plot. A second conclusion is that three temperatures are all that are needed in a shelf-life study. Finally it suggests that more than seven data points are needed to get more confidence.

V. SHELF-LIFE PLOTS

A second approach to the effect of temperature on food quality is to construct shelf-life plots. This is extremely useful, especially when little data is available to get rate constants or when only the time to reach a certain level of quality change has been determined. It can be shown that for any reaction order,

$$k = \frac{\Delta A}{\theta_s} \tag{23}$$

where

$$\Delta A = \text{amount lost at time } \theta_s$$

$$= A_0 - A_s \text{ for zero order}$$

$$= \ln \frac{A_0}{A_s} \text{ for first order}$$

since

$$\ln k = \ln \Delta A - \ln \theta_s = \ln k_0 - \frac{E_A}{R_T} \tag{24}$$

or

$$\ln k = \ln \Delta A + \ln (1/\theta_s) \tag{25}$$

thus

$$-\ln \theta_s = \ln \frac{1}{\theta_s} = -\ln \Delta A + \ln k_0 - \frac{E_A}{RT} \tag{26}$$

where θ_s is the shelf life for a certain quality change. Thus, as seen in Fig. 8a, a plot of log θ_s versus $1/T$ will have the slope of $+E_A/R$ as in the Arrhenius situation. In addition, if only a small temperature range is used (less than $\pm 20°C$), there is little error in plotting log θ_s versus T directly and assuming a straight line as in Fig. 8b (Labuza, 1982a). The same

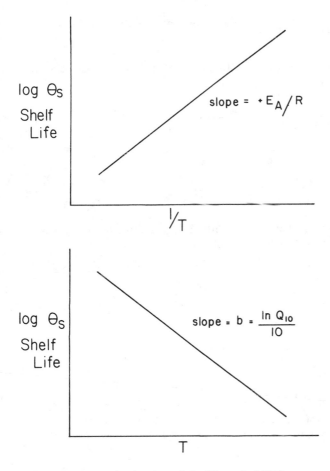

FIGURE 8 (a) Arrhenius plot of log shelf life versus reciprocal absolute temperature showing a slope of $+E_A/R$. (b) Pseudo Arrhenius plot (shelf-life plot) of log shelf life versus actual temperature showing a slope of b.

statistical evaluations could be used to derive the shelf-life plot as were done for the Arrhenius plot with the same problems. The equation for the shelf-life plot is

$$\theta_s = \theta_0 e^{-bT} \tag{27}$$

where

   T = temperature (°C or K)

   θ_0 = shelf life at T = 0°C or K° depending on temperature base

b = slope of ln θ_s vs T graph and would have positive value by definition

From this second plot then, the effect on shelf life of a 10°C increase in temperature can be derived, where

$$Q_{10} = \frac{\text{shelf life at T}}{\text{shelf life at T + 10°C}} = \frac{\text{rate constant at T + 10°C}}{\text{rate constant at T}} \tag{28}$$

It can also be shown that

$$Q_{10} = e^{10b} \tag{29}$$

The Q_{10} is not constant but depends on both the E_A and the temperature, as shown in Eq. (30), since b is a function of E_A

$$\log Q_{10} = \frac{2.187\ E_A}{T(T + 10)} \tag{30}$$

or

$$Q_{10} = \exp\left(\frac{5.036\ E_A}{T(T + 10)}\right) \tag{31}$$

where T is in Kelvin. In going back to the previous statistical analysis of the data, one could use the confidence limits in E_A to get the variation in Q_{10} or b, but again common sense might suggest that this would not be very

TABLE 7 Q_{10} Values for Browning of Nonhygroscopic Whey (25 to 35°C range)

Analytical method	Average Q_{10}	Q_{10} upper	Q_{10} lower
1	5.09	9.54	2.71
2	5.06	6.14	4.18
3	5.48	23.05	1.31
4	4.89	10.56	2.28
5	4.90	5.70	4.21
6	5.31	—	—

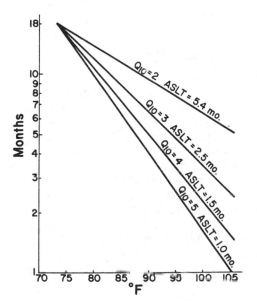

FIGURE 9   Hypothetical shelf-life plot for various Q_{10}'s passing through a shelf life of 18 months at 73°F.   Accelerated shelf-life times (ASLT) are those required at 40°C (104°F).

meaningful because of the large errors involved, and thus, the slope of the best straight line might be used solely to get the Q_{10} or b.   Using the previous data, Table 7 shows the average Q_{10} and 95% confidence limits based on the Table 5 calculations.   As with the rate constants, the values based on the averages are all quite similar.   The smallest range was found for method 5 which used each data point to calculate E_A.   Method 2, based on the 95% confidence limit extremes, gives a slightly larger range.   This again suggests that either of these two methods could be used to establish the basic kinetic parameters.

    An example of the use of the Q_{10} values and their importance is in the construction of a shelf-life plot for preplanning of a test.   For example, suppose the desired shelf life of a food were 18 months at 73°F.   In constructing Fig. 9, one could draw lines on the shelf-life plot passing through 18 months/73°F with slopes of different Q_{10}'s.   Where the lines cross at 104°F (40°C) would give an estimate of how long the test time (ASLT) must be to ensure the desired shelf life.   Thus, as seen in Fig. 9, the test time would be as little as 1 month if the Q_{10} were 5, or as much as 5.4 months if the Q_{10} were 2.   The confidence limits in Q_{10} are then apparent in the test design.   Labuza (1980, 1982a) has published E_A values for various food quality-loss mechanisms which can be used to determine Q_{10}'s, and thus, these plots.

VI. STORAGE UNDER UNSTEADY-STATE
TEMPERATURE CONDITIONS

A. General Variable Storage Temperature Distribution

In the preceding discussion, it has been assumed that the temperature of
the system under study has remained constant. However, foods rarely
experience constant temperatures. In the course of processing, distribution, and storage, food temperature fluctuations will occur and affect the
rates of deterioration as well as the accumulated extent of deterioration.
In a series of time-temperature-tolerance (TTT) studies described by
Guadagni (1968) and Olson (1968), the shelf life of frozen foods was observed to be related to the series of temperatures at which they were
stored. In these studies, shelf life was expressed by a measure of consumer acceptability designated as "high quality life" (HQL). These studies
utilized a linear relationship between log of HQL in days versus temperature, which is represented by Fig. 8b. This plot was then used to predict
days of quality left for a food undergoing any temperature distribution.
These studies used a simple ratio method to estimate the fraction of shelf
life used up by dividing the storage distribution into small time segments
denoted as $\Delta\theta_i$, as seen in Fig. 10.

For a given time segment $\Delta\theta_i$ and the average temperature T_i for that
time, the fraction of shelf life consumed was calculated from the ratio of
this time divided by the total shelf life θ_s the product would have it held at
temperature T_i. In equation form, this becomes

$$f_{con} = \left(\frac{\Delta\theta_i}{\theta_{s_i}}\right)_{T_i} \tag{32}$$

where

f_{con} = fraction of shelf life consumed

$\Delta\theta_i$ = time stored at temperature T_i (days)

θ_{s_i} = value of shelf life (days) at temperature T_i determined from
shelf-life plot (example of Fig. 8b)

For a series of different time-temperature intervals, the total of HQL
consumed is described by

$$\sum f_{con} = \sum \left(\frac{\Delta\theta_i}{\theta_{s_i}}\right)_{T_i} \tag{33}$$

This equation implicitly assumes a zero-order reaction since as noted before,

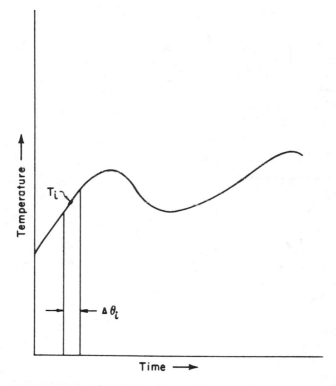

FIGURE 10 Hypothetical variable time/temperature distribution showing time increment $\Delta\theta_i$.

$$\frac{dA}{d\theta} = k = \frac{A_0 - A_s}{\theta_s} \tag{34}$$

In integral form,

$$\int_{A_0}^{A} \frac{dA}{A_0 - A_s} = \frac{A_0 - A}{A_0 - A_s} = \int \frac{d\theta}{\theta_s} = \sum \frac{\Delta\theta_i}{\theta_s} \tag{35}$$

The left-hand side denotes the change in A (or A_0 - A) divided by the total possible change in A (or A_0 - A_s). The right-hand side is the same as Eq. (33) when $d\theta$ is assumed to be $\Delta\theta_i$. Substituting for θ_s from the shelf-life equation gives

$$f_{con} = \int \frac{d\theta}{\theta_0 e^{-bT}} = \int \frac{e^{+bT} d\theta}{\theta_0} \tag{36}$$

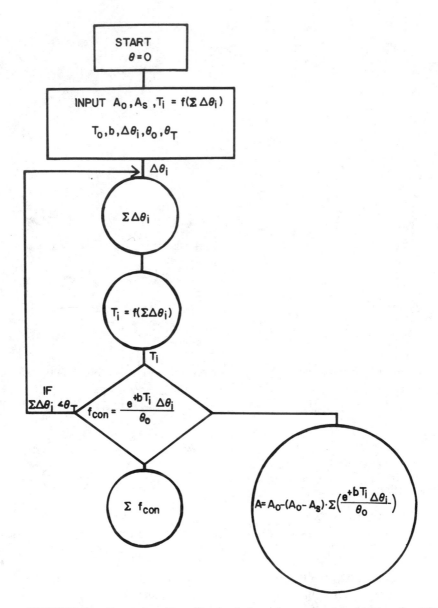

FIGURE 11 Computer iteration technique for analysis of either fraction of or total quality change as a function of time.

Thus, if a function is known for the variation in temperature with time
[i.e., $T = f(\theta)$], then Eq. (36) can be integrated, otherwise a computer iterative technique can be used as denoted in Fig. 11. Once the total fraction of
shelf life consumed has been evaluated, the time of shelf life left at any
temperature T_i is

$$\theta_{left} = (\theta_s)_{T_i}\left(1 - \sum f_{con}\right) \tag{37}$$

Gutschmidt (1974) found these equations to be very appropriate in estimating
the quality loss occurring during distribution of frozen chicken. It must be
noted, however, that these equations are appropriate only for pseudo-zero-
order reactions or first-order reactions in which the quality change is so
small that zero order can be applied.

When applying these equations to solve for the amount left of a given
quality factor (such as a nutrient) at any given time, the same equations are
applicable as long as the integrated form of Eq. (28) is used, which is

$$A = A_0 - \sum_{i=0}^{end} (k_i\theta_i)_{T_i} = A_0 - (A_0 - A_s)\int \frac{e^{+bT}}{\theta_0} d\theta \tag{38}$$

If at end of shelf life the value of A_s is 0, then

$$A = A_0\left(1 - \sum f_{con}\right) \tag{39}$$

It must be remembered that in the above equations (where θ_s was substituted for), the value of θ_s is the time to reach quality level A_s.

A similar equation can be developed for a first-order reaction which
takes the form

$$\frac{\ln (A_0/A)}{\ln (A_o/A_s)} = \int \frac{e^{+bT}}{\theta_0} d\theta \tag{40}$$

Again once a function for the temperature-time distribution is known, Eq.
(40) can be integrated or an iterative procedure used to sum up shelf life
consumed using the same format as described earlier. The amount of
quality left with time is

$$A = A_0 \exp\left[-\left(\ln \frac{A_s}{A_0}\right)\int \frac{e^{+bT}}{\theta_0}\right] \tag{41}$$

B. Arrhenius Form for Temperature Function

A different approach which is as applicable and more general is to use the Arrhenius relation for k rather than the shelf-life equations. For zero order this becomes, when $A_S = 0$,

$$A = A_0\left(1 - \int k_0 e^{-E_A/RT}\, d\theta\right) \tag{42}$$

For first order this takes the form

$$A = A_0 \exp\left(-\int k_0 e^{-E_A/RT}\, d\theta\right) \tag{43}$$

The equations are equivalent, respectively, to Eqs. (36) and (41). Since the integrals in these equations have no definite solution for any function of time/temperature, a computer iterative procedure, as in Fig. 11, would have to be used to solve for the extent of deterioration with time. This type of approach has been used by Lund (1977) to determine nutrient quality loss during canning.

In all the above equations, the test of fit, or more importantly the confidence limits on the predicted quality value left after any distribution, will depend on the variability in the values of either b and θ_0 or E_A and k_0; the former are functions of the latter. As was seen in Tables 5 and 6, with such large 95% confidence limits, especially on k_0, application of statistics may not be very meaningful. One technique could be used in which it is assumed that the lines (ln θ_S versus T or ln k versus 1/T) go through the mean value of the intercept with slopes according to the 95% confidence limits of b or E_A.

Several devices have been built and tested which are chemical zero-order integrators of time/temperature effects (Blixt and Tiru, 1976; Schubert, 1977). These can be placed on food packages and will indicate the extent of shelf life or quality lost if the device has a temperature sensitivity similar to that of the food. Several studies have shown that these devices are not as reliable as had been hoped for (Byrne, 1976; Hayakawa and Wong, 1974; Hu, 1972; Kramer and Farquhar, 1976).

VII. REGULAR SEQUENTIAL TIME/
        TEMPERATURE FUNCTIONS

A. General Rate Approach

While irregular temperature fluctuations can readily occur during distribution and storage, regular sequential temperature fluctuations can also occur. The diurnal-nocturnal change can result in a regular fluctuating temperature profile, such as for food stored in boxcars or warehouses.

In a study of commercial warehouses, Monroe, Brighton, and Bendix (1949) examined the average temperature of various geographical regions, warehouses, and the inside of cans stored within the warehouses. This study found that the warehouse temperatures were predictably cyclic, and nearly sinusoidal within the cans.

The original relationships derived to describe the effect of periodic temperature fluctuations on reaction rates were presented by Hicks (1944). Hicks used an exponential function for the rate of reaction as a function of temperature which is really a zero-order representation of the shelf-life equation presented earlier in rate form.

$$-\frac{dA}{d\theta} = R_0 e^{bT} \tag{44}$$

where

$-\dfrac{dA}{d\theta}$ = rate of loss of quantity A per unit time = θ_s^{-1}

R_0 = rate at temperature $T_0 = \theta_0^{-1}$

b = slope of ln (rate) or log shelf life versus temperature plot

T = temperatute of system

Hicks wrote a sine wave function for the temperature as a function of time θ as

$$T = T_m + a_0 \sin 2\pi\theta \tag{45}$$

where

T = temperature at time θ

T_m = mean temperature of the system = $\dfrac{T_{upper} + T_{lower}}{2}$

a_0 = amplitude of sine wave (as seen in Fig. 12)

Thus, by substituting Eq. (45) into Eq. (44), Hicks found that the change in the amount of A as a function of time was

$$A_0 - A_{sine} = \Delta A_{sine} = \int_0^\theta R_0 e^{b(T_m + a_0 \sin 2\pi\theta)} d\theta \tag{46}$$

Note that this is exactly the same equation as Eq. (38) or (39). Since there was no exact solution for this, Hicks found that using the ratio of the amount

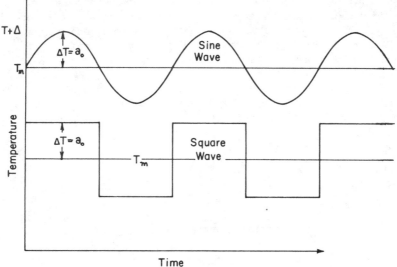

FIGURE 12   Regular sequential sine- and square-wave temperature distributions.

lost in the fluctuating sequence versus that lost at constant temperature, a
Bessel function solution for the integral results with the form

$$\frac{\Delta A_{sine}}{\Delta A_{con}} = \frac{A_0 - A_{sine}}{A_0 - A_{con}} = \Gamma_{sine} = 1 + \frac{(a_0 b)^2}{2^2} + \frac{(a_0 b)^4}{2^2 4^2} + \frac{(a_0 b)^6}{2^2 4^2 6^2} + \cdots \tag{47}$$

In both equations, A_{sine} refers to the amount left after experiencing a sine
wave, while A_{con} is the amount left at constant temperature T_m.

These equations show that the amount of loss during a fluctuating temperature sequence is greater than that predicted under constant conditions
of the arithmetic mean temperature T_m, since the Γ_{sine} is always greater
than 1. The constant temperature to which the fluctuating sequence is comparable is denoted as the effective temperature T_{eff}. This effective temperature is derived from the difference from the mean temperature of the
system by

$$T_{eff} - T_m = \Delta T_{eff} = \frac{1}{b} \ln \Gamma_{sine} = \frac{10}{\ln Q_{10}} \ln \Gamma_{sine} \tag{48}$$

Schwimmer, Ingraham, and Hughes (1955) extended Hicks' theory by examining systems which had undergone one of three types of regular fluctuations, sine wave, square wave, and ramp or saw-toothed (spike) waves. Using similar zero-order kinetics, Schwimmer found that for a square wave following the pattern shown in Fig. 12,

$$\frac{\Delta A_{sq}}{\Delta A_{con}} = \frac{A_0 - A_{sq}}{A_0 - A_{con}} = \Gamma_{square} = \frac{1}{2}(e^{ba_0} + e^{-ba_0}) \tag{49}$$

where A_{sq} is the quality left after any period in the square wave. Powers et al. (1965) published tables of Γ_{sine} and Γ_{square} for zero-order reactions as a function of Q_{10} and a_0, the amplitude. This was done using a simple computer program to solve Eqs. (47) and (49) for a large set of values. It can also be shown mathematically, since b is a function of Q_{10}, that

$$\Gamma_{sq} = \frac{1}{2}(Q_{10}^{a_0/10} + Q_{10}^{-a_0/10}) \tag{50}$$

As with the sine wave function, Γ_{sq} is always greater than 1; so there is always greater loss in a fluctuating temperature condition. For example, if Q_{10} were 2 and $a_0 = 10°C$, then

$$\Gamma_{sq} = \frac{1}{2}(2^1 + 2^{-1}) = 1.25$$

Thus, there is a 25% greater loss as compared to storage at the mean temperature. If the fluctuation were only 5°C, then $\Gamma_{sq} = 1.06$, or a 6% greater loss. If the analytical procedure has a ±10% variation, then one would not be able to see that the fluctuation gives a faster rate. Schwimmer, Ingraham, and Hughes (1955) also showed that for the square wave the effect of the temperature function took the same form as that for the sine wave, where

$$T_{eff} = T_m = \Delta T_{eff} = \frac{1}{b} \ln \Gamma_{sq} \tag{51}$$

Thus, with a Q_{10} of 2 (or b = 0.069) and $\Gamma_{sq} = 1.06$, the effective temperature difference is 0.85°C. If the constant temperature storage room varied by ±1°C, this would mask the effect of a ±5°C cycle.

Practical application of the Hicks (1944) and Schwimmer, Ingraham, and Hughes (1955) theories was presented in a series of studies on microbiological growth and enzyme denaturation during fluctuating temperature regimes (Powers et al., 1965; Wu, Eitenmiller, and Powers, 1974, 1975a, b). No evaluation using statistical procedures was made. While these studies demonstrate the increased rate of reaction over mean temperatures due to fluctuation, Labuza (1979) has noted a basic error in these studies which is attributable to assumptions made by Hicks (1944) and later by Schwimmer, Ingraham, and Hughes (1955).

The implicit assumptions in the above solutions are

1. The Arrhenius relation is not assumed by Eq. (44); thus, the equation is applicable only over a small temperature range due to the temperature dependence of b and Q_{10}. This does not create too much of an error for most food storage tests.

2. The reaction rate for the quality loss follows pseudo-zero-order kinetics. The aforementioned studies by Powers and Wu can be shown to involve first-order kinetics that were improperly analyzed using the zero-order solution (Labuza, 1979).

For first-order reactions, it was shown by Labuza (1979) that the Γ-function value remained the same for either sine-wave or square-wave fluctuations; however, the Δ term on the left-hand side of either Eq. (47) or (49) is

$$\frac{\Delta A_s}{\Delta A_{con}} = \frac{\ln (A_s/A_0)}{[\ln (A_{con}/A_0)]_{constant}} \tag{52}$$

B. Arrhenius Rate Approach

It can be shown (Labuza, 1979), through rigorous solutions using the general rate Eq. (2) and the Arrhenius relation, that all of the above solutions are exact integrals.

1. Square Wave, Zero Order

As before for zero order,

$$A_0 - A_{sq} = \int_0^\theta k \, d\theta \tag{53}$$

If for any total time period (θ) one-half the time is at $T_m + a_0$ and one-half at $T_m - a_0$, then by definition

$$A_{sq} = A_0 - \int_0^{\theta_T/2} k_{T_m+a_0} \, d\theta + \int_0^{\theta_T/2} k_{T_m-a_0} \, d\theta \tag{54}$$

Integrating at each constant temperature gives

$$A_{sq} = A_0 - \frac{\theta_T}{2} \left[(k_{T_m+a_0} + k_{T_m-a_0}) \right] \tag{55}$$

since by the Arrhenius relation,

$$Q_A = \frac{k_{T+a}}{k_T} = Q_{10}^{(a/10)(T+10)/(T+a)} \tag{56}$$

then substituting in

$$A_{sq} = A_0 - \frac{k_{T_m} T}{2}\left(Q_{10}^{(a_0/10)(T_m+10)/(T_m+a_0)} + Q_{10}^{-(a_0/10)(T_m+10)/(T_m-a_0)} \right) \tag{57}$$

On examining Eq. (50), Eq. (57) is approximately

$$A_{sq} - A_0 = (k_{T_m} \Gamma_{sq}) \theta_T = A_0 - k_{eff} \theta_T \tag{58}$$

where

$$k_{eff} = k_{T_m} \Gamma_{sq} \tag{59}$$

$$\Gamma_{sq} = \frac{1}{2}\left(Q_{10}^{+(a_0/10)(T+10)/(T+a_0)} + Q_{10}^{-(a_0/10)(T\ 10)(T-a_0)} \right) \tag{60}$$
$$\text{(note T is in Kelvin)}$$

The difference is the correction ratio in the exponents which corrects the Q_{10} for the fact that it is a function of temperature while E_A is not. Thus, using Eq. (60) for Γ_{sq} corrects for the variation in Q_{10}. The errors in projection of the amount left after a square-wave cycle would be based on the errors in k_{T_m} and in the E_A or Q_{10} values. Thus, the test of the predictability of the equation could be based on how close the predicted A_{sq} comes to the actual measured value. A more rigorous solution using E_A instead of Q_{10}'s in Eq. (56) could be done if desired but would not give much more accuracy. The solution is

$$A_{sq} = A_0 - \frac{k_0}{2} \theta_T \left(e^{-E_A/R(T_m+a_0)} + e^{-E_A/R(T_m-a_0)} \right) \tag{61}$$

This is an exact solution, since k is constant for each half of the square wave; thus, a calculator rather than a computer can be used to solve the problem.

2. Square Wave, First Order

For a first-order reaction,

$$A = A_0 \exp\left(\int k \, d\theta\right) \tag{62}$$

Thus, using the same steps as above, then

$$A_{sq} = A_0 \exp\left(-k_{T_m} \Gamma_{sq} \theta_T\right) \tag{63}$$

or

$$A_{sq} = A_0 \exp\left(-k_{eff} \theta_T\right) \tag{64}$$

where Γ_{sq} and k_{eff} are defined as in Eqs. (59) and (60). In addition, using the Arrhenius relation, then

$$A_{sq} = A_0 \exp\left[-\frac{k_0}{2} \theta_T \left(e^{-E_A/R(T_m+a_0)} + e^{-E_A/R(T_m-a_0)}\right)\right] \tag{65}$$

3. Sine-Wave Solutions

Utilizing the Arrhenius relation and Eq. (45) for a sine wave for zero-order kinetics, the solution becomes

$$A = A_0 - \int k_0 \exp\left(\frac{-E_A}{R(T_m + a_0 \sin 2\pi\theta)}\right) d\theta \tag{66}$$

There are no known functions for an exact solution to Eq. (66). Thus, an iterative computer procedure such as in Fig. 11 would have to be used. A similar function would exist for first-order reactions.

4. Other Functions

Recently Schubert (1977) derived, strictly from analytical functions, equations for ΔT_{eff} for different sequential temperature situations. These solutions are

Spike wave

$$\Delta T_{eff} = \frac{1}{b} \ln \frac{\sinh \partial 0}{\partial 0} \tag{67}$$

Square wave

$$\Delta T_{eff} = \frac{1}{b} \ln (\cosh \partial b) \tag{68}$$

Sine wave

$$\Delta T_{eff} = \partial [1 - \exp (0.27 \partial b)] \tag{69}$$

No tests for the validity of the equations exist in the literature.
For the spike wave using similar procedures, it can be shown that

$$\Gamma_{spike} = \frac{Q_{10}^{(a_0/10)(T+10)/(T+a_0)} + Q_{10}^{-(a_0/10)(T+10)/(T-a_0)}}{2n \ln Q_{10}} \tag{70}$$

where

$$n = \frac{a_0}{10} \frac{T + 10}{T + a_0} \tag{71}$$

For product undergoing a linear rise or fall following Eq. (72),

$$T = T_i + m\theta \tag{72}$$

where

T_i = temperature at $\theta = 0$

m = degrees rise or fall per unit time

Then for zero order using the shelf–life–equation method,

$$A_{lin} = A_0 - \int_0^\theta \frac{e^{b(T_i + m\theta)}}{\theta_0} d\theta \tag{73}$$

$$A_{lin} = A_0 - \frac{e^{bT_i}}{bm\theta_0} (e^{bm\theta} - 1) \tag{74}$$

No tests exist for this situation in storage of foods. It should be noted that
a similar equation for first–order reactions could be derived. As can be
seen, since an exact solution exists there is no need for a computer program. However, if the Arrhenius relation is used instead of the shelf–life

equation, then there are no exact solutions, and a computer program similar to solving Eq. (66) must be used.

5. Solution Example

Table 8 contains data for the browning of whey powder held in a square-wave function. Using least-squares analysis, the equation for extent of browning, assuming some constant effective temperature, is

$$B = (1.897 \pm 1.005) + (0.694 \pm 0.023) \times 10^2 \theta \qquad (75)$$

The r^2 value is 0.998. In comparison to Table 5, it can be seen that the rate constant of 0.694 falls between 35 and 45°C (0.288 at 35°C and 1.439 at 45°C) as it should. In addition, the confidence limits are such that the values do not overlap; thus, the fluctuating condition is significantly different from the mean temperature of 35°C.

To compare the actual extent of quality change to what would be predicted, Eq. (57) can be used with $a_0 = 10°C$ and $Q_{10} = 5.06$ with a range of 4.18 to 6.14 using method 2 from Table 7. Thus, for these Q_{10} values, the extent of degradation with time is shown in Table 9 in comparison with the

TABLE 8   Extent of Browning in Nonhygroscopic Whey Held at a_w 0.44 and a Square-Wave Function of 25/45°C with a Period of 5 Days at Each Temperature

Day	Sample weight (g)	Optical density OD × 10^2		Browning value (OD/g solid) × 10^2	
0	3.0376	5.2	4.9	1.7	1.6
	3.0412	5.2	5.2	1.7	1.7
	3.0402	5.5	5.6	1.8	1.9
10	3.0386	27.3	26.1	9.3	8.9
20	3.0354	45.5	44.6	15.5	15.2
30	3.0372	69.9	71.1	23.9	24.3
40	3.0379	83.5	85.1	28.5	29.0
50	3.0382	110.0	111.8	37.9	38.2
60	3.0356	122.9	122.0	42.1	41.7
70	3.0398	149.0	150.2	50.9	51.3
80	3.0412	167.3	168.5	57.1	57.5

TABLE 9   Comparison of Actual to Predicted Extent of Browning for a
25/45°C Square-Wave Storage Condition[a]

Days	Actual	Regression Eq. (75)	Browning value ($\times 10^2$) at a Q_{10} of		
			4.18	5.06	6.14
0	1.7	1.897	1.7	1.7	1.73
10	9.1	8.8	8.1	9.3	10.8
20	15.3	15.8	14.4	16.8	19.8
30	23.1	22.7	20.7	24.4	28.9
40	28.8	42.6	27.1	31.9	37.9
50	38.1	36.6	33.4	39.4	47.0
60	41.9	43.5	39.7	46.9	56.0
70	51.1	50.5	46.1	54.6	65.1
80	57.3	57.4	52.7	62.1	74.1

[a] $k_{T_m} = (0.288 \times 10^{-2})$ OD/day g solids.

actual measured value and the regression values. In this case only the
average value of k_{T_m} was used. As noted in Table 5, this also has a confidence limit so that each of the calculated values, using Eq. (57), has a
variation of about ±6%. The calculations of Table 9 show that use of the
average Q_{10} gives browning values quite close to the actual values or the
values from the regression line, thus indicating the validity of the mathematical procedures. Simple computer programs or calculator programs
could be written to complete the matrix of values using the confidence limits
on both k_{T_m} and Q_{10}. It should be noted that the calculated effective rate
constants by use of Eq. (59) are

Actual from regression $k_{eff} = 0.694 \pm 0.023$

$Q_{10} = 4.18$ $\Gamma_{sq} = 2.18$ $k_{eff} = 0.628 \pm 0.036$

$Q_{10} = 5.06$ $\Gamma_{sq} = 2.62$ $k_{eff} = 0.754 \pm 0.043$

$Q_{10} = 6.14$ $\Gamma_{sq} = 3.14$ $k_{eff} = 0.904 \pm 0.052$

As seen above, the average Q_{10} of 5.06 gives the best estimate of the actual effective rate constant as measured from the least-squares regression value.

The third test that can be used compares the effective temperature as determined by interpolation of the rate constant on the Arrhenius plot (see Fig. 6) with the use of Eq. (51) or (68). The results of this comparison are

$$Q_{10} = 4.18 \qquad T_{eff} = 40.45°C$$

$$Q_{10} = 5.06 \qquad T_{eff} = 40.94°C \qquad \Bigg\} \qquad \text{Equation (51)}$$

$$Q_{10} = 6.14 \qquad T_{eff} = 41.31°C$$

$$Q_{10} = 5.06 \qquad T_{eff} = 37.67°C \qquad \text{Equation (68)}$$

$$\begin{array}{c} \text{Interpolation} \\ \text{on} \\ \text{Arrhenius} \\ \text{plot} \end{array} \qquad \begin{array}{l} T_{eff} = 40.35°C \\ (\text{range } 40.13 \text{ to } 40.57) \end{array}$$

In this case, for interpolation, only the mean values of E_A and $\ln k_0$ were used from method 2 in Table 5. As seen again, the theoretical equations based on kinetics give a good comparison to the actual value, while Shubert's Eq. (68) does not.

Other tests of these equations, for both zero-order and first-order reactions undergoing various square-wave distributions, have been published by Kamman and Labuza (1981), Labuza and Saltmarch (1982), and Labuza, Bohnsack, and Kim (1982) including studies of protein quality loss, thiamin loss, and browning. In almost all cases the predictions were very good in comparison to actual changes. No literature exists, however, for application to sensory testing and for sine-wave distributions. However, if the predictions work for square waves, they should be applicable to any reasonable temperature distribution. It would behoove the industrial food scientist to put these functions into simple routine computer programs along with a simulated shelf-life temperature distribution function in order to make some predictions of expected quality losses. As a final note, in all of these equations moisture content was assumed to be constant. If the external relative humidity changes with time and if proper engineering equations are used to predict gain or loss, this can be coupled with the temperature distribution to accurately predict quality change. Karel, Mizrahi, and Labuza (1971) and Mizrahi, Labuza, and Karel (1970) present extensive computer iterative procedures for doing this, and Labuza and Contreras-Medellin (1981) have discussed the usefulness of such solutions. These methods have been applied to storage-life predictions of various food products (Mizrahi and Karel, 1977; 1978; Quast, Karel, and Rand, 1972; Quast and Karel, 1972) as well as to drugs (Yang and Roy, 1980; Veillard et al., 1979).

REFERENCES

Amdur, I. and Hammes, G. (1966). Chemical Kinetics. McGraw-Hill, New York.

Blixt, K. and Tiru, M. (1976). An enzymatic time/temperature device for monitoring the handling of perishable commodities, Develop. Biol. Stand. 36:237.

Byrne, C. H. (1976). Temperature indicators: The state of the art, Food Technol. 30(6):66.

Daniels, F. and Alberty, R. A. (1975). Physical Chemistry, 4th ed. John Wiley and Sons, Inc., New York.

Eyring, H. (1935). Activated complex and the absolute rate of reactions, J. Chem. Phys. 3:107.

Freund, J. (1967). Modern Elementary Statistics, 3d ed. Prentice-Hall, New York.

Guadagni, D. G. (1968). Cold storage life of frozen fruits and vegetables as a function of time and temperature, in Low Temperature Biology of Foodstuffs. Recent Advances in Food Science, Vol. 4 (J. Hawthorne, ed.). Pergamon Press, New York.

Gutschmidt, J. (1974). The storage life of frozen chicken with regard to the temperature in the cold chain. Lebensm.-Wiss. u. Technol. 7:137.

Hayakawa K. and Y. Wong (1974). Performance of Frozen Food indicators subjected to time variable temperatures ASHRAE J. 16(4):44.

Hicks, E. W. (1944). Note on the estimation of the effect of diurnal temperature fluctuation on reaction rates in stored foodstuffs and other materials. J. Coun. Sci. Ind. Res. (Australia) 17:111.

Hu K. (1972). Time-temperature indicating system writes status of product shelf life Food Technol 26(8):56.

Kamman, J. F. and Labuza, T. P. (1981). Kinetics of thiamin and riboflavin loss in pasta as a function of constant and variable storage conditions. J. Food Sci. 46:1457.

Karel, M., Mizrahi, S., and Labuza, T. P. (1971). Computer prediction of food storage, Modern Pkg. August.

Kramer A. and Farquhar. J. (1976). Testing of time-temperature indicating and defrost devices Food Technol 30(2):53.

Labuza, T. P., Tsuyuki, H., and Karel, M. (1969). Kinetics of linoleate oxidation in model systems, J. Amer. Oil Chem. Soc. 46(8):409.

Labuza, T. P., Tannenbaum, S. R., and Karel, M. (1970). Water content and stability of low-moisture and intermediate-moisture foods, Food Technol. 24(5):543.

Labuza, T. P. (1971). Kinetics of lipid oxidation in foods, CRC Crit. Rev. Food Technol. 2:335.

Labuza, T. P., Warren, R. M., and Warmbier, H. C. (1977). The physical aspects with respect to water and nonenzymatic browning, in Protein Crosslinking (M. Friedman, ed.). Plenum Press, New York.

Labuza, T. P. and Kreisman, L. (1978). Open shelf life dating of foods. Office of Technology Assessment Contract OTA-C-78-001. Washington, D. C.

Labuza, T. P. (1979). A theoretical comparison of losses in foods under fluctuating temperature sequences, J. Food Sci. 44:1162.

Labuza, T. P. (1980). Temperature/entropy/enthalpy compensation in food reactions, Food Technol. 34(2):67.

Labuza, T. P. (1982a). Shelf Life of Foods. Food and Nutrition Press, Westport, Conn.

Labuza, T. P. (1982b). Moisture gain or loss from packaged foods, Food Technol. 36(4):92.

Labuza, T. P. and Contreras-Medellin, R. (1981). Prediction of moisture protection requirements for foods, Cereal Foods World. 26(7):335.

Labuza, T. P., Bohnsack, K., and Kim, N. N. (1982). Kinetics of protein quality change in egg noodles stored under constant and fluctuating temperatures. Cereal Chem. 59:142.

Labuza, T. P. and Saltmarch, M. (1981). The kinetics of the nonenzymatic browning reaction as affected by water in foods, in Water Activity: Influences on Food Quality (L. Rockland and G. F. Stewart, eds.) Academic Press, New York.

Labuza, T. P. and Saltmarch, M. (1982). Kinetics of browning and protein quality loss in whey powder during steady state and nonsteady state storage conditions, J. Food Sci. 47:92.

Lund, D. B. (1977). Design of thermal processes for maximizing nutrient retention, Food Technol. 31(2):71.

Mizrahi, S., Labuza, T. P., and Karel, M. (1970). Computer aided prediction of browning in dehydrated cabbage, J. Food Sci. 35:799.

Mizrahi, S. and Karel, M. (1977). Moisture transfer in a packaged product in isothermal storage, J. Food Proc. Preserv. 1:225.

Mizrahi, S. and Karel, M. (1978). Evaluation of kinetic models for reactions in moisture sensitive products using dynamic storage conditions, J. Food Sci. 43:750.

Monroe, K. H., Brighton, K. W., and Bendix, G. H. (1949). The nutritive value of canned foods. XXVIII. Some studies of commercial warehouse temperatures with reference to stability of vitamins in canned foods, Food Technol. 2:292.

Olson, R. L. (1968). Objective tests for frozen food quality. In Low Temperature Biology of Foodstuffs. Recent Advances in Food Science, Vol. 4 (J. Hawthorne, ed.). Pergamon Press, New York.

Pope, D. G. (1980). Accelerated stability testing for prediction of drug stability, Drug Cos. Ind. 127(5):54.

Powers, J. J., Lukasziwicz, W., Wheeler, R., and Dornsetter, T. P. (1965). Chemical and microbial activity ratio under square wave and sinusoidal temperature fluctuation, J. Food Sci. 30:520.

Quast, D., Karel, M., and Rand, W. (1972). Development of a mathematical model for oxidation of potato chips as a function of oxygen, extent of oxidation and ERH, J. Food Sci. 37:673.

Quast, D. and Karel, M. (1972). Effects of environmental factors on the oxidation of potato chips, J. Food Sci. 37:584.

Ragnarsson, J. O. and Labuza, T. P. (1977). Accelerated shelf life testing for oxidative rancidity in foods. A review, Food Chem. 2:291.

Schubert, H. (1977). Criteria for the application of TTT indicators to quality control of deep frozen foods. 11F/11R Commissions C1/C2, p. 104.

Schwimmer, S., Ingraham, L. L., and Hughes, H. M. (1955). Temperature tolerance for frozen food processing. Effective temperature in thermally fluctuating systems, Ind. Eng. Chem. 27(6):1149.

Silivius, J. R. and McElhaney, R. (1981). Nonlinear Arrhenius plots and analysis of reaction in biological membranes, J. Theor. Biol. 88:135.

Sontag, T. A., Pangborn, R. M., and Little, A. C. (1981). Potential fallacy of correlating hedonic responses with physical and chemical measurements, J. Food Sci. 46:583.

Van Arsdel (1957). The time temperature-tolerance of frozen foods. I. Introduction—The problem and the attack, Food Technol. 11:28.

Van Arsdel and Guadagni, D. G. (1959). Time-temperature-tolerance of frozen foods. XV. Methods of using temperature histories to estimate changes in frozen food quality, Food Technol. 13(2):14.

Veillard, M., Benetejac, R., Duchene, D., and Cartensen, J. (1979). Moisture transfer tests in blister package testing, Drug Develop. Ind. Pharm. 5(3):227.

Wolf, J. C., Thompson, D. R., and Reineccius, G. A. (1977). Initial losses of available lysine in model systems, J. Food Sci. 42(6):1540.

Wu, A. C. M., Eitenmiller, R. R., and Powers, J. J. (1974). Effect of fluctuating temperature on the stability and activity of invertase, J. Food Sci. 39:1179.

Wu, A. C. M., Eitenmiller, R. R., and Powers, J. J. (1975a). Responses of chymotrypsin and lysozyme under fluctuating temperature treatments, J. Food Sci. 40:840.

Wu, A. C. M., Eitenmiller, R. R., and Powers, J. J. (1975b). Effect of fluctuating temperature treatment on milk coagulation and inactivation of soybean trypsin inhibitors, J. Food Sci. 40:1171.

Yang, W. and Roy, S. (1980). Projection of tentative expiry dating from one point accelerated stability testing, Drug Develop. Ind. Pharm. 6(6):591.

5
Quantitative Analysis and Simulation of Food Quality Losses During Processing and Storage

MARCUS KAREL Massachusetts Institute of Technology, Cambridge, Massachusetts

I. STATEMENT OF THE PROBLEM

Quantitative analysis of quality changes in any product during its processing, storage, and preparation requires certain initial assumptions, as well as a set of real or assumed data. The analysis can only be as valid as the assumptions.

The first, and absolutely necessary, assumption is that there exists a parameter or a set of parameters which can be measured and which correlate with "quality" of the product. The existence of this assumption is particularly worth stressing for the case of <u>food</u> quality, which is often described in nonquantitative terms (e.g., fresh vs. stale, wholesome vs. spoiled, flavorful vs. rancid). The usual approach in food technology is to derive a quantitative parameter correlating with quality. This may be a single chemical entity whose loss or increase correlates with quality. Examples include ascorbic acid loss in stored frozen vegetables (Guadagni, 1968), increase in trimethylamine in stored frozen fish (Haard et al., 1975), and increase in malondialdehyde and related compounds in oxidation-sensitive foods (Grosch, 1975). The parameter may be based on physical measurements, such as in the case of mechanical properties related to texture (Elton, 1969).

Even when a single measurement does not adequately describe quality, it is possible to derive a single "score" which is a quality parameter obtained by some weighted statistical analysis of a set of microbiological, physical, chemical, or organoleptic tests. A well-known example of this approach is the scoring system for bread, which includes physical measurements (e.g., bread volume) as well as organoleptic, subjective evaluation (Matz, 1960). We shall designate such a primary, or derived, parameter with the symbol Q.

The quantitative analysis of quality changes requires further that we describe the changes in Q as a result of processing, storage, and preparation conditions in a manner which allows prediction or simulation of these changes. An approach which has been evolved by several researchers over the last 30 years is to obtain differential equations relating rate of change of quality to significant processing or storage conditions, as shown in the generalized equation

$$\frac{dQ}{dt} = f(E_i, F_j) \tag{1}$$

where

E_i = environmental factor ($i = 1, \ldots, n$)

F_j = composition factor ($j = 1, \ldots, m$)

The effects of some of these factors can be expressed relatively easily in a functional relationship that is similar for many deterioration reactions

and food systems. Others are more complicated and unique in their behavior and must be derived separately for each product and food system
investigated.

It is usually convenient to consider compositional factors outside the
food as environmental factors. Thus the absolute or relative humidity
surrounding the food would be listed as one of the "E" factors.

In order to analyze the quality change, we also require the knowledge
of how each of the pertinent variables changes with time:

$$E_i = f(t) \tag{2}$$

$$F_j = f(t) \tag{3}$$

These may be simple (e.g., E_i = constant, as in constant temperature
storage) or may take the form of complex differential, partial differential,
or parametric equations.

Combination of Eqs. (1) to (3) allows, given an algorithm for their
solution, the prediction and/or simulation of quality changes in any process.
Relatively simple problems, primarily related to storage-induced quality
changes in moisture-sensitive foods, were analyzed and solved in the above
manner as early as the 1940s (Oswin, 1954; Heiss, 1956). In my doctoral
thesis (Karel, 1960) I formulated several differential equations relating
oxidation-related changes in foods stored in oxygen-permeable containers.
The analytical solution of these equations, however, proved awkward, and
the solution of these and other complex problems was shown to be tackled
best by numerical methods, using computers (Mizrahi, Labuza, and Karel,
1970; Saguy and Karel, 1980). The explosive growth in availability of computer hardware and software was instrumental in allowing us to solve even
highly complex quality-loss-related equations. In my opinion the rate-
limiting steps are now the analytical methodology for experimental assessment of quality changes and the availability of realistic models of quality
deterioration, Eq. (1). Once these models are available the mathematical
methodology for their incorporation in a simulation scheme is relatively
easy.

It is appropriate, therefore, to consider what generalizations can be
made about the effects of compositional and environmental factors on quality
changes.

II. DEPENDENCE ON ENVIRONMENTAL FACTORS
AND ON FOOD COMPOSITION

Chapter 4 deals very extensively with effects of temperature on food deterioration and also mentions briefly the effect of water activity. We consider it desirable, nevertheless, to consider both of these very important
factors in our discussion.

A. Temperature

It is commonly assumed, and often found valid, that the temperature dependence of quality deterioration follows the Arrhenius relation:

$$k = k_0 e^{-E/RT} \tag{4}$$

where

k = rate constant in a given quality loss

k_0 = constant, independent of temperature

E = activation energy

T = absolute temperature

When the other variables are unchanging, the Arrhenius relation is usually applicable, and its use constitutes the soundest approach to modeling temperature dependence. The activation energy (E) does, however, vary with concentration and other composition factors. Furthermore, when the reaction mechanism changes with temperature, the activation energy may be variable.

It should be fairly obvious that Eq. (4) is unlikely to apply when quality loss is due to different reactions as temperature changes or when the dominant quality-loss reaction is affected by subsidiary reactions with different temperature coefficients.

Functions other than the Arrhenius relation have been occasionally suggested as suitable for correlation of food stability data. Kwolek and Bookwalter (1971) considered, among others, the following forms of temperature dependence:

Linear

$$f(T) = c + bT \tag{5}$$

Exponential

$$f(T) = cT^b \tag{6}$$

Hyperbolic

$$f(T) = \frac{c}{b - T} \tag{7}$$

They found that the hyperbolic model and the Arrhenius equation gave the best correlation and could be applied over the widest range of temperatures when tested with data selected from three published stability studies, which included both organoleptic and chemical food quality indexes.

Over a limited temperature range, however, a linear approximation may be used, as Olley and Ratkowsky (1973) demonstrated for various refrigerated foods. These studies led them to the conclusion that in the temperature range of 0 to 6°C, a linear approximation was adequate for prediction of the temperature effect on a variety of deterioration reactions in refrigerated fish, beef, and poultry. Furthermore, they claimed that a single proportionality constant is adequate for the many diverse reactions reported in the literature for these items.

B. Water Activity and Water Content

Water is only one of the many components of food. It is, however, worth considering separately because water has a profound effect on physical, chemical, and biological processes in foods. In physical changes it is the role of water as plasticizer of polymeric components and as solvent for the low-molecular-weight compounds that is of key importance (Tsouroflis, Flink, and Karel, 1976). Microbiological processes are affected by water primarily because of the osmotic stress set up between cell interior and its surroundings in food (Brown, 1976).

Chemical reactions are influenced in a more complex manner. Water can act in one of the following roles:

1. Solvent for reactants and/or products

2. Reactant (in hydrolytic reactions)

3. Reaction product (in condensation reactions)

4. Modifier of activities of catalysts and inhibitors

Another important reason for considering water as not only a compositional but also an environmental factor is that water is volatile and is ubiquitous in the environment. All materials, including foods, tend to equilibrate with water in the environment. The condition of equilibrium is, of course, the equality of chemical potential of water in various phases (e.g., in and out of microbial cells, in food, and in air surrounding it). At constant temperature chemical potentials are equal when activities are equal. For this reason it is considered that the appropriate property of water to be correlated with rates of quality loss is water activity, which can be defined adequately for our purposes by Eq. (8) (Karel, 1975; Troller and Christian, 1978; Rockland and Stewart, 1981).

$$a \equiv \left(\frac{p_f}{p^0} \right)_T \tag{8}$$

where

a = water activity

p_f = partial pressure of water in food

p^0 = vapor pressure of water at temperature T

A number of relations have been suggested to describe the dependence of reaction rates on moisture content (or on water activity). Some authors who studied browning of vegetables reported an exponential relation between rate of browning and moisture content (Hendel et al., 1955; Legault et al., 1951).

Oxidation of potato chips was investigated by Quast and Karel (1972), who reported that oxidation was related to the function $a^{-1/2}$, where a is the water activity. Sometimes a linear relation between m and reaction rate or between a and reaction rate provides a satisfactory approximation of the behavior of nutrient retention in dry foods. Beetner et al. (1974) found that retention of thiamin and riboflavin in extrusion is affected by several variables (including moisture content), each of which could be approximated by a linear relationship that was valid within the narrow range of experimental values studied. Labuza (1972) reviewed studies on nutrient losses during dehydration and storage and suggested that the data of several investigators who studied retention of ascorbic acid imply a linear relation between the logarithm of the reaction rate constant and water activity. Wanninger (1972) proposed a model in which the reaction rate constant is directly proportional to water concentration.

Villota, Saguy, and Karel (1980) analyzed literature data on shelf life of stored dehydrated vegetable products. They classified the literature data into four main categories, as follows:

1. Products stored in air and failing due to off flavors

2. Similar products stored in absence of air

3. Products stored in air and failing due to changes in color

4. Products similar to those under 3 but stored in absence of air

They were able to correlate the data using a general mathematical model given by

$$\ln t_f = \alpha_1 + \alpha_2 \frac{E}{R} \frac{1}{T} + \alpha_3 (m - BET) \tag{9}$$

where

t_f = time of failure, days

T = temperature, K

m = moisture, g H_2O/g sample

BET = monolayer moisture content, g H_2O/g sample

α_1, α_2, α_3 = constants

E = activation energy, cal/mol

R = molar gas constant, 1.987 cal/(K mol)

The fit was excellent for several commodities in which flavor was the quality parameter subject to failure. When color was the cause of failure, in some cases the correlation was less perfect.

C. Oxygen

When the effect of oxygen is considered, it is often simply a question of total amount available for reaction with food components. If this amount is limited to a level that causes no significant effect in the food, and if there is no potential for additional oxygen coming into contact with the food, then the rate of quality loss is irrelevant. In other cases, the total amount of oxygen potentially able to react with nutrients is, in fact, significant, and consideration must be given to the effect of concentration (or partial pressure) of oxygen on the rate. A general equation form often found useful in correlating reaction rates with oxygen concentration is given in

$$R = \frac{[O_2]}{\alpha_4 + \alpha_5 [O_2]} \qquad (10)$$

where

R = reaction rate

$[O_2]$ = oxygen concentration

α_4, α_5 = constants

Note that when $\alpha_5[O_2]$ is very much less than α_4, the rate is linear with oxygen concentration, and when $\alpha_5 [O_2]$ is very much greater than α_4, the rate becomes independent of concentration. Both of these extremes can occur in the same product, depending on temperature, surface-to-volume ratio, oxygen pressure, and extent of reaction.

D. Compositional Factors Other Than Oxygen and Water

Quality of foods and the rate of change of quality in processing and storage depend obviously on the chemical composition. In many cases it is possible to correlate quality loss with the loss of a particular component, such as a vitamin or a pigment. In these cases it is possible to relate the rate of change of quality to the concentration C_i of the given component and to other reactants, if any, as shown in

$$\frac{dC_i}{dt} = kC_1^{n_1} C_2^{n_2} \cdots$$
(11)

The reaction has an overall order n (n = $\Sigma\, n_i$).

The rate constant k thus has the units of C^{1-n} per unit time. The value of n can be a fraction or a whole number. The order is strictly an empirical concept. However, when the stoichiometric equation represents the mechanism, the reaction order and the molecularity have the same value.

A zero-order reaction behavior has been observed for nonenzymatic browning in many products, for example, fruit products (Karel and Nickerson, 1964; Resnick and Chirife, 1979; Saguy, Kopelman, and Mizrahi, 1979b), vegetables (Hendel et al., 1955; Mizrahi, Labuza, and Karel, 1970), dry milk (Flink et al., 1974; Labuza, Mizrahi, and Karel, 1972), freeze-dried model systems containing carbohydrates, lipids, and proteins (Karel and Labuza, 1968), and ascorbic acid loss (Deng et al., 1978).

A first-order reaction was applicable in many food systems such as loss of ascorbic acid (Karel and Nickerson, 1964; Labuza, 1972; Laing, Schleuter, and Labuza, 1978; Lee et al., 1977; Nagy and Smoot, 1977; Reimer and Karel, 1978; Saguy, Kopelman, and Mizrahi, 1979a; Wanninger, 1972), color loss (Chou and Breene, 1972; Saguy, Kopelman, and Mizrahi, 1978), and softening of texture during heat processing (Bourne, 1976; Nichlas and Pflug, 1961; Paulus and Saguy, 1980).

Different reaction orders may be obtained, depending on the food systems and the quality factors involved. Enzyme-catalyzed reactions that follow Michaelis-Menton kinetics usually represent mixed-order reactions. The same mixed-order reaction is expected in lipid oxidation. If, however, the reaction order is not known from theory, it can be approximated by a numerical method.

III. EXAMPLES OF PREDICTION AND/OR
      SIMULATION OF QUALITY LOSSES IN
      PROCESSING AND STORAGE

As indicated previously, the simulation of quality changes requires the specification of the dependence of the designated quality measure on environmental and compositional factors [Eq. (1)], the specification of the time course of change of these factors [Eqs. (2) and (3)], and a method for analytical or numerical solution of these equations. We shall present a few examples of this approach.

A. Moisture Increase in a Packaged Product Stored
   in a Permeable Container at Constant Temperature
   and Humidity

We shall illustrate the quantitative approach on this very simple example, which occurs in many practical food packaging problems. We assume the moisture content to be the quality factor (e.g., moisture content must be maintained within some specified limits). Equation (1) becomes Eq. (12) expressing the rate of change of moisture content.

$$\frac{dm}{dt} = K_1(p_e - p)s^{-1} \tag{12}$$

where

   m = moisture content, g H_2O/g solids

   s = weight of solids in package, g

   t = time, days

   p_e = partial pressure of water outside package, torr

   p = partial pressure of water inside package, torr

   K_1 = overall water permeance constant for the package, g H_2O/
        (day torr)

If the assumption is made that the temperature of storage and the humidity of environment remain constant and that K_1 is independent of p, we obtain

$$\frac{dm}{dt} = K_2 - K_3 a \tag{13}$$

where

   $K_2 = K_1 p_e s^{-1}$

   p^0 = vapor pressure of H_2O at storage temperature

   $K_3 = K_1 p^0 s^{-1}$

   a = water activity inside the package

   The solution of the problem now requires the knowledge of a relation between m and a. Linearity is often assumed (an assumption often justified for relatively narrow ranges of m involved in pertinent storage times) as shown in

$$m = \sigma a + \rho \tag{14}$$

where σ, β = constant (g H_2O/g solids). We can write

$$\frac{dm}{dt} = \left(K_2 + \frac{K_3 \beta}{\sigma} \right) - \frac{K_3}{c} m \tag{13a}$$

This equation can easily be integrated and provides m as a simple function of time:

$$m = \left[m_i - \left(\frac{K_2 \sigma}{K_3} + \beta \right) \right] \exp \left[- \frac{K_3}{\sigma} t \right] + \frac{K_2 \sigma}{K_3} + \beta \tag{15}$$

where

m_i = initial moisture content

t = storage time

B. Prediction of Changes in a Quality Factor Dependent in Its Stability on Moisture Content

The previous example leads immediately to the simple and quite frequent problem of quality loss in moisture-sensitive foods. Consider, for instance, the case of flavor deterioration in a dry product stored under the simple conditions assumed in the example of Sec. III, A, and assume that the deterioration of quality depends on moisture content as given by Eq. (9).
Equation (1) can then be reformulated as

$$- \frac{dQ}{dt} = K_4 (m - BET) \tag{16}$$

Over a reasonably narrow range of moisture changes, we can assume that moisture changes as given by

$$m = m_i + K_5 t \tag{17}$$

We obtain then

$$-dQ = (K_6 + K_7 t) \, dt \tag{18}$$

where

$K_6 = K_4 m_i - K_4 BET$
$K_7 = K_4 K_5$

This equation is readily integrated to give

$$Q = Q_i - K_6 t - \frac{1}{2} K_7 t^2 \tag{19}$$

where Q_i is the initial quality.

The above two examples are of sufficient simplicity to allow analytical solution of the differential equations involved. In many cases this approach is quite feasible and appropriate (Karel, 1975). As the complexity of the problems increases, however, a more appropriate approach is the use of computer-assisted iteration techniques. Examples of this approach are given below.

C. Computer-Aided Prediction of Extent of Browning in Dehydrated Cabbage

The situation analyzed here is based on Mizrahi, Labuza, and Karel (1970) and is similar to the previous two examples in the following respects:

1. Temperature and humidity of storage are constant.

2. K_1 is a constant.

3. There is a single quality factor, namely absence of excessive browning.

However, Mizrahi, Labuza, and Karel (1970) found that the rate of browning, which is independent of time, depends on moisture content in a complex manner:

$$r \equiv \frac{dB}{dt} = K_8 \left[1 + \sin \left(-\frac{\pi}{2} + \frac{m\pi}{18} \right) \right]^n \tag{20}$$

where

B = browning in Klett units

n = exponent

K_8 = constant

In addition, the dependence of m on a was also considered nonlinear and given by

$$a = \frac{K_9 + m}{K_{10} + m} \tag{21}$$

where K_9, K_{10} = constant

Additional complications considered by Mizrahi, Labuza, and Karel (1970) were due to the generation of moisture by the browning reaction. These factors, however, shall be ignored here in order to introduce the iteration approach in as simple an example as possible. Figure 1 shows a flow chart of the computer iteration. The chart may be simply explained as follows: "Input" consists of constants and initial conditions, which have been explained previously, and of the chosen iteration interval Δt, which

FIGURE 1   Simulation of browning in a stored, dried vegatable.   (From Mizrahi, Labuza, and Karel, 1970.   Reprinted from Journal of Food Science Vol. 35, p. 799 ©1970 by Institute of Food Technologists.)

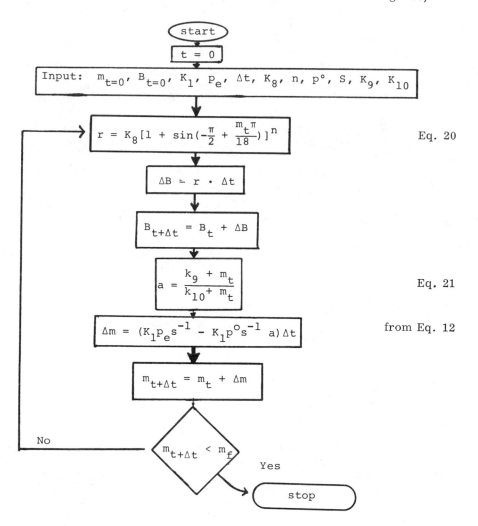

start

$t = 0$

Input: $m_{t=0}$, $B_{t=0}$, K_1, p_e, Δt, K_8, n, p°, S, K_9, K_{10}

$$r = K_8 [1 + \sin(-\frac{\pi}{2} + \frac{m_t \pi}{18})]^n$$

Eq. 20

$$\Delta B = r \cdot \Delta t$$

$$B_{t+\Delta t} = B_t + \Delta B$$

$$a = \frac{k_9 + m_t}{k_{10} + m_t}$$

Eq. 21

$$\Delta m = (K_1 p_e s^{-1} - K_1 p^\circ s^{-1} a) \Delta t$$

from Eq. 12

$$m_{t+\Delta t} = m_t + \Delta m$$

$m_{t+\Delta t} < m_f$

No

Yes

stop

the work of Mizrahi, Labuza, and Karel (1970) was 0.1 day. The iteration is then conducted for as many Δt intervals as desired (a cutoff value for t, m, or B must be provided; otherwise, the impossible instruction is given to the computer to continue the iteration indefinitely). In Fig. 1 subscripts t and t + Δt refer to values at the beginning and end of interval Δt, respectively.

Rate of browning at the beginning of the interval Δt is determined using Eq. (20). Increase in browning ΔB is added to B_t to obtain $B_{t+\Delta t}$. Water activity in the package is then calculated for time t using Eq. (21), and this activity is used to calculate the moisture increase in interval Δt, using the integrated form of Eq. (12). This amount is added to m_t to obtain $m_{t+\Delta t}$.

The new values of m, B, and t replace the previous ones, and the iteration is repeated.

The example chosen here to illustrate the computer iteration method is still extremely simple. In fact, Mizrahi, Labuza, and Karel (1970) found that the function r = f(m) could be expressed by functions simpler than Eq. (20) (at some loss of precision) and the above problem could be solved analytically. However, the important feature of this example which we wish to convey is the use of differential equations, which are then integrated numerically by iteration using an arbitrary number of time intervals. This method allows a number of variations. In our examples the time intervals are equal throughout the iteration, and the same functions describing the various parameters are valid throughout the iteration. However, it is very easy to use, if desirable, iterations in which Δt changes (e.g., as $\Delta B/\Delta t$ increases, Δt can be decreased to give iterations at approximately equal intervals of the reaction). We could also specify different isotherms [Eq. (21)] for different moisture-content ranges.

The use of iterative simulation of quality losses in processing and storage was found by our group to be a very versatile tool, and a necessary step on the way to process optimization (Mishkin, Karel, and Saguy, 1982). The limiting factor in its application is the development of appropriate models [Eqs. (2) and (3)] (Saguy and Karel, 1980).

It should be noted also that numerical integration, iteration, and printing out of results no longer require the use of expensive computers. Many, if not most, problems of quality prediction can be solved using programmable calculators. For instance, the Texas Instruments SR 58/59 series calculators have built-in software for numerical integration using Simpson's rule. Runge-Kutta methods for solution of one or two first-order differential equations are readily programmable on these calculators (Yakush, 1979).

D. More Complex Problems in Quality Loss
   During Storage

The general approach demonstrated above for simple shelf-life simulations can be readily extended to more complex problems. Some examples include

1. Simulation of quality losses under fluctuating storage conditions

2. Simultaneous functioning of two deteriorative mechanisms (Quast and Karel, 1972)

3. Quality changes in a food packaged in a material with barrier properties changing as the moisture content of the food changes in storage (Karel, Mizrahi, and Labuza, 1971)

4. Loss of vitamin content in a food stored in a permeable container (Riemer and Karel, 1978)

Many other examples could be cited.

E. Simulation of Quality Changes During Processing

Recently, interest has increased in prediction and simulation of quality changes during processes, such as thermal processing (Lund, 1973; Saguy, Kopelman, and Mizrahi, 1979a,b; Saguy and Karel, 1979; Teixeira et al., 1969), extrusion (Lorenz, Jansen, and Harper, 1980), cooking (Ohlsson, 1980a,b), and dehydration (Aguilera et al., 1975; Saguy et al., 1978).

We shall illustrate the application to dehydration using a realistic but still simplified example based on the work published by Aguilera et al. (1975). The techniques are similar to those used by Aguilera et al. (1975), but we have simplified a few assumptions for this illustration. Consider the air dehydration of potato, and assume that nonenzymatic browning, as measured by optical density, is the relevant quality factor. The rate of browning is a function of both temperature and moisture content of the potato:

$$r \equiv \frac{dB}{dt} = f(m,\ T) \tag{22}$$

where B is the browning in arbitrary units.

This function does not have to be expressed in the form of a single equation. Aguilera et al. (1975) used the data of Hendel et al. (1955) which yielded the following equations, valid for moisture contents above 1% dry basis.

$$\ln r_{T=50°C} = K_{11} - K_{12} \ln m \qquad \text{for } m > 20\% \tag{23a}$$

$$\ln r_{T=50°C} = K_{13} + K_{12} \ln m \qquad \text{for } 1\% < m \le 20\% \tag{23b}$$

$$\ln \left(\frac{r_T}{r_{T=50°C}} \right) = \frac{-E}{R} \left(\frac{1}{323} - \frac{1}{T} \right) \tag{24}$$

$$E = 26 \qquad \text{for } m > 16\% \tag{25}$$

$$E = 42 - m \qquad \text{for } m \le 16\% \tag{25a}$$

where

m = moisture content (% dry basis)

R = the gas constant [kcal/(mol K)]
T = temperature (K)

E = activation energy (kcal/mol)

We assume further that the drying occurs at constant air condition in a single "falling rate period" and that the browning can be analyzed as a quality factor, the deterioration of which depends on the average moisture in the dried material at any time. We shall also consider the average temperature to be the determining factor, but this assumption is actually unnecessary because temperature gradients occurring in air-dried materials are negligible (Karel, 1974). The conditions of the air must be specified in terms of T_{DB}, T_{WB}, and m_e (dry-bulb and wet-bulb temperatures and equilibrium moisture content of the product to be dried, respectively).

We follow Aguilera et al. (1975) in assuming the drying of potato slices in which unidimensional diffusion is controlling the rate, and under these conditions the average moisture content may be approximated by

$$\frac{m - m_e}{m_i - m_e} = \frac{8}{\pi^2} e^{-\pi^2 Dt/4L^2} \tag{27}$$

where

m_e = equilibrium moisture content

m_i = initial moisture content

D = diffusion coefficient, cm^2/sec

L = half-thickness of the slab, cm

t = drying time, sec

We assume further that the temperature of the slab (T_S) is given by

$$\ln (T_{DB} - T_S) = \ln (T_{DB} - T_{WB}) - \beta t \tag{28}$$

where

T_{DB} = dry-bulb temperature of air

T_{WB} = wet-bulb temperature of air

β = constant given by Eq. (29)

$$\beta = \frac{\Delta[\ln (m/m_i)]}{\Delta t} \tag{29}$$

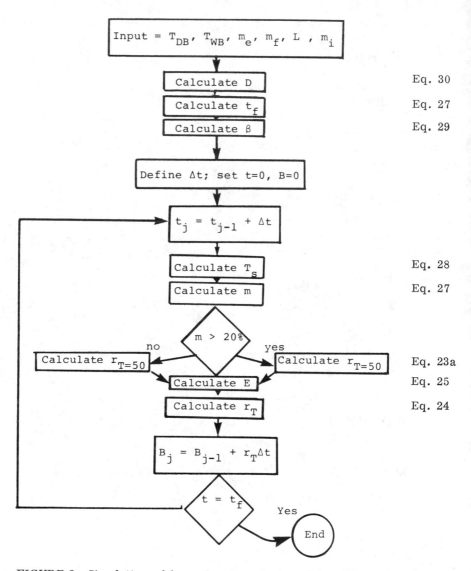

FIGURE 2   Simulation of browning in a drying slab.   (From Aguilera et al., 1975.)

The diffusion coefficient D is considered to be independent of m, but a function of T as shown in

$$D = 1.67 \times 10^{-5} e^{-7500/RT_{DB}} \tag{30}$$

where D is in centimeters squared per second.

The above set of equations can now be used to calculate browning as shown in Fig. 2.

The flow chart shows the following steps:

1. Input contains information about air conditions (T_{DB}, T_{WB}, and m_e), the initial moisture (m_i) and the final moisture content desired (m_f), and the slab half-thickness (L).

2. D is calculated using the specified T_{DB} and Eq. (30).

3. Total drying time (t_f) is calculated using the specified values of m_i, m_e, m_f, and L and the calculated D.

4. The value of β is calculated using Eq. (29).

5. The iteration begins using an interval (Δt) of predetermined size. [This could be specified as a fraction of t_f, e.g., $\Delta t = 0.01t_f$]. Arbitrarily, t is set equal to 0, and B is set equal to 0.

6. The temperature T_S is determined using Eq. (28).

7. m is calculated using Eq. (27).

8. The rate of browning is determined at each t_j as a function of m and T_S using Eqs. (23a) or (23b), (25), and (24).

9. The extent of browning in interval Δt is added to previous extent.

10. The iteration is repeated until $t = t_f$.

Our example is still relatively simple. More complex conditions of drying have been simulated successfully. The study on which we base our example considered, for instance, variation of moisture and of browning with position in the slab (Aguilera et al., 1975). In another study Villotta and Karel (1980b) simulated ascorbic acid retention during drying of a model system in which the complexity of conditions was much greater (i.e., shrinkage, consideration of three-dimensional diffusion, complex dependence of reaction rate on m and T). The basic premise of the simulations, namely, the coupling of mass- and heat-transport relations with models of kinetic behavior of quality, is well illustrated by the simple example we cited.

ACKNOWLEDGMENT

This work was supported in part by the National Science Foundation under Grant ENG-7824342 from the Division of Engineering.

REFERENCES

Aguilera, J. M., Chirife, J., Flink, J. M., and Karel, M. (1975). Lebensm.-Wiss. u.-Technol. 8:128.

Beetner, G., Tsao, T., Frey, A., and Harper, J. (1974). J. Food Sci. 39:207.

Bourne, M. C. (1976). Texture of fruits and vegetables, in Rheology and Texture in Food Quality (J. M. DeMan, P. W. Voisey, V. F. Rasper, and D. W. Stanley, eds.). Chap. 8, Avi, Westport, Conn.

Brown, A. D. (1976). Bact. Revs. 40:803.

Chou, H. and Breene, W. M. (1972). J. Food Sci. 37:66.

Deng, J. C., Watson, M., Bates, R. P., and Schroeder, E. (1978). J. Food Sci. 43:457.

Elton, G. A. H. (1969). Bakers Digest 43(3):24.

Flink, J. M., Hawkes, J., Chen, H., and Wong, E. (1974). J. Food Sci. 39:1244.

Guadagni, D. G. (1968). In Low Temperature Biology of Foodstuffs (J. Hawthorne, ed.). Pergamon Press, Oxford, p. 399.

Grosch, W. (1975). Z. Lebensm. Unters.-Forsch. 157:70.

Haard, N. F., Martins, I., Newbury, R., and Botta, R. (1979). J. Inst. Can. Sci. Technol. Aliment. 12(2):84.

Heiss, R. (1956). Verpackung feuchtigkeitsempfindlicher Gueter, Springer-Verlag, Berlin.

Hendel, C. E., V. G. Silveira, and W. O. Harrington (1955). Food Technol. 9:433.

Karel, M. (1960). Some effects of water and of oxygen on rates of reactions of food components. Ph.D. Thesis, MIT, Cambridge, Mass.

Karel, M. and Nickerson, J. T. R. (1964). Food Technol. 18:104.

Karel, M. and Labuza, T. P. (1968). J. Agr. Food Chem. 16:717.

Karel, M., Mizrahi, S., and Labuza, T. P. (1971). Modern Pack. 44(8): 54-58.

Karel, M. (1974). Fundamentals of dehydration processes, in Advances in Preconcentration and Dehydration of Foods (A. Spicer, ed.). Applied Science, London, pp. 45-94.

Karel, M. (1975). Protective packaging of foods, in Physical Principles of Food Preservation (M. Karel, ed.). Marcel Dekker, New York, p. 399.

Kwolek, W. F. and Bookwalter, G. N. (1971). Food Technol. 25(10):51.

Labuza, T. P. (1972). CRC Crit. Rev. Food Technol. 3(2):217.

Labuza, T. P., Mizrahi, S., and Karel, M. (1972). Trans. ASAE 15: 150.

Laing, B. M., Schlueter, D. L., and Labuza, T. P. (1978). J. Food Sci. 43:1440.

Lee, Y. C., Kirk, J. R., Bedford, C. L., and Heldman, D. R. (1977). J. Food Sci. 42:640.

Legault, R. R., Hendel, C. E., Talburt, W. F., and Pool, M. F. (1951). Food Technol. 5:417.

Lorenz, K., Jansen, G. R., and Harper, J. (1980). Cereal Foods World 25(4):161.

Lund, D. B. (1973). Food Technol. 27:16.

Matz, S. A. (1960). Bakery Technology and Engineering, Avi, Westport, Conn.

Mishkin, M., Karel, M., and I. Saguy (1982). Food Technol. 36(7):101.

Mizrahi, S., Labuza, T. P., and Karel, M. (1970). J. Food Sci. 35:804.

Nagy, S. and Smoot, J. M. (1977). J. Agr. Food Chem. 25:135.

Nicholas, R. C. and Pflug, I. J. (1961). Food Technol. 16(2):104.

Ohlsson, T. (1980a). J. Food Sci. 45:836.

Ohlsson, T. (1980b). J. Food Sci. 45:848.

Olley, J. and Ratkowsky, D. A. (1973). Food Technol. Australia 25(2):66.

Paulus, K., and I. Saguy (1980). J. Food Sci. 45:239.

Oswin, C. R. (1954). Protective Wrappings, Cam, London.

Quast, D. G. and Karel, M. (1972) J. Food Sci. 37:584.

Quast, D. G., Karel, M., and Rand, W. M. (1972). J. Food Sci. 37.070.

Quast, D. G., and Karel, M. (1973). Mod. Pack. 46(3):50.

Resnick, S. and Chirife, G. (1979). J. Food Sci. 44:601.

Reimer, J. and Karel, M. (1978). J. Food Process. Preserv. 1:293.

Rockland, L. and Steward, G. (eds.) (1981). Water Activity Influences on Food Quality, Academic Press, New York.

Saguy, I., Mizrahi, S., Villota, R., and Karel, M. (1978). J. Food Sci. 43:1861.

Saguy, I., Kopelman, I. J., and Mizrahi, S. (1978). J. Agr. Food Chem. 26:360.

Saguy, I., Kopelman, I. J., and Mizrahi, S. (1979a). J. Food Process. Eng. 2:213.

Saguy, I., Kopelman, I. J., and Mizrahi, S. (1979b). J. Food Process. Preserv. 2:175.

Saguy, I. and Karel, M. (1979). J. Food Sci. 44:1485.

Saguy, I. and Karel, M. (1980). Food Technol. 34(2):78.

Villota, R. and M. Karel (1980a). J. Food Process. Preserv. 4:111-134.

Villota, R. and M. Karel (1980b). J. Food Process. Preserv. 4:141-159.

Villota, R., Saguy, I., and Karel, M. (1980). J. Food Sci. 45:398-399, 401.

Teixeira, A. A., Dixon, J. R., Zahradnik, J. W., and Zinsmeister, G. E. (1969). J. Food Sci. 33:845.

Troller, J. and Christian, J. (1978). Water Activity and Food, Academic Press, New York.

Tsouroflis, S., Flink, J. M., and Karel, M. (1976). J. Sci. Food Agr. 27:509.

Wanninger, L. A., Jr. (1972). Food Technol. 26(6):42.

Yakush, S. A. (1979). Computer Programs in Biomed. 9:103.

6

Heat Transfer and Related Topics

DARYL B. LUND and JOHN P. NORBACK   University of Wisconsin,
Madison, Wisconsin

I. INTRODUCTION

Numerical techniques have found ready application particularly in solving heat-transfer problems associated with food processing. Generally steady-state problems are straightforward, and analytical solutions exist for the linear differential equations which describe rates of heat transfer. On the other hand, unsteady-state heat-transfer problems are more difficult, and numerical solutions are often the only resort to obtaining the solution. In this chapter, we shall make some general observations on the application of numerical techniques to solve the partial differential equations encountered in unsteady-state heat transfer.

By similarity arguments these same techniques are applied to mass-transfer problems. Furthermore, it should be emphasized that in solving these types of problems, one should first attempt to find an analytical solution. The diffusion equation (thermal or mass) with constant diffusivity has been solved analytically for numerous initial and boundary conditions, e.g.,

Carslaw and Jaeger (1959), Crank (1956), and Luikov (1968). Luikov (1968) also presents some solutions of the diffusion equation for variable diffusivity. Although this condition is extremely important in mass transfer during drying (cf. Schoeber, 1976), it is not of major significance in heat transfer (that is, thermal diffusivity is relatively insensitive to temperature). For both heat transfer and mass transfer, however, the assumption of isotropic behavior should be verified experimentally.

There are two basic unsteady-state processes worthy of comment. The first is unsteady-state heat transfer without phase change or significant occurrence of reactions or physical phenomena with latent heats. This is generally encountered in heating or cooling within the limits of -10 to 100°C or under pressure conditions prohibiting vaporization. This bound is dictated by the fact that water is the primary solvent for food systems, and phase changes are restricted to approximately this range. A slightly expanded range is realized with the presence of solutes and their effect on freezing point and boiling point.

Generally for unsteady-state heat transfer without phase change, the accompanying heats of reaction are small compared to sensible heat. Thus they generally can be ignored. Furthermore, latent heat associated with physical phenomena such as starch gelatinization, protein denaturation, or gelation is sufficiently small compared to sensible heat changes; so they also can be ignored.

The second basic unsteady-state process occurs with a significant associated latent heat. In the case of pure heat transfer, the latent heat of fusion or vaporization is incorporated with freezing and boiling. In the case of drying, the heat-transfer equation is coupled to the mass-transfer equation.

It should be pointed out that the numerical solutions to these heat-transfer problems may be used for several purposes. First the solution can be used to predict the behavior of state variables in the system (e.g., temperature or concentration as a function of time and position). Second, the solution could be used for parameter estimation using a nonlinear statistical procedure (e.g., the modified Levenberg-Marquardt procedure; Brown and Dennis, 1972). In this case the optimum value for a parameter or set of parameters if found which minimizes the sum of squares of differences between the predicted and experimental results. Finally these numerical solutions are essential for optimization or maximization problems where heat transfer is only one component of the system. For example, numerical solutions of heat transfer have been applied for the optimization of thermal processes for nutrient retention (Teixeira et al., 1969).

II. NUMERICAL TECHNIQUES

To illustrate the potential of a computer to generate useful information for systems involving unsteady-state heat transfer consider two mathematical expressions which can be discretely modeled:

$$\frac{\partial T}{\partial t} = \frac{\partial}{\partial x}\left(a\frac{\partial T}{\partial x}\right) \tag{1}$$

$$\frac{\partial T}{\partial t} = \frac{\partial}{\partial x}\left(a\frac{\partial T}{\partial x}\right) + L(t) \tag{2}$$

boundary and initial condition are given by

1. At $x = X$, $T = $ constant or $q = hdT$.

2. At $t = 0$, $T(x, t) = T_0$ or $T(x, t) = T_i(x, t)$.

3. At $x = X/2$, $dT/dx = 0$.

In some circumstances Eq. (1) may be modeled using Fourier series (Green-span, 1961). But these methods do not generalize easily to problems described by Eq. (2). Thus, in most problems encountered in practice, discrete methods are required.

While many schemes are possible for discretizing physical systems to model heat penetration and corresponding system states, the machine is essentially blind to two types of difficulties which may arise. The first consideration is that the machine generated model may not be stable, and the second is that the model may not be accurate.

Stability of a discrete approximation method may depend strictly on the step size chosen for the approximation or on machine's capacity to represent the identifying values for the state of a system or both. Any time the machine is in overflow because of performing the calculations required by the discretization of the system, the system is not stable. If the source of the instability is not due to errors in coding, then the system itself must be examined to determine the source of the instability. For many nonlinear systems, including those considered here, analyzing for the source of instability may not be possible. If instability occurs, a usual response is to decrease step sizes to cope with it. This may lead to accuracy difficulties which will be discussed later.

Some stability information is available, however. Solutions to Eqs. (1) and (2) are said to possess the "max-min property" (Friedman, 1964). A solution has the max-min property if over every bounded simply connected region, the solution takes on its maximum and minimum values for the region on the boundary of the region. A numerical solution to Eqs. (1) or (2) which satisfies this property is said to be "physically reasonable," which Greenspan (1974) has used to define a stable solution. He developed an appealing, intuitive argument to demonstrate that if h and k are the step sizes for x and t, respectively, in a forward-difference discretization of Eq. (1) and $\lambda = k/h^2$, then the system is stable if $0 = \lambda = 1/2$. This argument applies only to the two-dimensional case, and higher-dimensional analogs are not generally available. Detailed analysis of stability and its relation to convergence may be found in Hildebrand (1968). Despite the difficulties which may be caused by stability problems, at least when overflow occurs, it is signaled by the machine. Numerical solutions which are inaccurate have no such advantage. Keeping in mind that to have any solution at all requires that the numerical system be stable, we now pursue the question of accuracy for numerical models of Eqs. (1) and (2).

It is possible to construct stable approximations to Eqs. (1) and (2) for all values of λ in a backward-difference discretization of these problems with respect to the variable t. This result is not a panacea, however, since to generate the successive approximate solutions requires solving a linear system of equations for each row determined by a discrete value for x in the numerical scheme. In cases where steep slopes require that the step size in the x direction is small, this leads to a large linear system to be solved for every discrete value of t. This difficulty can be compounded in higher dimensions by steps associated with the other variables required.

If the numerical model is stable and the algorithm used is known to converge, then accuracy of the approximation depends upon the characteristics of the machinery used for computation and on the computing time expended to achieve the approximation. The backward-difference scheme, while making it possible to choose any step size for x and t, may lead to accuracy problems because of accumulation of error during a large number of steps. To a certain extent the machine and the algorithm are in competition: the algorithm requires a number of steps to converge, and at each step inaccuracies accumulate. Balancing these two trends depends upon the rate of convergence for the numerical scheme chosen and the characteristics of the machine used.

Many of the algorithms developed to model Eq. (1) can be generalized and applied to Eq. (2) in a straightforward fashion when the resulting system is mildly nonlinear. In many of these cases (Greenspan, 1974) computational requirements for these generalized systems are similar, or increased in a minor way. The requirement that L(t) is mildly nonlinear leads to algorithms of similar convergence characteristics for systems of the forms of Eqs. (1) and (2). Stability of the generalized systems remain about the same. Many instances in food processing do not meet this "mildly nonlinear" requirement, however. The initial-value, step-ahead procedures which work well on relatively shallow gradients require that the results from solving a linear system at one point in space be used to generate a linear system for an adjacent point. A similar recursive relation connects the linear system from one instant in discrete time to the next. Where pronounced nonlinearities occur, the impact of truncation errors and round-off errors in the machine accumulate beyond control. These effects can be so pronounced that they obscure convergence as well as provide inaccurate models.

For nonlinear systems a method which avoids the step-by-step error accumulation is required. Such a method has been developed and may be applied (Greenspan, 1974; Carassa and Parter, 1970) in circumstances where the extra work required does not become too burdensome. This method, sometimes called a "boundary-value technique," depends upon knowing the solution to the problem expressed by Eq. (2) in a truncated region and applying this information to solving the unsteady-state problem. This can be accomplished by using a Taylor's series expansion to approximate the operator

$$\frac{\partial T}{\partial t} - \frac{\partial}{\partial x} \left(\alpha \frac{\partial T}{\partial x} \right)$$

and generating a (large) linear system of equations, depending on backward- and forward-difference expressions with respect to time. If a system of the form of Eq. (2) has a solution, the Taylor expansions can be integrated into one large linear system which can be solved for an approximate solution to the overall problem. To solve such a large system may require special methods such as an application of Newton's method or Monte-Carlo techniques.

Some special adjustments of these methods may be required to accommodate the geometry of containers or equipment. Choice of backward-step methods, for example, may conveniently surmount computational difficulties when numerical schemes are applied near the center of cylindrical containers. Another strategy which may be of use over sharp, short gradients is to vary the step size for this region. For particular problems, piecing together approximations with different step sizes in different regions may provide a computationally tractible way to a discrete model.

Application of a machine to approximate numerically systems of the form of Eqs. (1) and (2) can yield useful results. Some caution must be maintained regarding the viability of these solutions, however. These numerical results may be intrinsically inaccurate and may be uneconomical in the sense of requiring too much computation. When the numerical solution is stable and accurate, then the food processor can achieve useful information by these methods.

A. Thermal Process Calculations

An example of the application of computers in food processing is the calculation of thermal processes that are applied to foods to extend the shelf life. Several models have been published and were reviewed by Hayakawa (1977). To illustrate the principles, the lethality-Fourier number method developed by Lenz and Lund (1977a, b) will be reviewed here.

For thermal death kinetics of microorganisms or, in fact, most heat-labile constituents in foods, first-order kinetics apply:

$$\frac{dx}{dt} = -kx \tag{3}$$

where

$$x = \frac{c}{c_0}$$

c = concentration at time t

c_0 = initial concentration at time t = 0

The reaction rate constant k is a function of temperature, and the Arrhenius equation may be used to describe the functional relation

$$k = k_r \exp\left[-\frac{E_a}{R_g}\left(\frac{1}{T} - \frac{1}{T_r}\right)\right] \tag{4}$$

where E_a is the Arrhenius activation energy (cal/mol); R_g is the gas law constant [1.987 cal/(mol K)] ; k_r is the reaction rate constant at a reference temperature T_r (min^{-1}).

Inserting Eq. (4) into Eq. (3) and integrating over process time gives

$$-\frac{\ln x}{k_r} = \int \exp\left[\frac{-E_a}{R_g}\left(\frac{1}{T} - \frac{1}{T_r}\right)\right] dt \tag{5}$$

To obtain a dimensionless solution, the Fourier number $(\alpha t/R^2)$ is substituted for time t yielding

$$L = \frac{-\alpha \ln x}{k_r R^2} = \int_0^t \exp\left[\frac{-E_a}{R_g}\left(\frac{1}{T} - \frac{1}{T_r}\right)\right] d\tau \tag{6}$$

where lethality number is defined by the left-hand side of Eq. (6), τ is the heating Fourier number, α is thermal diffusivity (cm^2/min), and R is the radius of the container (cm). The only Fourier-number-dependent variable in Eq. (6) is the temperature T, which is also a function of position. For heating a conduction-heating product in a can, the functional relationship between T and Fourier number is the solution of the unsteady-state heat conduction equation for a finite cylinder.

$$\frac{\partial T}{\partial t} = \alpha\left[\frac{1}{r}\frac{\partial}{\partial r}\left(r\frac{\partial T}{\partial r}\right) + \frac{\partial^2 T}{\partial z^2}\right] \tag{7}$$

This solution is given by Carslaw and Jaeger (1959) as

$$\frac{T_s - T}{T_s - T_i}\Bigg\} = \frac{4}{\pi}\sum_{n=0}^{\infty}\frac{(-1)^n}{2n+1}\exp\left[-\left(n+\frac{1}{2}\right)^2\pi^2\gamma^2\tau\right]\cos\left[\left(n+\frac{1}{2}\right)\pi\xi\right]\Bigg\}$$

$$x\Bigg\{2\sum_{k=1}^{\infty}\frac{J_0(\beta_k\rho)}{\beta_k J_1(\beta_k)}\exp\left[-\beta_k^2 t\right]\Bigg\} \tag{8}$$

where T_s is the heating-medium temperature and β are the roots of $J_0(\beta) = 0$.

For the temperature profile during cooling, following an initial period of heating, the unsteady-state heat-conduction equation for finite cylinders was solved by separation of variables. The boundary conditions are (1) symmetry at the center of the containers, (2) negligible resistance to heat transfer at the container surfaces, and (3) the initial temperature profile, given as the temperature profile at the end of heating.

These boundary conditions can be written

1. $\dfrac{\partial T}{\partial r} = 0$ at $r = 0$

2. $\dfrac{\partial T}{\partial z} = 0$ at $z = 0$

3. $T = T_w$ at $r = R$

4. $T = T_w$ at $z = \ell$

5. $T = T_s - (T_s - T_i)\theta_1$ at $t = 0$

where T_w is the cooling-water temperature (°F) and θ_1 is given by the right-hand side of Eq. (7).

If Eq. (6) is dimensionalized and the solution is assumed to be of the form

$$T = R(r)Z(z)B(t)$$

where R, Z, and B are only functions of the corresponding variables, then the resulting solution is

$$\frac{T - T_w}{T_s - T_i} = \left(4 \sum_{j=1}^{\infty} \sum_{i=0}^{\infty} \frac{(-1)^i \exp\left\{ -\left[\gamma^2 (i + 1/2)^2 \pi^2 + \beta_j^2 \right] \tau_w J_0(\beta_j \rho) \right\}}{(i + 1/2)\pi b_j J_1(b_j)} \right)$$

$$\times \cos\left[\left(i + \frac{1}{2} \right)\pi\xi \right] \left\{ \left(\frac{T_s - T_w}{T_s - T_i} \right) - \exp\left\{ -\left[\beta_j^2 + \left(i + \frac{1}{2} \right)^2 \pi^2 \gamma^2 \right] \tau \right\} \right\}$$

$$\tag{9}$$

where τ_w is the cooling Fourier number. The infinite series in Eqs. (8) and (9) were terminated when additional terms contributed less than 0.01% to the unaccomplished temperature.

Equation (6) was solved numerically on an Univac 1108 computer incorporating Eq. (8) or (9) with $\rho = \xi = 0$ (center) as the temperature profile. Using Simpson's rule, the integral of Eq. (6) was evaluated using Fourier increments of 0.02 units. This increment size was found to produce results which were within 0.1% of results when smaller increments were used. This corresponds to 4-min increments in a 307 x 306 can when the thermal diffusivity is 0.117 cm^2/min.

Equations (6) to (9) use R (can radius) as the characteristic dimension in defining the Fourier number. This was chosen because for most cans, R is the smaller dimension. However, the squat cans $(R = \ell)$, the smaller dimension ℓ is the controlling factor in heat penetration, and ℓ must be substituted for R in determining τ in Eqs. (6) to (9). Under these conditions it was found that the Fourier increment size of 0.02 units could also be used to integrate Eq. (6).

The lethality-Fourier number method for calculating lethality at a point can be rearranged so that retention $(x = c/c_0)$ at a point is calculated. Retentions calculated at a series of points can then be volume averaged. For conduction-heating foods, the density is approximately constant and volume averaging is equivalent to mass averaging. The average retention is then incorporated into an average lethality number (L) to facilitate presentation of results.

From the previous developments, the lethality number (L) for heating and cooling at each point is

$$L = \frac{-\alpha \ln x}{k_r R^2} = \int_0^\tau \exp\left[\frac{-E_a}{R_g}\left(\frac{1}{T_h} - \frac{1}{T_r}\right)\right] d\tau$$

$$+ \int_\tau^{\tau+0.7} \exp\left[\frac{-E_a}{R_g}\left(\frac{1}{T_c} - \frac{1}{T_r}\right)\right] d\tau \qquad (10)$$

where T_h is the temperature during heating at the position identified by ξ and ρ (normalized length Z/ℓ and normalized radius r/R, respectively) and T_c is the temperature during cooling at the same position.

The integral for heating is evaluated over Fourier heating time (τ), and the integral for cooling is evaluated from the end of heating to $\tau + 0.7$. This upper limit on cooling time was chosen since for most situations encountered in thermal processing, the temperature at time $\tau + 0.7$ for all points in the container will be sufficiently low so that thermal destruction does not occur (Lenz and Lund, 1977a). Equations (8) and (9) give the temperature profiles as a function of time and position for heating and cooling, respectively. Equation (10) can be rearranged to give

$$x = \exp\left(\frac{-k_r R^2}{\alpha} L\right) = \exp(-b_\tau L) \qquad (11)$$

where $b_t = k_r R^2 / \alpha$. To obtain an average retention ($\bar{x}$), Eq. (11) is volume averaged to give

$$\bar{x} = \frac{\int_V x \, dV}{\int_V dV} = \frac{\int_V \exp(-b_\tau L) \, dV}{\int_V dV} \tag{12}$$

Since the volume of the container is a constant (V_c), the volume integrals can be normalized, and $\bar{x}$ can be incorporated in an average lethality number (L) to yield

$$\bar{L} = \frac{-\alpha \ln \bar{x}}{k_r R^2} = \frac{-\alpha}{k_r R^2} \ln \int_0^1 \exp(-b_\tau L) \, d\underline{V} \tag{13}$$

where $\underline{V} = V/V_c$. If the container is a finite cylinder (such as a can), the volume integral becomes

$$\bar{L} = \frac{-\alpha}{k_r R^2} \ln \int_{-1}^1 \int_0^1 \exp(-b_\tau L) \rho \, d\rho \, d\xi \tag{14}$$

This integral can be evaluated by dividing the can into small-volume elements and evaluating the lethality at the center of each element. This method requires at least 100 elements and is the basis of the finite difference method of Teixeira et al. (1969). With this many volume elements, the computational time becomes excessive, and the method is expensive. However, Timbers and Hayakawa (1967) and Hayakawa (1969) reduced the number of elements by using a weighted-residual technique. This technique involves rewriting the volume integral as

$$\int_{-1}^1 \int_0^1 \exp(-b_\tau L) \rho \, d\rho \, d\xi = \sum_{i=1}^n W_i \sum_{j=1}^m W_j \exp(-b_\tau L_{ij}) \tag{15}$$

where L_{ij} is evaluated at certain points (ξ_i, ρ_j). The weighting factors (W_i, W_j) and points (ξ_i, ρ_j) can be found in Abramowitz and Stegun (1964) for various numbers of points (n, m). For this application, it was found that 24 points (n = 6, m = 4) were adequate to obtain the value of the integral within 0.1% of that when more points were used. Furthermore, this can be reduced to 12 points since the integral is symmetric in the ξ direction. The final result is

$$\bar{L} = \frac{-\alpha}{k_r R^2} \ln \left[2 \sum_{i=1}^3 W_i \sum_{j=1}^4 W_j \exp(-b_\tau L_{ij}) \right] \tag{16}$$

The program listing for calculating L and $\overline{L}$ is given in Lenz (1977) and is available from Lund.

III. CONCLUSIONS

Computing machines have greatly increased the capability of engineers to describe food processing systems quantitatively. Numerical solutions to the differential and integral equations which model food processing allow us to predict the behavior of state variables (e.g., concentration and temperature) of the system. Furthermore, these equations can be used to estimate parameters of the system and to assess the sensitivity of the state variable to these parameters. Finally, these quantitative expressions describing food systems are essential for optimization. This is one area in which there is the potential for much greater application of the techniques described in this and other chapters in this book.

Applying numerical techniques is not as simple as it appears on the surface. As we have pointed out, there are problems with stability and accuracy of solutions. In both areas there is the opportunity for research and application to problems in food processing.

IV. APPENDIX

Calculation of thermal processes applied to foods is an excellent example of use of computers in the food industries. The development cited in the text of this chapter is based on a first-order kinetic model which exhibits a temperature dependence described by the Arrhenius equation and conduction-heating products for which the analytical solution to the differential equation for heat transfer is known. Although this model is theoretically rigorous, the industry uses a more empirically derived model based on the Ball or formula method.

Hayakawa (1977) reviewed the calculation methods and published codes based on the semiempirical methods. The programs can be used to calculate heating time if the endpoint concentration of the target factor (microorganism, spore, or enzyme) is known or to calculate the final concentration if the thermal process is given for both center point and mass average. To compile the program, 158,000 cores are required, and 78,000 cores are required for computation.

A very useful application of computers is for optimization of processes. Plant (1970) published two codes in his Ph.D. thesis which may be useful in food applications. The thesis presents the theoretical aspects of an iterative procedure that can be used to compute optimal controls. The suggested procedure can be used to compute time-optimal, fuel-optimal, and energy-optimal controls for a class of linear time-invariant systems using a digital or a hybrid computer. The maximum principle of Pontryagin is used to reduce the optimization problem to a two-point boundary-value problem. A linearization technique, similar to Newton's method, is used to determine a new guess at each iteration; in this manner, the optimal control sequence that forces any given initial state to a target hypersphere about the origin of the state space can be generated with a modest amount of computer usage. Two FORTRAN programs are given; one for the time-optimal case and the other for the fuel-optimal case with a fixed response time. Experimental results which illustrate the usefulness of these programs are included.

ACKNOWLEDGMENT

This is a contribution from the College of Agricultural and Life Sciences, University of Wisconsin, Madison.

REFERENCES

Abramowitz, M. and Stegun, I. A. (eds.) (1964). Handbook of Mathematical Functions. National Bureau of Standards, Applied Mathematics Series 55.

Brown, K. M. and Dennis, L. E. (1972). Derivative free analogues of the Levenberg-Marquardt and Gauss Algorithms for nonlinear least squares approximations, Numerische Mathematik 18:289.

Carassa, A. and Parter, S. V. (1970). An Analysis of "Boundary Value Technique" for parabolic problems, Math. Comp. 110:315-340.

Carslaw, H. S. and Jaeger, J. C. (1959). Conduction of Heat in Solids, 2d ed. Clarendon Press, Oxford.

Crank, J. (1956). The Mathematics of Diffusion. University Press, Oxford.

Friedman, A. (1964). Partial Differential Equations of the Parabolic Type. Prentice-Hall, Englewood Cliffs, N.J.

Greenspan, D. (1961). Introduction to Partial Differential Equations. McGraw-Hall, New York.

Greenspan, D. (1974). Discrete numerical methods in physics and engineering, in Mathematics in Science and Engineering, Vol. 107. Academic, Press, New York.

Hayakawa, K. (1969). New parameters for calculating mass average sterilizing values to estimate nutrients in thermally conductive food, Can. Inst. Food Technol. J. 2:165-172.

Hayakawa, K. (1977). Mathematical methods for estimating proper thermal processes and their computer implementation, Advan. Food Res. 23:76-141.

Hildebrand, F. B. (1968). Finite Differential Equations and Simulations. Prentice-Hall, Englewood Cliffs, N.J.

Lenz, M. E. (1977). The lethality-Fourier number method. Its use in estimating confidence intervals of the lethality or process time of a thermal process and in optimizing thermal processes for quality retention. Ph.D. Thesis. Department of Food Science, University of Wisconsin, Madison.

Lenz, M. K. and Lund, D. B. (1977a). The lethality-Fourier number method. Experimental verification of a model for calculating temperature profiles and lethality in conduction-heating canned foods, J. Food Sci. 42:989-996, 1001.

Lenz, M. K. and Lund, D. B. (1977b). The lethality-Fourier number method. Experimental verfication of a model for calculating average quality factor retention in conduction-heating canned foods, J. Food Sci. 42:997-1001.

Luikov, A. V. (1968). Analytical Heat Diffusion Theory. Academic Press, New York.

Plant, J. B. (1970). An iterative procedure for the computation of optimal controls. Ph.D. Thesis, Massachusetts Institute of Technology, Cambridge, Mass.

Schoeber, W. J. A. H. (1976). Regular regimes in sorption processes, Ph.D. Thesis. Eindhoven Technical University, The Netherlands.

Teixeira, A. A., Dixon, J. R., Zahradik, J. W., and Zinsmeister, G. E. (1969). Computer optimization of nutrient retention in the thermal processing of conduction-heated foods, Food Technol. 23:845-850.

Timbers, G. E. and Hayakawa, K. (1967). Mass average sterilizing value for thermal processes, Food Technol. 21:1069, 1072-1073.

7

Linear Programming and Its Implementation

FILMORE E. BENDER and AMIHUD KRAMER* University of Maryland, College Park, Maryland

I. INTRODUCTION

Although linear programming problems can be solved without the use of computers, the ability to solve a linear programming problem repeatedly at a nominal cost is the key to the effective utilization of linear programming in the food industry. This chapter begins with a simple least-cost feed-mix problem which is solved graphically in order to illustrate the underlying logic of linear programming. This same problem is then solved using the simplex algorithm. The graphic and simplex solutions are then related to

---

*Deceased

one another. The last half of this chapter is devoted to setting up a small but reasonably realistic least-cost mayonnaise blend problem.

It is important to note that the linear programming framework is a useful way to organize information and keep track of the arithmetic of the problem. The computer merely relieves the individual of the computational burden. In the final analysis it is the responsibility of the user to define what can and cannot be done, what constraints are relevant, when the analysis is complete, and when it is time to implement a specific action.

II. THE FEED-MIX PROBLEM

Consider the need to mix a quantity of feed that meets the following specifications:

1. Contains at least 3 lb of fat

2. Contains no more than 5 lb of fiber

3. Contains at least 2 lb of protein

Assume that there are two ingredients (A, B) avilable that can be mixed to produce the needed feed. Table 1 shows the characteristics of these ingredients, including their cost per pound. Figure 1 shows a graph that illustrates all of the combinations of A and B that satisfy our three conditions concerning fat, fiber, and protein.

A. Graphic Solution

This figure was developed in the following manner. We began by calculating the quantity of A needed to achieve the minimum fat. Using only ingredient A, we would need 7.5 lb of A:

$$7.5 \text{ lb A} = 3 \text{ lb fat} \div 0.4 \ \frac{\text{lb fat}}{\text{lb A}}$$

TABLE 1   Characteristics of Two Feed Ingredients

One pound of A	One pound of B
Contains 0.4 lb fat	Contains 0.2 lb fat
Contains 0.4 lb fiber	Contains 0.5 lb fiber
Contains 0.2 lb protein	Contains 0.3 lb protein
Costs $0.30	Costs $0.20

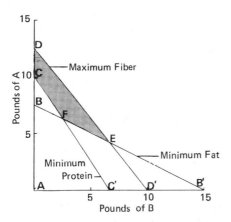

FIGURE 1 The combinations of ingredients A and B that meet the fat, fiber, and protein constraints.

On the other hand, to achieve the needed fat using only ingredient B would require 15 lb of B:

$$15 \text{ lb B} = 3 \text{ lb fat} \div 0.2 \ \frac{\text{lb fat}}{\text{lb B}}$$

The line BB' in Fig. 1 connects these two points and shows all of the possible combinations of A and B that just meet the requirement of 3 lb of fat. Since fat is a minimum restriction, the area above and to the right of this boundary is the region of acceptable solutions, i.e., combinations of A and B that yield 3 lb of fat or more.

A maximum limitation on fiber means that no more than 12.5 lb of A can be permitted:

$$12.5 \text{ lb A} = 5 \text{ lb fiber} \div 0.4 \ \frac{\text{lb fiber}}{\text{lb A}}$$

Similarly, the maximum limitation on fiber means that no more than 10 lb of B can be permitted:

$$10 \text{ lb B} = 5 \text{ lb fiber} \div 0.5 \ \frac{\text{lb fiber}}{\text{lb B}}$$

The line DD' in Fig. 1 connects these two points and shows all of the possible combinations of A and B that just match the maximum fiber permitted in the mix. Since fiber represents a maximum restriction, the area to the left of and below the boundary DD' is the region of acceptable solutions, i.e., combinations of A and B that contain 5 lb of fiber or less.

The protein restriction is a minimum requirement. If only A is utilized, a minimum of 10 lb must be used in order to achieve the needed protein:

$$10 \text{ lb A} = 2 \text{ lb protein} \div 0.2 \frac{\text{lb protein}}{\text{lb A}}$$

On the other hand, if only ingredient B is used, a minimum of 6.66 lb of B must be used in order to achieve the needed protein.

$$6.66 \text{ lb B} = 2 \text{ lb protein} \div 0.3 \frac{\text{lb protein}}{\text{lb B}}$$

The line CC' in Fig. 1 connects these two points and shows all of the possible combinations of A and B that just meet the requirement of 2 lb of protein. Since protein is a minimum restriction, the area above and to the right of this boundary is the region of acceptable solutions, i.e., combinations of A and B that provide 2 lb of protein or more.

The shaded area in Fig. 1 labeled CDEF is the feasible region defined by these three restrictions. Only points within this area (CDEF) satisfy all three restrictions.

Only corner points of a feasible region are of interest to us. Table 2 shows the cost at the four corners. In addition, the fat, fiber, and protein content of the four proposed rations indicate that all four solutions meet the three restrictions. Since all four solutions meet the stated requirements, solution F is the optimal answer. It is the least-cost solution to our original problem. Any other point which satisfies the requirements of fat, fiber, and protein will cost as much or more than the $2.38 of point F.

We can view the search for a least-cost solution in terms of a family of isocost curves. This would be a family of lines where each line represents combinations of A and B with a given cost. In our case, ingredient A costs 30¢/lb. Ingredient B costs 20¢/lb. We could purchase 1.5 lb of B for the same amount of money as it takes to purchase 1 lb of A.

TABLE 2   Calculation of the Costs at the Corners of Fig. 1

Point	A (lb)	B (lb)	Fat (lb)	Fiber (lb)	Protein (lb)	Cost (dollar)
C	10	0	4.0	4.0	2.0	3.00
D	12.5	0	5.0	5.0	2.5	3.75
E	4.17	6.67	3.0	5.0	2.8	2.59
F	6.25	2.5	3.0	3.75	2.0	2.38

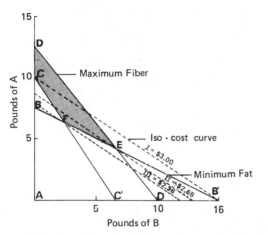

FIGURE 2   A family of isocost curves developed in searching for a least-cost solution.

Figure 2 shows the feasible region CDEF with three parallel isocost curves. The lowest of these (i.e., III – $2.38) passes through the point F. This confirms the optimal solution determined in Table 2. In practice, we would rarely solve a linear programming problem graphically. However, it is useful to look at graphic solutions of simple problems in order to develop an intuitive sense of how linear programming works. Only relatively simple problems may be conveniently solved using graphs.

B. Simplex Solution

The simplex is an algorithm developed to solve linear programming problems. It can be viewed as a systematic search procedure. Starting from a convenient initial point, the simplex searches from one corner to the next until an optimal solution is found. The Appendix to this chapter provides a summary statement of the steps involved in using the simplex algorithm. These steps can be followed by a clerk. Alternatively, a simple computer program can be written to follow these steps. However, virtually all computers have a standard linear programming package which has been prepared by the manufacturer. The Appendix to this chapter shows a typical data input format.

In order to set up a linear programming model, we must define three types of activities: (1) real activities denote those items which represent some physical action that we can take; (2) disposal activities denote those items which keep track of any slack in the system; and (3) "artificial activities," which are not always needed, provide a means of starting our search for an optimal from a convenient and known location.

TABLE 3   Initial Tableau of the Least-Cost Ration Problem

C_j	Supply or Activity Level B	Disposal Activities			Real Activities		Artificial Activities	
		$0\ \dfrac{\text{¢ profit}}{\text{lb excess fat}}$	$0\ \dfrac{\text{¢ profit}}{\text{lb unused fiber}}$	$0\ \dfrac{\text{¢ profit}}{\text{lb excess protein}}$	$-30\ \dfrac{\text{¢ profit}}{\text{lb } A}$	$-20\ \dfrac{\text{¢ profit}}{\text{lb } B}$	$-m\ \dfrac{\text{¢ profit}}{\text{unit } Q_1}$	$-m\ \dfrac{\text{¢ profit}}{\text{unit } Q_2}$
		Fat x_1	Fiber x_2	Protein x_3	A x_4	B x_5	Q_1 x_6	Q_2 x_7
$-m\ \dfrac{\text{¢ profit}}{\text{unit } Q_1}$	3 lb fat	$-1\ \dfrac{\text{lb fat}}{\text{lb excess fat}}$	$0\ \dfrac{\text{lb fat}}{\text{lb unused fiber}}$	$0\ \dfrac{\text{lb fat}}{\text{lb excess protein}}$	$.4\ \dfrac{\text{lb fat}}{\text{lb } A}$	$.2\ \dfrac{\text{lb fat}}{\text{lb } B}$	$1\ \dfrac{\text{lb fat}}{\text{unit } Q_1}$	$0\ \dfrac{\text{lb fat}}{\text{unit } Q_2}$
$0\ \dfrac{\text{¢ profit}}{\text{lb unused fiber}}$	5 lb fiber	$0\ \dfrac{\text{lb fiber}}{\text{lb excess fat}}$	$1\ \dfrac{\text{lb fiber}}{\text{lb unused fiber}}$	$0\ \dfrac{\text{lb fiber}}{\text{lb excess protein}}$	$.4\ \dfrac{\text{lb fiber}}{\text{lb } A}$	$.5\ \dfrac{\text{lb fiber}}{\text{lb } B}$	$0\ \dfrac{\text{lb fiber}}{\text{unit } Q_1}$	$0\ \dfrac{\text{lb fiber}}{\text{unit } Q_2}$
$-m\ \dfrac{\text{¢ profit}}{\text{unit } Q_2}$	2 lb prot.	$0\ \dfrac{\text{lb protein}}{\text{lb excess fat}}$	$0\ \dfrac{\text{lb protein}}{\text{lb unused fiber}}$	$-1\ \dfrac{\text{lb protein}}{\text{lb excess protein}}$	$.2\ \dfrac{\text{lb protein}}{\text{lb } A}$	$.3\ \dfrac{\text{lb protein}}{\text{lb } B}$	$0\ \dfrac{\text{lb protein}}{\text{unit } Q_1}$	$1\ \dfrac{\text{lb protein}}{\text{unit } Q_2}$
Z_j	$-5m$ ¢ profit	$m\ \dfrac{\text{¢ profit}}{\text{lb excess fat}}$	$0\ \dfrac{\text{¢ profit}}{\text{lb unused fiber}}$	$m\ \dfrac{\text{¢ profit}}{\text{lb excess protein}}$	$-.6m\ \dfrac{\text{¢ profit}}{\text{lb } A}$	$-.5m\ \dfrac{\text{¢ profit}}{\text{lb } B}$	$-m\ \dfrac{\text{¢ profit}}{\text{unit } Q_1}$	$-m\ \dfrac{\text{¢ profit}}{\text{unit } Q_2}$
$Z_j - C_j$	$-5m$ ¢ profit	$m\ \dfrac{\text{¢ profit}}{\text{lb excess fat}}$	$0\ \dfrac{\text{¢ profit}}{\text{lb unused fiber}}$	$m\ \dfrac{\text{¢ profit}}{\text{lb excess protein}}$	$-.6m + 30\ \dfrac{\text{¢ profit}}{\text{lb } A}$	$-.5m + 20\ \dfrac{\text{¢ profit}}{\text{lb } B}$	$0\ \dfrac{\text{¢ profit}}{\text{unit } Q_1}$	$0\ \dfrac{\text{¢ profit}}{\text{unit } Q_2}$

Table 3 shows the initial tableau for this linear programming problem. The initial tableau is a summary statement of all of the relevant information. Table 3 shows all of the coefficients with their appropriate dimensions. In practice the dimensions are omitted. The initial tableau and complete solution to this problem are shown in Table 4.

The problem that we are solving can be stated compactly: Minimize

$$\Pi = 30x_4 + 20x_5$$

subject to

$$0.4x_4 + 0.2x_5 \geq 3$$
$$0.4x_1 + 0.5x_5 \leq 5$$
$$0.2x_4 + 0.3x_5 \geq 2$$

where

$$x_4 = \text{pounds of A}$$
$$x_5 = \text{pounds of B}$$

Since it is usually more convenient to maximize the objective function, we could state the same problem as follows: Maximize

$$\Pi = -30x_4 - 20x_5$$

subject to

$$0.4x_4 + 0.2x_5 \geq 3$$
$$0.4x_4 + 0.5x_5 \leq 5$$
$$0.2x_4 + 0.3x_5 \geq 2$$

Working with inequalities is often cumbersome. The initial tableau utilizes disposal activities to convert inequalities to equalities. In addition, it is usually convenient to start at some easily identifiable location such as point A in Fig. 1, where we are using 0 lb of A and 0 lb of B. In order to accomplish this, the initial tableau uses artificial activities.

The initial tableau (of Tables 3 and 4) starts our search for an optimal solution from point A of Fig. 1. However, point A is outside of the feasible region. The artificial activities (Q_1 and Q_2) are arbitrary devices to satisfy the minimum fat and protein restrictions until we can modify our plan by bringing in a positive quantity of A and B. In order to ensure that the artificial activities are removed from the plan (i.e., become nonbasic), we impose a very great cost on a unit of an artificial activity. The convention is to use -m as the cost, where m is greater than any other cost in the problem. Any cost figure containing any fraction of m is considered greater

TABLE 4 Simplex Solution of the Least-Cost Ration Problem

C_j	Supply or Activity Level B	Disposal Activities — 0 Fat x_1	0 Fiber x_2	0 Protein x_3	Real Activities — -30 A x_4	-20 B x_5	Artificial Activities — $-m$ Q_1 x_6	$-m$ Q_2 x_7	
$-m$ Q_1 (Min Fat)	3	-1	0	0	.4	.2	1	0	Initial Tableau
0	5	0	1	0	.4	.5	0	0	
$-m$ Q_2 (Min Protein)	2	0	0	-1	.2	.3	0	1	
Z_j	-5m	m	0	m	-.6m	-.5m	-m	-m	
$Z_j - C_j$	-5m	m	0	m	-.6m + 30	-.5m + 20	0	0	
-30 A	7.5	-2.5	0	0	1	.5	2.5	0	Second Iteration
0 Maximum Fiber	2	1	1	0	0	.3	-1	0	
$-m$ Q_2 (Min Protein)	.5	.5	0	-1	0	.2	-.5	1	
Z_j	-.5m	-.5m	0	m	0	-.2m	1.5m	-m	
$Z_j - C_j$	-225	+75	0	m	0	+5	+75	0	
-30 A	10	0	0	-5	1	1.5	0	5	Third Iteration
0 Maximum Fiber	1	0	1	2	0	-.1	0	-2	
0 Fat	1	1	0	-2	0	.4	-1	2	
Z_j							m	m	
$Z_j - C_j$	-300	0	0	150	0	-25	+150	-150	
-30 A	6.25	-3.75	0	2.5	1	0	3.75	-2.5	Fourth Iteration
0 Maximum Fiber	1.25	.25	1	1.5	0	0	-.25	-1.5	
-20 B	2.5	2.5	0	-5	0	1	-2.5	5	
Z_j							m	m	
$Z_j - C_j$	-237.5	62.5	0	25	0	0	+87.5	+200	

156

than any cost figure that does not contain m. We could use an arbitrarily high cost (e.g., -999999) to accomplish this same goal.

The first row of the initial tableau of Tables 3 and 4 is the C_j row (i.e., the cost row or objective function). This row indicates the cost or revenue that we receive from one unit of each activity. Here our goal is cost minimization. However, if we retain the convention that revenues are positive and that costs are negative, we will still maximize the objective function (Π). As a result, in order to minimize cost, our goal will be to maximize the following objective function:

$$\Pi = 0x_1 + 0x_2 + 0x_3 - 30x_4 - 20x_5 - mx_6 - mx_7$$

Since we cannot have negative amounts of activities, the linear program will maximize Π by making x_4, x_5, x_6, and x_7 as small as possible.

The next row is one of our constraints. It keeps track of the minimum quantity of fat needed for the final mix. Initially, we stated that the final mix must contain at least 3 lb of fat. Table 1 stated that 1 lb of A contained 0.4 lb of fat and that 1 lb of B contained 0.2 lb of fat. We used this information to draw our first boundary (BB') in Fig. 1. The boundary showed the region defined by the inequality

$$0.4x_4 + 0.2x_5 \geq 3$$

In the case of a minimum restriction, the simplex procedure eliminates inequalities by converting them to equalities through the creation of disposal (slack) activities and artificial activities. The disposal activity keeps track of how much more fat than the specified minimum is in the mixture. The artificial activity simply permits the fat minimum to be satisfied with something that does not physically exist. (Another way of looking at an artificial activity is to view it as an accounting device which keeps track of how much of a minimum remains unsatisfied.) Thus, our inequality is changed to the following equality:

$$-1x_1 + 0.4x_4 + 0.2x_5 + 1x_6 = 3$$

or

$$0.4x_4 + 0.2x_5 + 1x_6 = 3 + 1x_1$$

that is, the fat provided by ingredient A ($0.4x_4$) plus the fat provided by ingredient B ($0.2x_5$) plus the fat provided by the artificial activity ($1x_6$, which will be 0 in our optimal solution) will equal the minimum fat required (3 lb) plus any excess above that minimum ($1x_1$).

The next row is our second constraint. It keeps track of the total amount of fiber in the mix. It states that no more than 5 lb of fiber are permitted in the final mix. A pound of A has 0.4 lb of fiber. A pound of B has 0.5 of fiber. We used this information to draw our second boundary (DD') in Fig. 1. The boundary showed the region defined by the following inequality:

$$0.4x_4 + 0.5x_5 \leq 5$$

Since this is a maximum restriction, we need add only the disposal activity to establish the following equality:

$$1x_2 + 0.4x_4 + 0.5x_5 = 5$$

That is, the pounds of fiber that we are permitted to put into the mix, but have not done so ($1x_2$), plus the pounds of fiber from ingredient A ($0.4x_4$) plus the pounds of fiber from ingredient B ($0.5x_5$) will equal the total pounds of fiber permitted (5).

The fourth row is our last constraint for this problem. This row relates to our requirement that the final mix must contain at least 2 lb of protein. Table 1 stated that 1 lb of A contained 0.2 lb of protein and that 1 lb of B contained 0.3 lb of protein. We used this information to draw our final boundary (CC') in Fig. 1. This boundary showed the region defined by the inequality

$$0.2x_4 + 0.3x_5 \geq 2$$

As we did in the case of the minimum requirement for fat, we convert this inequality to an equality by adding a disposal activity (which keeps track of excess protein) and an artificial activity (which keeps track of how much of the minimum is unsatisfied). Thus, our inequality is changed to the following equality:

$$-1x_3 + 0.2x_4 + 0.3x_5 + 1x_7 = 2$$

or

$$0.2x_4 + 0.3x_5 + 1x_7 = 2 + 1x_3$$

That is, the protein provided by ingredient A ($0.2x_4$) plus the protein provided by ingredient B ($0.3x_5$) plus the protein provided by the artificial activity ($1x_7$, which will be zero in our optimal solution) will equal the minimum protein required (2 lb) plus any excess above the minimum ($1x_3$).

The next row is the Z_j row. The Z_j row represents the cost of each activity in terms of the activities in basis, i.e., in terms of the activities in the current plan. It should be noted that Table 4 shows the C_j value of each activity as a row across the top of the table. It also shows the C_j values of each of the activities in each basis as a column at the left-hand edge of the table.

In ordinary practice, the Z_j row is only calculated for the initial tableau. Once the Z_j row is calculated, the Z_j - C_j row can be calculated, and no further use is made of the Z_j row.

The Z_j row is calculated by multiplying each element in a column by the corresponding C_j for each row and summing these products. The calculations are as follows:

1. For the supply or B column,

$$-m \times 3 = -3m$$
$$0 \times 5 = 0$$
$$-m \times 2 = \underline{-2m}$$
$$-5m$$

As a result, -5m is entered as the Z_j value for the supply column.

2. For the disposal activities,

   a. Fat

$$-m \times -1 = m$$
$$0 \times 0 = 0$$
$$-m \times 0 = \underline{0}$$
$$m$$

   b. Fiber

$$-m \times 0 = 0$$
$$0 \times 1 = 0$$
$$-m \times 0 = \underline{0}$$
$$0$$

   c. Protein

$$-m \times 0 = 0$$
$$0 \times 0 = 0$$
$$-m \times -1 = \underline{m}$$
$$m$$

These values for the disposal activities are entered in the Z_j row for their respective columns.

3. For the real activities,

    a. For A,

$$-m \times 0.4 = -0.4m$$
$$0 \times 0.4 = 0$$
$$-m \times 0.2 = \underline{-0.2m}$$
$$-0.6m$$

    b. For B,

$$-m \times 0.2 = -0.2m$$
$$0 \times 0.5 = 0$$
$$-m \times 0.3 = \underline{-0.3m}$$
$$-0.5m$$

These values for the real activities are entered in the Z_j row for their respective columns.

4. For the artificial activities,

    a. For Q_1,

$$-m \times 1 = -m$$
$$0 \times 0 = 0$$
$$-m \times 0 = \underline{0}$$
$$-m$$

    b. For Q_2,

$$-m \times 0 = 0$$
$$0 \times 0 = 0$$
$$-m \times 1 = \underline{-m}$$
$$-m$$

These values for the artificial activities are entered in the Z_j row for their respective columns.

With the Z_j row completed, it is possible to calculate the $Z_j - C_j$ row. For each column, subtract the C_j value from the corresponding Z_j value. For the supply or B column, $C_j = 0$ by definition. So that $Z_j - C_j = -5m - 0 = -5m$.

1. For the disposal activities, the $Z_j - C_j$ values are

   a. For fat,

   $$m - 0 = m$$

   b. For fiber,

   $$0 - 0 = 0$$

   c. For protein,

   $$m - 0 = m$$

2. For the real activities, the $Z_j - C_j$ values are

   a. For ingredient A,

   $$-0.6m - (-30) = -0.6m + 30$$

   b. For ingredient B,

   $$-0.5m - (-20) = -0.5m + 20$$

3. For the artificial activities, the $Z_j - C_j$ values are

   a. For Q_1,

   $$-m - (-m) = 0$$

   b. For Q_2,

   $$-m - (-m) = 0$$

Upon entering these values in the $Z_j - C_j$ row for their respective columns, we have completed our initial tableau and are ready to use the simplex procedure to solve this linear programming problem.

The initial tableau as shown in Table 4 describes our problem in a systematic way and enables us to search for an optimal solution. The initial tableau describes point A (i.e., the origin) of Fig. 1. It says that we are outside of the feasible region. We are using 0 lb of ingredient A and 0 lb of ingredient B. We are meeting our minimum fat requirement with 3 units of Q_1 (an artificial activity) and our minimum protein requirement with 2 units of Q_2 (an artificial activity). We are incurring an infinitely large cost (-5m) for this plan.

In order to search for an improved plan, we examine the $Z_j - C_j$ row for the most negative value. Remember that m stands for more than any other number in the problem. As a result, we concentrate our attention on columns with negative values of m. The real activity A has the most negative $Z_j - C_j$ with the value $-0.6m + 30$. We can improve our plan by moving in the direction of increasing quantities A, that is, moving along the vertical axis toward point B in Fig. 1.

Having chosen to move in the A direction (i.e., A is the incoming column), we must determine the maximum distance that we can move before we reach a corner. We determine this distance by dividing the values in the supply column by the rate at which the ingredient satisfies needs (minimum constraints) or uses available resources (maximum constraints).

For fat,

$$7.5 \text{ lb A} = 3 \text{ lb fat} \div 0.4 \frac{\text{lb fat}}{\text{lb A}}$$

For fiber,

$$12.5 \text{ lb A} = 5 \text{ lb fiber} \div 0.4 \frac{\text{lb fiber}}{\text{lb A}}$$

For protein,

$$10 \text{ lb A} = 2 \text{ lb protein} \div 0.2 \frac{\text{lb protein}}{\text{lb A}}$$

These equations indicate that fat is the most restricting constraint. The simplex proceeds by producing a new plan which uses as much of ingredient A as possible. The artificial activity Q_1 leaves the plan (basis) and is replaced by A. This is accomplished by dividing each element in the fat (Q_1) row by the fat-A coefficient. (This intersection of the outgoing row and the incoming column is the pivot point).

The newly defined A row in the second iteration (Table 4) is determined as follows:

$$7.5 \text{ lb A} = 3 \text{ lb fat} \div 0.4 \frac{\text{lb of fat}}{\text{lb A}}$$

$$-2.5 \frac{\text{lb A}}{\text{lb excess fat}} = -1 \frac{\text{lb fat}}{\text{lb excess fat}} \div 0.4 \frac{\text{lb fat}}{\text{lb A}}$$

$$0 \frac{\text{lb A}}{\text{lb unused fiber}} = 0 \frac{\text{lb fat}}{\text{lb unused fiber}} \div 0.4 \frac{\text{lb fat}}{\text{lb A}}$$

$$0 \frac{\text{lb A}}{\text{lb excess protein}} = 0 \frac{\text{lb fat}}{\text{lb excess protein}} \div 0.4 \frac{\text{lb fat}}{\text{lb A}}$$

$$1 \frac{\text{lb A}}{\text{lb A}} = 0.4 \frac{\text{lb of fat}}{\text{lb A}} \div 0.4 \frac{\text{lb fat}}{\text{lb A}}$$

$$0.5 \frac{\text{lb A}}{\text{lb B}} = 0.2 \frac{\text{lb fat}}{\text{lb A}} \div 0.4 \frac{\text{lb fat}}{\text{lb A}}$$

$$2.5 \frac{\text{lb A}}{\text{lb Q}_1} = 1 \frac{\text{lb fat}}{\text{lb Q}_1} \div 0.4 \frac{\text{lb fat}}{\text{lb A}}$$

$$0 \frac{\text{lb A}}{\text{lb Q}_2} = 0 \frac{\text{lb fat}}{\text{lb Q}_2} \div 0.4 \frac{\text{lb fat}}{\text{lb A}}$$

With this newly defined A row and the initial tableau, the simplex procedure defines the second iteration. It should be noted that since the Z_j
row of the initial tableau was needed only to define our $Z_j - C_j$ row (i.e.,
our decision row), the Z_j row does not appear in future iterations.

The new elements are determined by adjusting the corresponding elements of the initial tableau. For the first column, these calculations are
as follows:

$$2 \text{ lb fiber} = 5 \text{ lb fiber} - \left(7.5 \text{ lb A} \times 0.4 \, \frac{\text{lb fiber}}{\text{lb A}}\right)$$

$$0.5 \text{ lb protein} = 2 \text{ lb protein} - \left(7.5 \text{ lb A} \times 0.2 \, \frac{\text{lb protein}}{\text{lb A}}\right)$$

Our $Z_j - C_j$ calculation is done in two parts by calculating the m component and the dollar component separately.

$$-0.5m\text{¢} = -5m\text{¢} - \left(7.5 \text{ lb A} \times -0.6m \, \frac{\text{¢}}{\text{lb A}}\right)$$

$$-225\text{¢} = 0\text{¢} - \left(7.5 \text{ lb A} \times 30 \, \frac{\text{¢}}{\text{lb A}}\right)$$

This yields a profit for the current plant of $-0.5m - 225$ cents. We
interpret a negative profit as a cost. The cost of this plan is still infinitely
large (i.e., there is an m component) because we are still outside of the
feasible region. The current plan corresponds to point B of Fig. 1.

The remaining columns are calculated in a similar manner:
For excess fat,

$$1 \frac{\text{lb fiber}}{\text{lb excess fat}} = 0 \frac{\text{lb fiber}}{\text{lb excess fat}} - \left(-2.5 \frac{\text{lb A}}{\text{lb excess fat}} \times 0.4 \, \frac{\text{lb fiber}}{\text{lb A}}\right)$$

$$0 \frac{\text{lb protein}}{\text{lb excess fat}} = 0 \frac{\text{lb protein}}{\text{lb excess fat}} - \left(-2.5 \frac{\text{lb A}}{\text{lb excess fat}} \times 0.2 \, \frac{\text{lb protein}}{\text{lb A}}\right)$$

Again, $Z_j - C_j$ has two components:

$$-0.5m \frac{\text{¢}}{\text{lb excess fat}} = m \frac{\text{¢}}{\text{lb excess fat}}$$
$$- \left(-2.5 \frac{\text{lb A}}{\text{lb excess fat}} \times -0.6m \, \frac{\text{¢}}{\text{lb A}}\right)$$

$$0.75 \frac{\text{¢}}{\text{lb excess fat}} = 0 \frac{\text{¢}}{\text{lb excess fat}}$$
$$- \left(-2.5 \frac{\text{lb A}}{\text{lb excess fat}} \times 30 \, \frac{\text{¢}}{\text{lb A}}\right)$$

$$= -0.5m + 75$$

For fiber, because the cell value for the A row in the fiber column in the second iteration is zero, the fiber column for the second iteration will be the same as for the initial tableau. The reader may check by following the pattern of calculations used for the other columns.

For excess protein, as in the case of the fiber column, the protein column is unchanged.

For A, because A is the activity which has just entered the basis (i.e., it is the incoming row for the second iteration), all of the remaining cell entries will be zeros. The reader may check by following the pattern of calculations used for the other columns.

For B,

$$0.3 \; \frac{\text{lb fiber}}{\text{lb B}} = 0.5 \; \frac{\text{lb fiber}}{\text{lb B}} - \left(0.5 \; \frac{\text{lb A}}{\text{lb B}} \times 0.4 \; \frac{\text{lb fiber}}{\text{lb A}} \right)$$

$$0.2 \; \frac{\text{lb protein}}{\text{lb B}} = 0.3 \; \frac{\text{lb protein}}{\text{lb B}} - \left(0.5 \; \frac{\text{lb A}}{\text{lb B}} \times 0.2 \; \frac{\text{lb protein}}{\text{lb A}} \right)$$

Again, $Z_j - C_j$ has two components:

$$-0.2m \; \frac{\text{¢}}{\text{lb B}} = -0.5m \; \frac{\text{¢}}{\text{lb B}} - \left(0.5 \; \frac{\text{lb A}}{\text{lb B}} \times -0.6m \; \frac{\text{¢}}{\text{lb A}} \right)$$

$$5 \; \frac{\text{¢}}{\text{lb B}} = 20 \; \frac{\text{¢}}{\text{lb B}} - \left(0.5 \; \frac{\text{lb A}}{\text{lb B}} \times 30 \; \frac{\text{¢}}{\text{lb A}} \right)$$

For Q_1, artificial activities are treated the same way that disposal and real activities are treated.

$$-1 \; \frac{\text{lb fiber}}{\text{lb Q}_1} = 0 \; \frac{\text{lb fiber}}{\text{lb Q}_1} - \left(2.5 \; \frac{\text{lb A}}{\text{lb Q}_1} \times 0.4 \; \frac{\text{lb fiber}}{\text{lb A}} \right)$$

$$-0.5 \; \frac{\text{lb protein}}{\text{lb Q}_1} = 0 \; \frac{\text{lb protein}}{\text{lb Q}_1} - \left(2.5 \; \frac{\text{lb A}}{\text{lb Q}_1} \times 0.2 \; \frac{\text{lb protein}}{\text{lb A}} \right)$$

Again, $Z_j - C_j$ has two components:

$$1.5m \; \frac{\text{¢}}{\text{lb Q}_1} = 0m \; \frac{\text{¢}}{\text{lb Q}_1} - \left(2.5 \; \frac{\text{¢}}{\text{lb Q}_1} \times -0.6m \; \frac{\text{¢}}{\text{lb A}} \right)$$

$$-75 \; \frac{\text{¢}}{\text{lb Q}_1} = 0 \; \frac{\text{¢}}{\text{lb Q}_1} - \left(2.5 \; \frac{\text{¢}}{\text{lb Q}_1} \times 30 \; \frac{\text{¢}}{\text{lb A}} \right)$$

For Q_2, as we have seen, since the cell entry of the A row in the second iteration of Table 4 is 0 for the Q_2 column, the Q_2 column is unchanged for the second iteration.

Upon completion of the second iteration, we continue our search for an optimal solution by searching the $Z_j - C_j$ row for the most negative entry. The most negative entry in the second iteration is for "excess fat," which is one of our disposal activities. However, its $Z_j - C_j$ value of $-0.5m + 75$ is the most negative because of our interpretation of m as infinitely large. Having made the decision to bring excess fat into the plan, the third iteration is calculated using the procedures that have been established.

The third iteration defines a plan that

1. Uses 10 lb of A

2. Leaves 1 lb of allowable fiber unused

3. Contains 1 lb of excess fat (i.e., 1 lb more than the required minimum)

4. Costs 300¢

The plan defined by the third iteration corresponds to point C of Fig. 1. Since the cost of the plan contains no m component, we know that the plan is feasible. That is, we can actually meet all of our constraints. However, our goal is not only a feasible plan but an optimal one. We search the $Z_j - C_j$ row of the third iteration to determine if any negative entries remain which would indicate that further cost savings exist. The most negative $Z_j - C_j$ is for the real activity B with a value of -25.

The fourth iteration is the optimal solution to our original problem, since no negative entries remain in the $Z_j - C_j$ row. It corresponds to point F of Fig. 1. It shows that the least-cost plan

1. Uses 6.25 lb of A

2. Leaves 1.25 lb of allowable fiber unused

3. Uses 2.5 lb of B

4. Costs 237.5¢

This is the least-cost plan, given our original problem that we obtain at least cost a mix that meets the following specifications:

1. Contains at least 3 lb of fat

2. Contains no more than 5 lb of fiber

3. Contains at least 2 lb of protein

In addition to the primal solution which tells us how much of A and B to use, the final iteration of Table 4 contains the dual solution which tells us how much these three constraints are costing on the margin. The shadow prices are the $Z_j - C_j$ entries for the disposal activities.

Fat = 62.5

Fiber = 0

Protein = 25

These shadow prices are interpreted as follows:

If we reduce the fat minimum by 1 lb (i.e., from 3 to 2 lb), we could save 62.5¢.

If we increase the fiber maximum by 1 lb (i.e., from 5 to 6 lb), we would not save anything. This appears reasonable when we reexamine Fig. 1. The least-cost solution is at point F. If we increase the maximum allowable fiber, we will move the line DD' up and to the right. This will make the feasible region larger but will not affect the current optimal solution.

If we reduce the protein minimum by 1 lb (i.e., from 2 to 1 lb), we could save 25¢.

Whether the restrictions on the mix should be changed is a question that only a nutritionist can answer. The linear program does the best it can within the boundaries that have been defined. It signals through the shadow prices how expensive some of these constraints are. A nutritionist using a computer and a linear program can solve many problems in a short period of time. By examining the dual solution to see which constraints are costing the most, a nutritionist can often make balanced tradeoffs that result in still lower cost solutions. This is the exploration of a trained nutritionist. A linear programming specialist or computer operator cannot make the needed nutritional judgments.

III.   ADDING A RESTRICTION ON THE WEIGHT
OF THE MIX

The reader undoubtedly noticed that the final mix of Table 4 resulted in a total of 8.75 lb (6.25 lb A plus 2.5 lb B). Ordinarily, it is desirable to have a linear program define a solution that is in some reasonable unit; for example, 10 lb.

If we wanted a total mix equal to 10 lb, we would need to add that constraint. Our problem would then be to determine the least-cost mix that meets the following specifications:

1.   Contains at least 3 lb of fat.

2.   Contains no more than 5 lb of fiber.

3.   Contains at least 2 lb of protein.

4.   Weighs exactly 10 lb.

Using the same two ingredients A and B that were described in Table 1, we can construct a graph showing our choices for this new problem. Figure

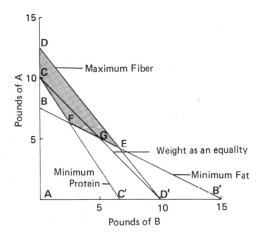

FIGURE 3   The addition of an equality as a constraint greatly reduces the feasible region.

3 shows these choices. The boundaries CC' (minimum protein), BB' (minimum fat) and DD' (maximum fiber) were our original three constraints. The shaded area CDEF was our original feasible region as shown in the construction of Fig. 1.

The requirement that the mix must weight exactly 10 lb is shown by the line CD'. Only points along the line CD' result in combinations of ingredients A and B that exactly equal 10 lb. As a result, our feasible region is reduced from the shaded area CDEF to the line segment CG. Only points along the line segment CG simultaneously satisfy all four constraints. Since only corner points are of interest, the least-cost solution will be either point C or point G. This small modification of our original problem illustrates two important points about linear programming problems.

1. Given an initial problem and its solution, if an additional constraint is imposed, the result must be to increase cost (or reduce profit) because adding a constraint reduces our choices. The only other possibility is that the constraint is trivial (e.g., in our case, that the total mix must weight no more than 15 lb).

2. Although adding a constraint reduces the choices available for a solution, if the added constraint is an equality, it greatly reduces the choices. A quick inspection of Fig. 3 confirms this situation. The addition of the weight equality resulted in a great reduction in the size of the feasible region. As a result, the authors are reluctant to use equality constraints where inequalities will suffice. In some instances such as food-blending problems, it is often essential to specify weight of the final mix as an equality.

TABLE 5 Simplex Solution of the Least–Cost Ration Problem Where Weight Is a Restriction

Column groupings: Disposal Activities = x_1, x_2, x_3; Real Activities = x_4, x_5; Artificial Activities = x_6, x_7, x_8.

c_j	Basis	Supply or Activity Level B	Fat x_1 (0)	Fiber x_2 (0)	Protein x_3 (0)	A x_4 (−30)	B x_5 (−20)	Q_1 x_6 (−m)	Q_2 x_7 (−m)	Q_3 x_8 (−m)	Iteration
$-m$	Q_1 (Min Fat)	3	−1	0	0	.4	.2	1	0	0	Initial Tableau
0	Fiber	5	0	1	0	.4	.5	0	0	0	
$-m$	Q_2 (Min Protein)	2	0	0	−1	.2	.3	0	1	0	
$-m$	Q_3 (Weight)	10	0	0	0	1	1	0	0	1	
	Z_j	$-15m$	m	0	m	$-1.6m$	$-1.5m$	$-m$	$-m$	$-m$	
	$Z_j - C_j$	$-15m$	m	0	m	$-1.6m + 30$	$-1.5m + 20$	0	0	0	
0	A	7.5	−2.5	0	0	1	.5	2.5	0	0	Second Iteration
0	Maximum Fiber	2	1	1	0	0	.3	−1	0	0	
$-m$	Q_2 (Min Protein)	.5	.5	0	−1	0	.2	−.5	1	0	
$-m$	Q_3 (Weight)	2.5	2.5	0	0	0	.5	−2.5	0	1	
	$Z_j - C_j$	$-3m - 225$	$-3m + 75$	0	m	0	$-.7m + 5$	$4m - 75$	0	0	
	A	10	0	0	−5	1	1.5	0	5	0	Third Iteration
	Maximum Fiber	1	0	1	2	0	−.1	0	−2	0	
	Fat	1	1	0	−2	0	.4	−1	2	0	
	Q_3 (Weight)	0	0	0	5	0	−.5	0	−5	1	
	$Z_j - C_j$	-300	0	0	$-5m + 150$	0	$.5m - 25$	m	$6m - 150$	0	
	A	10	0	0	0	1	1	0	0	1	Fourth Iteration
	Maximum Fiber	1	0	1	0	0	.1	0	0	−.4	
	Fat	1	1	0	0	0	.2	−1	0	.4	
	Protein	0	0	0	1	0	−.1	0	−1	.2	
	$Z_j - C_j$	-300	0	0	0	0	-10	m	m	$m - 30$	
	A	5	−5	0	0	1	0	5	0	−1	Fifth Iteration
	Maximum Fiber	.5	−.5	1	0	0	0	.5	0	−.6	
	B	5	5	0	0	0	1	−5	0	2	
	Protein	.5	.5	0	1	0	0	−.5	−1	.4	
	$Z_j - C_j$	250	50	0	0	0	0	$m - 50$	m	$m - 10$	

Table 5 shows the linear programming solution to our problem which is graphed in Fig. 3. The weight restriction has resulted in an additional row. As in the case with minimum restrictions, we need to add an artificial activity in order to meet the equality initially (i.e., when A = 0 and B = 0 and we are outside of the feasible region). Unlike our previous inequalities which we converted to equalities through the introduction of disposal activities, our weight restriction is an equality, and no slack activity is introduced.

The weight row in the initial tableau of Table 5 can be read as follows:

$$1x_4 + 1x_5 + 1x_8 = 10$$

where x_8 as an artificial activity with an infinitely high cost (-m) will be equal to zero in the optimal solution.

The fifth iteration of Table 5 contains the optimal solution to our new problem. The primal solution states that we should

1. Use 5 lb of A

2. Leave unused 0.5 lb of allowable fiber

3. Use 5 lb of B

4. Exceed minimum protein by 0.5 lb

5. Incur a cost of 250¢

The dual solution indicates that if we could reduce the fat minimum by 1 lb, we could reduce the cost of our feed by 50¢. There would be no saving from increasing allowable fiber or reducing minimum protein.

The graphic and simplex solution of these two simple problems have illustrated the basic logic of linear programming. The next section illustrates how a small but reasonably realistic problem might be approached.

IV. A LEAST-COST MAYONNAISE BLEND

Consider the case of a mayonnaise manufacturer who wishes to produce a new economy product. Initially, he decides to limit his formula to oil, yolk, salt, vinegar, mustard, and water, and places the following constraints on these ingredients:

1. The amount of oil can vary from a low of 65% to a high of 80%, and 25% of the yolk is considered to be oil.

2. The amount of yolk can vary from 6.5% to 8.0%.

3. Salt content cannot exceed 0.8%. Since salted yolks are used in preparing the product, one-tenth of the yolk is counted as salt.

4. The vinegar is one-tenth acid, and the acid content can vary from 0.2 to 0.5% of the total weight of the product.

5. All of the vinegar and half of the yolk is moisture, and the moisture content can vary from a low of 12% to a high of 18%.

6. The content of mustard flour can vary from a low of 0.25% to a high of 1.0%.

7. The moisture content must be less than or equal to 50 times the acid content.

8. The oil content must be less than or equal to 12 times the yolk content.

The cost of each of the ingredients is as follows:

Ingredient	Symbol	Cost ($/lb)
Oil	X_1	0.58
Yolk	X_2	0.93
Salt	X_3	0.03
Vinegar	X_4	0.26
Mustard	X_5	0.71
Water	X_6	0.00

As we work through this problem, we will write our constraints as equalities or inequalities without identifying the disposal or artificial activities needed for use with the simplex algorithm. A problem of this size would normally be solved on the computer and, as shown in the Appendix to this chapter, most computer algorithms work directly from the equality or inequality as we will develop it here.

Most books which illustrate linear programming as a management tool imply that someone in the firm identifies a problem, writes out the relevant activities and constraints (i.e., the entire initial tableau) and has it subsequently solved on a computer and immediately implemented by the firm. Fortunately, it never happens that way. The process can best be viewed as an evolutionary one where management poses a problem. The linear programming model and the computer keep track of the arithmetic. Management reviews the computer solution and identifies constraints or activities that were initially omitted. The problem is redefined and resolved. Management examines the new solution, and so on until management feels that the solution is practical and superior. It is not possible to convey to the reader, within the confines of a book, this interactive process between management and the computer. The following simple illustration may help.

Suppose that our mayonnaise problem had been stated only in terms of the ingredients. We would then identify our objective function as minimize

$$58x_1 + 93x_2 + 3x_3 + 26x_4 + 71x_5 + 0x_6$$

Since we are developing a blend, it would probably be convenient to require that 100 lb be mixed in order to simplify the interpretation of our results since the answer would come out in percentages. Consequently, we would define the following weight constraint:

$$1x_1 + 1x_2 + 1x_3 + 1x_4 + 1x_5 + 1x_6 = 100$$

With an objective function and a single constraint, we have a valid linear programming problem. The least-cost solution to this problem is to put 100 lb of water into a barrel and label it mayonnaise. Management will immediately perceive that this solution has a number of difficulties not the least of which will be consumer complaints and the possibility of a series of lawsuits. At this point, management might point out that minimum quantities of oil and yolk are required. Given the new solution, management might point out that for flavor considerations, certain quantities of mustard and acid are needed. And so this process continues until management is satisfied that the blended mayonnaise meets the needs of the firm. The linear programming model assures that no lower cost formulation exists given the ingredients and prices that were identified.

The constraints for this problem, in addition to the weight constraint just given, would be as follows:

1. Oil constraints. The minimum oil content is 65%. Since we are blending a 100 lb mix, we can state this as a minimum of 65 lb of oil in the mix. One lb of oil contains 1 lb of oil. One lb of yolk contains 0.25 lb of oil. Our constraint can be written as

$$1x_1 + 0.25x_2 \geq 65$$

Using similar reasoning our maximum oil constraint can be written as

$$1x_1 + 0.25x_2 \leq 80$$

2. Yolk constraints. Only yolk contains yolk. Therefore, our yolk constraints can be written as

$$1x_2 \geq 6.5 \qquad 1x_2 \leq 8.0$$

3. Salt constraint.

$$0.1x_2 + 1x_3 \leq 0.8$$

4. Acid constraints.

$$0.1x_4 \geq 0.2 \qquad 0.1x_4 \leq 0.5$$

5. Moisture constraints.

$$0.5x_2 + 1x_4 + 1x_6 \geq 12$$

$$0.5x_2 + 1x_4 + 1x_6 \leq 18$$

6. Mustard constraints.

$$1x_5 \geq 0.25 \qquad 1x_5 \leq 1.0$$

The first six constraints examined were relatively straightforward. Even without linear programming, it would be possible to keep track of these constraints. The two remaining constraints are ratios. They are more difficult to comprehend. However, it is the ability of the linear programming model to deal with them that makes it such a useful management tool in the food industry.

    7. Moisture-to-acid ratios. The specification for this mayonnaise requires the following:

Moisture ≤ 50(acid)

Half of yolk, all of vinegar and all of water are moisture.

$$0.5x_2 + 1x_4 + 1x_6 \leq 50(\text{acid})$$

Vinegar is the sole source of acid and is 10% acid.

$$0.5x_2 + 1x_4 + 1x_6 \leq 50(0.1x_4)$$

or

$$0.5x_2 + 1x_4 + 1x_6 \leq 5x_4$$

TABLE 6 Initial Tableau for Least-Cost Mayonnaise Blend

Item	Oil x_1	Yolk x_2	Salt x_3	Vinegar x_4	Mustard x_5	Water x_6	Right-hand side
Cost	-0.58	-0.93	-0.3	-0.26	-0.71	0	
Weight	1	1	1	1	1	1	$= 100$
Min oil	1	0.25					≥ 65
Max oil	1	0.25					≤ 80
Min yolk		1					≥ 6.5
Max yolk		1					≤ 8.0
Max salt		0.1	1				≤ 0.8
Max acid				0.1			≤ 0.5
Min acid				0.1			≥ 0.2
Min moisture		0.5		1		1	≥ 12
Max moisture		0.5		1		1	≤ 18
Moisture/acid		0.5		-4		1	≤ 0
Oil/yolk	1	-11.75					≤ 0
Min mustard					1		≥ 0.25
Max mustard					1		≤ 1.0

TABLE 7   Primal Solution to Least-Cost
Mayonnaise Blend Problem

Activity	Level
Oil	78.280
Salt	.134
Yolk	6.667
Vinegar	3.600
Mustard	.250
Water	11.069
Total mayonnaise	100.000
Total cost	$ 52.72

Since we cannot have an unknown (in this case, $5x_4$) on the right-hand side of the inequality, we rewrite the expression as

$$0.5x_2 - 4x_4 + 1x_6 \leqslant 0$$

8. Oil-to-yolk ratio. In a manner similar to that used for constraint 7,

Oil $\leqslant 12$(yolk)

$$1x_1 + .25x_2 \leqslant 12(1x_2)$$
$$1x_1 + .25x_2 \leqslant 12x_2$$
$$1x_1 - 11.75x_2 \leqslant 0$$

By examining each constraint separately, the problem can be broken down into manageable components. Table 6 shows the initial tableau for this problem. The solution is summarized in Table 7.

Given the solution in Table 7 management may decide that whole egg should be considered as a potential ingredient in addition to yolk. In addition, several types of oil may be included in the same matrix (e.g., cotton-seed oil, soybean oil, coconut oil) in order to compete in the blend as prices change. As a result of taste panel data, additional constraints may be added, for example, a sugar-to-acid ratio or a salt-to-sugar ratio. In other words, the solution shown in Table 7 is merely a good starting point in the development of a least-cost mayonnaise formulation which will be changed from time to time as ingredient prices change.

V. SUMMARY

This chapter illustrated the logic of linear programming as a tool which can assist decision making. The final problem examined was small but reasonably realistic. If management views linear programming as a useful device to keep track of relationships and choices and views the computer as a means of performing the necessary arithmetic calculations quickly and accurately, then management will have greatly enhanced its ability to evaluate alternatives and make profitable decisions.

VII. APPENDIX

A. Using the Simplex Algorithm to Solve Linear Programming Problems

In the text, a simple linear programming problem with minimum constraints is solved using the simplex algorithm in Table 4. A similar problem with minimum constraints and an equality constraint is solved in Table 5. Each step of the necessary calculations is shown in the discussion of Tables 4 and 5. This section provides a summary listing of these steps.

    1. Define the initial tableau.

    a. List the activities (i.e., the things that can be done) as columns.

    b. List the constraints (i.e., the factors which limit the levels that the activities can take) as rows.

    c. List the supply of resources, the level of equality constraints, and the level of the requirements imposed as the B column which keeps track of these amounts.

    d. (1) For each maximum constraint, define a disposal activity which consists of a 1 in the affected row and 0 elsewhere.
       (2) For each minimum constraint, define a disposal activity which consists of a -1 in the affected row and 0 elsewhere.
       (3) No disposal activity should be defined for an equality constraint.
       (4) For each minimum constraint and each equality constraint, define an artificial activity which is composed of a 1 in the affected row and 0 elsewhere.

    e. List the cost or revenue for each activity across the top of the initial tableau as the C_j row. List revenues as positive values. List costs as negative values. The disposal activities have zero C_j values. For the artificial activities, enter a C_j value of $-m$, where this is interpreted as an infinitely large cost.

    f. Fill in the cell values for the real activities. If an activity decreases the B value (i.e., the right-hand side), the cell value is

positive. If an activity increases the B value, the cell value is
negative.

(1) For maximum constraints, a positive value indicates that the
    activity uses the resource; a negative value indicates that the
    activity augments the resource.

(2) For minimum constraints, a positive value indicates that the
    activity satisfies the requirement; a negative value indicates
    that the activity increases the requirement.

(3) For equality constraints, a positive value indicates that the
    activity satisfies the equality; a negative value indicates that
    the activity disturbs the equality.

g. (1) Define the Z_j row. For each row, list the C_j value as a new
       column to the left of the initial tableau. For rows that are
       maximum constraints, the C_j is 0 because it is the disposal
       activity that is in basis. For minimum or equality constraints,
       the C_j is -m because it is the artificial activity that is in basis.

       For each column (including the B column), the Z_j value is
       determined by multiplying each cell value by the C_j for that row.
       The C_j's will be either 0 or -m. The sum of these products is
       the Z_j value for that column.

   (2) Define the $Z_j - C_j$ or decision row. In the first column (B), the
       value for $Z_j - C_j$ in the initial tableau will be the Z_j value cal-
       culated in the previous step. For the remaining columns (i.e.,
       disposal, real, and artificial activities), the $Z_j - C_j$ values are
       calculated by subtracting C_j for each column from the Z_j for
       each column.

The initial tableau is now complete. The Z_j row is no longer needed.

2. Examine the $Z_j - C_j$ row for all disposal, real, and artificial activities. If all entries are zero or positive, the problem is solved, and we
have an optimal solution. If one or more $Z_j - C_j$ values are negative, we
choose the most negative. Remember that m is greater than any other number which does not contain m. In addition, if an artificial activity shows a
negative $Z_j - C_j$ value, an error has been made.

The column with the most negative $Z_j - C_j$ value is our incoming column. That is, it is an activity that will enter our plan and replace something that we are currently doing. Since we want to bring in as much of the
incoming activity as possible, the next step is to determine the most limiting constraint.

3. For each cell in the B column, divide by the corresponding cell of
the incoming activity. This division represents the use of resources or the
satisfaction of requirements by the incoming activity. Consequently, we do
not do this division for the $Z_j - C_j$ cells. In addition, a cell entry of zero
in our incoming column indicates that this activity uses none of this resource
or satisfies none of this requirement. Our quotient is infinite. By the same
token, a negative cell entry means that the incoming column does not use

the resource but rather adds to it, or in the case of a requirement, it does not satisfy the requirement but rather increases it. Again, our quotient is infinite.

Having proceeded with our division, one row at a time, we choose the row with the smallest quotient as the outgoing row. That is, the incoming column will replace the outgoing row in our plan.

4. The intersection of the incoming column and the outgoing row is the pivot point. The new incoming row for the next iteration is defined by dividing the outgoing row, one cell at a time, by the pivot point. This redefines our old row in terms of our new entering activity.

5. The remaining cells of the new iteration are defined by using the incoming column (which was chosen in step 2) and the incoming row (which was obtained by step 4) to update the cell values of the previous iteration.

The updating process redefines each activity in terms of the changed plan that results from bringing in the incoming row. With our new iteration complete, we return to step 2 and continue until the problem is solved.

B. Computerized Linear Programming Algorithms

In the main text of this chapter, we used the simplex algorithm to solve linear programming problems. We did this so that the reader would have a clearer conception of how linear programming solves complex problems. In addition, presenting the simplex algorithm permits the reader to solve relatively small linear programming problems without resorting to the use of computers.

However, access to a computer is necessary for any practical use of linear programming. Fortunately, nearly all computers have a linear programming package available. Even more important, most large-scale computers use identical input formats. Once an individual becomes familiar with using a computer format, it is relatively easy to switch from one brand of computer to another.

Tables 8 and 9 show the coded information for defining the two simple linear programming problems solved in the chapter. As can be seen in these examples, only the structural matrix has to be entered into the computer. The disposal and artificial activities are created by the computer algorithm as a result of the row identification provided by the user. These examples were coded according to the instructions provided by UNIVAC for its FMPS package which are summarized here:

1. Row Names Format

The row names are comprised of eight alphabetic or numeric characters punched in card columns 5 to 12. Of course, each row name must be distinct.

Column 3 of the row identification cards contains an indicator which identifies whether the row is (1) an objective function row or objective function change vector for parametric programming, (2) an equation

178

TABLE 8 Computer Data to Solve the Least-Cost Ration Problem Shown in Table 4

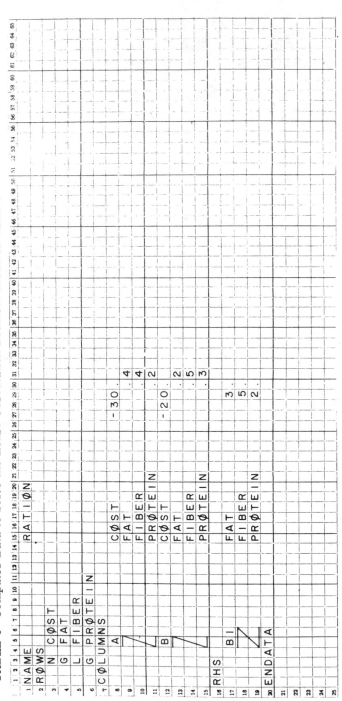

TABLE 9 Computer Data to Solve the Modified Least-Cost Ration Problem Shown in Table 5

```
NAME          REV-RATIØ
RØWS
 N  CØST
 G  FAT
 L  FIBER
 G  PRØTEIN
 E  WEIGHT
CØLUMNS
    A   CØST     -30.4    FAT       .4
        FIBER      .2     PRØTEIN
        WEIGHT    -20.2
    B   CØST      -20.1   FAT       .2
        FIBER      .5     PRØTEIN
        WEIGHT     1.
RHS
    BI  FAT        3.     FIBER     5.
        PRØTEIN    2.     WEIGHT   10.
ENDATA
```

constraint (one that must be satisfied as an equality by the solution), (3) an "equal or less than equal" inequality constraint, or (4) a "greater than or equal" inequality constraint. The character to be punched in each case is as follows:

	BCD character punched in column 3	Meaning of character
(1)	N	Row is an objective function or change vector row.
(2)	E	Row is an equality constraint.
(3)	L	Row is an "equal or less than equal" inequality constraint (i.e., maximum).
(4)	G	Row is a "greater than or equal" inequality constraint (i.e., minimum).

Blanks are considered characters and are not suppressed.

2. Matrix Elements Format

The nonzero elements are punched one per card in the following format:

Card columns	Contents
1-4	Blank
5-12	Column name
15-22	Row name
25-36	The value for this element (i.e., the a_{ij})
37-80	Blank

The columns can be in any order, but all the elements of one column must be together. The order of rows within a given column is arbitrary. In any case, the rows will be ordered to match the order of the row identification cards.

3. Right-Hand-Side Format

The right-hand-side elements must have a column name. Here we have used the name B1. The format for the right-hand-side's cards is identical to that of the matrix elements cards above.

ACKNOWLEDGMENTS

The authors wish to express their appreciation to AVI Publishing Co. (P.O. Box 831, Westport, Conn.) for permission to incorporate into this chapter all figures and tables (except Tables 6 to 8), which were previously published in Bender, Kramer, and Kahan (1976), <u>Systems Analysis for the Food Industry</u> (pp. 47, 49, 51, 52, 63, 65, 92, 93). In addition, the authors wish to acknowledge the assistance of Susan Blatchley Epstein in the development of the examples used.

REFERENCES

Bender, F. E., Kramer, A., and Kahan, G. (1976). <u>Systems Analysis for the Food Industry</u>. AVI, Westport, Conn.

Beneke, R. R. and Winterboer, R. (1973). <u>Linear Programming Applications to Agriculture</u>. Iowa State University Press, Ames, Iowa.

Dantzig, G. B. (1963). <u>Linear Programming and Extensions</u>. Princeton University Press, Princeton, N. J.

Heady, E. O. and Candler, W. (1958). <u>Linear Programming Methods</u>. Iowa State University Press, Ames, Iowa.

8

Production Control, PERT, and Transportation Problems

FILMORE E. BENDER and AMIHUD KRAMER* University of Maryland, College Park, Maryland

I. INTRODUCTION

There are many aids which provide management with a framework for solving seemingly complex problems rather simply. This chapter examines three general areas:

1. Production control, where our primary concern is with tools and techniques designed to assist management in the evaluation of both simple and complex tradeoffs.

*Deceased

2. Program evaluation and review technique (PERT), where our primary concern is the establishment of a clear statement of the tasks which must be accomplished in order to complete a time-dimensioned project (e.g., the building of a new factory, the development of a new food product or the courting of a young woman with marriage as the final activity in the project).

3. Transportation problems where several techniques are examined. The common element among these techniques is that they deal with some facet of a firm's transportation problems.

All of the techniques discussed in this chapter require repetitive calculations. With the current availability of programmable pocket calculators, not to mention desk-top computers designed for home and office use, many of the procedures presented in this chapter can be implemented by the reader without resorting to large-scale computers. However, for some of the algorithms presented, it would be preferable to use available commercially developed computer programs.

II. PRODUCTION CONTROL

There are various aspects to production control, all of which require managerial judgment. The following procedures provide a framework for examining certain classes of production problems.

A. Assignment Problems

One of the most common questions asked within an organization is Who will do what job? or Which factory will supply which warehouse? These are assignment problems. One of the responsibilities of management is to assign an individual to a specific task or to assign the output of factory A to warehouse X, etc. The following examples illustrate a simple repetitive algorithm which can be of considerable assistance when faced with assignment problems.

1. Labor: Selecting the Right Person for the Right Job

Whenever there are a number of tasks to perform and certain personnel to perform them, the question arises as to how to assign to each person the work which is most suitable to her or him. If the assignment is to be done purely on the basis of trial and error, the probability of obtaining a best solution is not high, because of the vast number of possibilities. If we should take only a small problem, such as assigning six men to six jobs, the number of possible permutations is six factorial, or $6! = 6 \times 5 \times 4 \times 3 \times 2 \times 1 = 720$. Obviously, it is impossible to try every combination. Fortunately, however, there are certain algorithms by which a best solution may be obtained rather quickly and easily. Simple methods apply only to situations where the number of tasks is equal to the number of men. Consider the case of a processor that has six positions to fill and has six people available for these positions. He gives each of the six men an opportunity to work at each job, and from this he obtains information in terms of minutes of labor required to accomplish each job by each man. He summarizes this information in the form of an "effectiveness matrix," as shown in Table 1. Columns A to F represent the six available men, and the rows 1 to 6 represent the tasks. The values in the matrix are minutes required to accomplish a task. Thus, man A requires 21 min to accomplish task 1, 19 min to accomplish task 2, etc.

Of course, our purpose is to accomplish all tasks in the least possible time. If we simply assigned a person to a job in the order they happen to be listed, that is person A to job 1, person B to job 2, etc., we would arrive at a total time of 163 min, as indicated in Table 2. We may rightly suspect that this is not the best solution, since there are no less than 720

TABLE 1   An "Algorithm" in the Form of an Effectiveness Matrix for the Optimal Selection of Six Men for Six Tasks

A		WORKERS						B			WORKERS			
	A	B	C	D	E	F			A	B	C	D	E	F
1	21	27	25	30	29	36	− 21	=	0	6	4	9	8	15
2	19	30	20	27	25	32	− 19	=	0	11	1	8	6	13
3	14	20	15	25	22	20	− 14	=	0	6	1	11	8	6
4	35	32	30	35	28	40	− 28	=	7	4	2	7	0	12
5	11	15	22	26	20	17	− 11	=	0	4	11	15	9	6
6	28	38	36	27	30	42	− 27	=	1	11	9	0	3	15

(TASKS — left side label for rows 1–6 of matrix A)

		C	− 0	− 4	− 1	− 0	− 0	− 6
	1 ✓	[0]	2	3	9	8	9
	2 ✓	✗	7	[0]	8	6	7
	3 ✓	✗	2	✗	11	8	[0]
	4	7	✗	1	7	[0]	6
	5 ✓	✗	[0]	10	15	9	✗
	6 ✓	1	7	8	[0]	3	9

(TASKS — left side label for rows 1–6 of matrix C)

TABLE 2   A First Assignment Solution
from Table 1a

Man	A task	Time (min)
A	1	21
B	2	30
C	3	15
D	4	35
E	5	20
F	6	42
		163

possible permutations of person-task combinations, and we have used no judgment in selecting the one above. We could probably do better by examining the entries in Table 1 carefully. Thus, for example, we might look at column A and select task 5 for person A, since he can perform this job in the least time. We could then follow with task 3 for person B, etc. If we should do this, we would achieve a substantial improvement as shown in Table 3.

However, to be certain of obtaining the best assignment, we proceed as follows:

We examine the entries in each row of Table 1A and subtract the smallest entry from each value in the row. Thus for example, we subtract 21 from row 1, so that the entries in row 1 of Table 1 are $21 - 21 = 0$, $27 - 21 = 6$, $25 - 21 = 4$, $30 - 21 = 9$, $29 - 21 = 8$, and $36 - 21 = 15$. We

TABLE 3   Revised Solution from
Table 1a

Man	A task	Time (min)
A	5	11
B	3	20
C	2	20
D	6	27
E	4	28
F	1	36
		142

TABLE 4 Optimal Assignment from
Table 1c

Man	A task	Time (min)
A	1	21
B	5	15
C	2	20
D	6	27
E	4	28
F	3	20
		131

proceed to subtract 19 from each entry in row 2, 14 from row 3, 28 from row 4, 11 from row 5, and 27 from row 6. The reduced values are shown in Table 1 Section A.

We now examine the entries in each column of Table 1, and subtract the smallest value in each column from all the values in the column. Thus we subtract 0, 4, 1, 0, 0, and 6 from columns A, B, C, D, E, and F, respectively. These reduced values are shown in Table 1 Section C.

We now examine each row in Table 1 in order, marking the 0 value in the row if there is only one there and crossing out any other 0's that may be in the same column in which the marked 0 is located. Thus in row 1, we block the 0 in column A and cross out all of the other 0's in column A, in this case located in rows 2, 3, and 5. When we now examine row 2, the only unmarked 0 is in column C, which we may now block, at the same time crossing out the 0 below it which is in row 3. We may now block the 0 in row 3 column F, since other 0's in this row were previously crossed out, at the same time crossing out the 0 in row 5 column F. In row 4 we find two unmarked 0's, so we cannot block either at this time. Proceeding to row 5, we block the 0 in column B and cross out the 0 above it. In row six we block the 0 in column D. We now have a blocked 0 in each row except E. We now examine the columns, and find a blocked 0 for A, B, C, D, and F. In column E we find only one 0 in row 4, which we now block, thereby completing the solution.

Assigning each man (A to F) the task for which he has a blocked 0, we arrive at the assignments shown in Table 4 and the time required to complete the tasks.

We may now be assured of the best solution, which is that these tasks cannot be done by these six men in less than the indicated 131 min.

At times the solution cannot be completed with this number of iteration; so additional manipulations are required. Furthermore, if the number of men is not equal to the number of tasks, the procedure is again different. Any problem of this kind may be solved as a linear programming problem.

The unit of measurement need not necessarily be one of time but could be of an aspect of quality or a quality-time index.

2. Material: What Goes Where?

In order to illustrate the general nature of the assignment problem, the same algorithm is used to examine a problem of assigning raw product.

Suppose six vining stations are scattered throughout a pea-growing area. A complete harvesting unit can supply enough pea vines to maintain full production at one vining station. It is necessary, therefore, to assign each of the fields being harvested at a particular time to one of six vining stations in such a way as to minimize transportation costs.

We first set up an effectiveness matrix, matrix A in Table 5, in which we list the fields vertically (1 to 6) and the vining stations horizontally (A to F). The values in the table are the number of tenths of a mile from field to vining station. It is assumed that the cost per mile is constant for the entire area. We now proceed to solve this problem by creating succesive matrices. In all problems of this sort, the second matrix, which in our example is matrix B, is prepared by subtracting the smallest element in each row from all of the elements in its row. The third matrix, which in our example is matrix C, is prepared by subtracting the smallest element in each column of matrix B from each element in its column. As illustrated in the person-jobs problem, an optimal solution may be obtained at this point. In most instances, subsequent matrices are required, prepared by an iterative technique which will be described following a more detailed discussion of the calculation of matrices B and C.

The smallest element in row 1 of matrix A is 13; hence, 13 is subtracted from every element in row 1; 15, the smallest element in row 2, is subtracted from every element in row 2. This procedure is continued for lines 3 through 6 in order to complete matrix B.

To compute matrix C, we locate the smallest element in column 1 of matrix B, which in our example is 0, and subtract it from every element in column 1. The smallest element in column 2 is also 0, and so once again 0 is subtracted from every element in column 2. This procedure is continued for columns 3 through 6 in order to complete matrix C.

We are now in a position to make our first attempt at assigning fields to vining stations. We reexamine row 1 of matrix C and find more than one 0 so that we cannot make an assignment in this row at this time. Moving to row 2, we block the one 0 found in this row which happens to be in column C, at the same time crossing out the other 0's in column C which happen to be in rows 3 and 4. As a result of this procedure, we now have no 0's in rows 3 and 4 and can therefore make no assignments for these rows at this time. In row 5, again we have more than one 0 and can make no assignment.

TABLE 5   Effectiveness Matrix for Minimizing Transportation Costs from Fields to Vining Stations

A	A	B	C	D	E	F	→ B	A	B	C	D	E	F
1	13	24	16	15	29	24	−13=	0	11	3	2	16	11
2	16	23	15	30	22	25	−15=	1	8	0	15	7	20
3	24	26	22	41	31	43	−22=	2	4	0	19	9	21
4	12	7	6	26	19	26	− 6=	6	1	0	20	13	20
5	7	5	6	17	11	16	− 5=	2	0	1	12	6	11
6	14	5	12	20	18	19	− 5=	9	0	7	15	13	14
								−0	−0	−0	−2	−6	−11

↓

D	A	B	C	D	E	F	← C	A	B	C	D	E	F
1	X	11	4	[0]	10	X		[0]	11	3	X	10	X
2	X	7	X	12	[0]	8	A	1	8	[0]	13	1	5
3	1	3	[0]	16	2	9	x x	2	4	X	17	3	10
4	5	[0]	X	17	6	8	x x	6	1	X	18	7	9
5	2	X	2	10	X	[0]		2	X	1	10	[0]	X
6	9	X	8	13	7	3	x	9	[0]	7	13	7	3
		x	x							x			

↓

E	A	B	C	D	E	F
1	X	12	5	[0]	10	X
2	X	8	1	12	[0]	8
3	[0]	3	X	15	1	8
4	4	X	[0]	16	5	7
5	2	1	3	10	X	[0]
6	8	[0]	8	12	6	2

In the last row, we find only one 0 in column B, which we block, and cross out the 0 above it in the same column.

We now examine column A of matrix C, and finding only one 0 at row 1, we block this item, at the same time crossing out the other 0's in row 1 at columns D and F. Columns B and C are already blocked. The one 0 in column D had been previously crossed out so that we can still make no assignment here. We can block the only 0 in column E, row 5, but at the same time we must cross out the only remaining 0 in column F, row 5.

Having completed this assignment procedure, an examination of the matrix reveals that rows 1, 2, 5, and 6 have assigned 0's, but rows 3 and 4 have no assigned 0's. In other words, we have not yet reached an optimal solution and will not do so until every row and every column has one assigned 0. It therefore becomes necessary to resort to an iterative procedure which will have the effect of putting more 0's into our next matrix.

The following steps constitute the iterative procedure.

1. Place a small x next to each row having no assigned 0's.

2. Place a small x under each column having 0's in rows which have just been checked.

3. Place a small x next to each row having an assigned 0 in a column which has been checked.

4. Repeat steps 2 and 3 until it is no longer possible to check rows or columns.

5. Draw lines through every unchecked row and every checked column.

6. Find the smallest element among those elements which have no lines drawn through them, and subtract that element from all of those elements which have no lines drawn through them. Place these new figures in the same relative position in a new matrix. Add this smallest element to all of those elements found at the intersection of two lines, and, as before, carry the sum over to the same relative position in a new matrix. Finally, without adding or subtracting anything, carry over those elements which are covered by only one line.

Applying this procedure to our example, we begin by checking rows for which we have no assignments. These are rows 3 and 4. Next we check all columns in which there are 0's in the checked rows. In this case, the only 0's in the unassigned rows of 3 and 4 are both in column C, which is there- fore the only column checked. We now examine the checked column for any assigned (blocked) 0's and check the rows in which these appear. In column we find one such assignment in row 2, which we check. Examining the three checked rows, we find no 0's in any additional columns and so our checking process is complete. We now draw lines through every unchecked row and every checked column, which as shown in matrix C are rows 1, 5, 6, and column C. These are the smallest number of lines that could have been drawn to pass through every 0 in the matrix.

When we reexamine matrix C, we find the lowest value not crossed by a line to be 1 (row 2, column A, also 2E, 4B). We construct matrix D by adding this value to every element found at the intersection of two lines.

Thus, the values in line 1 of matrix D of Table 5 remain the same as in matrix C, except for the value of 3 which is increased to 4. The values of row 2 of matrix C, on the other hand, are all reduced by 1, since they are uncrossed, except for the crossed 0 in column C which remains at 0. We continue on in a similar manner in order to complete matrix D.

Once again we attempt to assign fields to vining stations. Examining rows 1 and 2 of matrix D, we cannot make any assignments because each row contains more than one 0. We may block the one 0 in row 3, at the same time crossing out the zero in column C, row 4 and row 2. Having crossed out the 0 at 4C, we may now block the 0 in 4B and cross out the 0's in 5B and 6B. Row 5 still has two 0's, thus no assignment can be made. Row 6 has none, and so we can begin to look at the columns. There are two 0's in column A, and columns B and C have been assigned. Column D has one 0; so D can be blocked and 1A and 1F crossed out. Column E has two 0's which precludes any assignments, but the one 0 in column F can be blocked, at the same time crossing out the 0 at 5E. The 0 at 2E may now be blocked, but the 0 at 2A must be crossed out. Thus, we find that we have improved the situation from matrix C by reducing the unassigned fields from 2 to 1, but since we still do not have a complete answer, we must iterate one more time.

We begin by checking the only unassigned row in matrix D, which is row 6. Since this row has only one 0 in column B, we check column B. The only assigned (blocked) value in column B is in row 4, and so row 4 gets checked. Row 4 has a 0 in column B, which is already checked, and also in column C, which we now check. Column C has an assignment in row 3 which we must now check, but row 3 has a 0 only in column C which had been checked previously; so our checking process is complete. We again draw lines through the unchecked rows, 1, 2, and 5 and checked columns B and C and find that the lowest unlined value is 1 (element 3A).

TABLE 6   Optimal Assignment from Table 5e

Field	Vining station	Mileage (from Matrix A)
1	D	15
2	E	22
3	A	25
4	C	6
5	F	16
6	B	5
		89

We proceed with construction of matrix E by subtracting this value of 1 from every unlined value in matrix D and adding 1 to every value which lies at an intersection of two lines. This time, we must go to row 6 before we find only one 0 which we can block in column B while crossing out 4B. Similarly, an examination of the columns reveals only column D with on 0 which is blocked while crossing out 1A and 1F. Column F can now be blocked at 2E, crossing out 2A. Thus, column A can be blocked at 3A. This necessitates crossing out 3C, thereby leaving the blocking of 4C as the complete solution.

The best solution, therefore, is the assignment of fields to vining stations as shown in Table 6.

Due to its special nature, the assignment problem can be solved readily by hand. Alternatively, a small computer program can be written to perform the necessary calculations. In either event, this algorithm can be used to solve assignments on a day-to-day basis since it simply represents a systematic and efficient search of the choices that exist.

B. Scheduling Problems

Many food processing procedures consist of the raw material being passed through a sequence of machine centers, each of which may perform its function on different lots of material for various periods of time. It therefore becomes very difficult to select an optimal order of lots or jobs to pass through a sequence of equipment in such an order that there will be a minimum of idle time.

A broad definition of sequencing would lead us to consider problems in which several tasks must be performed in order to accomplish a job, and it is necessary to determine the order in which the tasks must be performed in order to minimize some factor, such as time. A definition of this sort soon leads us into considering different types of sequencing problems. For example, the development of a new food product is a job in which many different tasks must be performed at the same time. Management will undoubtedly want to minimize development time which leads directly to a sequencing problem. PERT (discussed later in this chapter) has been design to solve problems of precisely this nature. As another example, it may be necessary to determine the route, from among several available alternative routes, which a perishable food product must take in order to minimize the distance traversed (and concurrently the time traveled) in moving from a particular origin to a particular destination. This problem is a sequencing problem because we have the option of moving our product through a number of different cities, and it is necessary to determine which sequence of cities will minimize the distance traveled. Such a problem can be solved using linear programming. Our intent in this section is to confine our attention to a group of sequencing problems found most often in production. Here we would like to determine the order in which jobs should be passed through a production line in order to maximize efficiency, i.e., minimize idle time, idle machine time, and overall production time.

The magnitude of a typical sequencing problem becomes painfully evident when we realize how quickly the number of sequences proliferate as we move away from the very simplest of sequencing problems. If three different products must be ground in the same grinder, there are 3! or 6 possible ways of sequencing the three products through the grinder. Products A, B, and C can be ordered in the following six ways:

1. ABC

2. ACB

3. BAC

4. BCA

5. CAB

6. CBA

If six products must be passed through the same grinder, there are 6! or 720 ways of sequencing the six products. If three products must pass through two grinders, and there exists no restriction regarding the grinder which is to be used first, there are $(3!)^2$ or 36 possible sequences for performing the job. Given six products and two grinders, we must choose between $(6!)^2$ or 518,400 possible sequences.

Generally speaking, we are dealing with problems involving the processing of n products on m machines, with or without additional products coming in for processing while the first n products are being processed. For the latter case, where there are no additional products coming in, mathematical solutions exist for only limited situations. For the former case, where products are coming in, there are no rigorous mathematical techniques. For those situations where there are no rigorous mathematical techniques available, heuristic methods or simulation techniques can be applied.

1. The Gantt Chart

Undoubtedly, one of the most efficient ways of displaying any kind of sequencing or scheduling information is by means of the Gantt, or bar, chart. Named after Henry L. Gantt, a man who made many contributions to scientific management early in the twentieth century, the Gantt chart survives as one of the mainstays of the production manager. An example of a Gantt chart displaying sequencing information is shown in Fig. 1. The usual feature of a Gantt chart is a time scale placed horizontally along the top or bottom and rows denoting resources required to accomplish a production job. In our case, the resources are three machines, and the chart

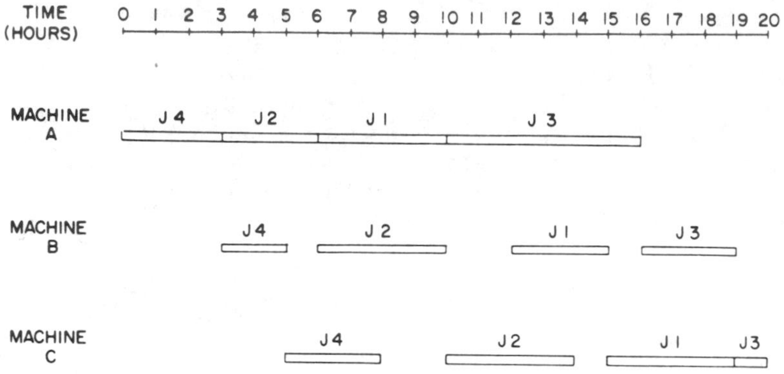

FIGURE 1   Gantt chart illustrating job sequencing.

indicates how these three machines are to perform four jobs. The analyst, having determined that the four jobs J1, J2, J3, and J4 should be performed in the sequence J4, J2, J1, and J3, can display this and other information by means of bars placed in the proper relative position. The chart in Fig. 1 shows that machine A will perform J4, J2, J1, and J3 in 16 hr and with no idle time. Machine B will perform the same four jobs in the same sequence but with a total of 4 hr of idle time between jobs. Machine C will perform comparably but with a total of 3 hr of idle time between jobs. The fact that each job must pass through one machine at a time is also illustrated, and this of course is the reason for the presence of idle time. Furthermore, the chart indicates that the total elapsed time from the beginning of the first job to the end of the last job is 20 hr.

2.   The Applicability of Sequencing Techniques

Sequencing methods are particularly applicable to the job shop. A job shop is a production operation consisting of several machines, where each machine performs a specific task, any two of which may or may not be alike. A machine shop is a good example of a job shop. Obviously, it follows that sequencing methods will be of less importance to the food processor than to many other industries simply because most food processing operations must follow a particular sequence. Rarely, will the food processor be afforded the luxury of sequencing operations according to the dictates of an algorithm. On the other hand, there are food processing operations which are performed in a particular sequence for no other reason than custom and habit, and not technological necessity. It may be true that substantial savings could be realized if these techniques are applied and age-old habits changed.

3.   Shortest-Processing-Time Rule

A simple rule that is often used by people concerned with sequencing is the shortest-processing-time (SPT) rule. This rule is illustrated by considering a situation where three jobs must be processed by one machine. Suppose the processing times for jobs 1, 2, and 3 are 15, 10, and 20 min, respectively. Following the SPT rule, we would perform job 2 first, job 1 second, and job 3 last because the processing time for job 2 is less than that of job 1, which in turn is less than that of job 3. The advantage of following this sequence is illustrated by the following:

Job	Waiting time (min)	Processing time (min)	Total elapsed time (min)
2	0	10	10
1	10	15	25
3	25	20	45
Sum	35	45	80
Average	35/3 = 11.67 min	45/3 = 15 min	80/3 = 26.67 min

Comparing the average waiting time (11.67 min), the average processing time (15 min), and the average total elapsed time (26.67 min) to figures obtained using other possible sequences amply illustrates the merits of the SPT rule. Consider the following alternative sequencings:

Job sequence	Average waiting time (min)	Average processing time (min)	Average total elapsed time (min)
2-1-3	11.67	15	26.67
2-3-1	13.33	15	28.33
1-2-3	13.33	15	28.33
1-3-2	16.67	15	31.67
3-2-1	16.67	15	31.67
3-1-2	18.33	15	33.33

The reader can see immediately that the average waiting time and average total elapsed time is less for sequence 2-1-3, obtained by following the SPT rule, than for any other possible sequence. Other rules of this type exist in sequencing theory, notably the first-come, first-served (FCFS) rule, and the probability-sequencing (PS) rule. For a detailed discussion of these rules, see Fabrycky, Ghane, and Torgensen (1972).

4. Graphic Solution Techniques

Some of the specific sequencing problems which have been studied, and the approach used to solve these problems are as follows:

1. Process n jobs on two machines all in the same order.

2. Process n jobs on three machines all in the same order, where the longest processing time on the second machine is less than or equal to the shortest processing time on both the first and third machines.

3. Process two jobs on m machines where the sequence of machines used for one job must not necessarily be the same as the sequence for the other job. A rather simple graphical technique can be used to solve this problem, but the solution is not necessarily optimal.

4. Process n jobs on m machines. This problem can be solved by a heuristic technique, which is subject to rather serious limitations, or by simulation, which is subject to no limitations but which does not necessarily give an optimal solution.

The following examples illustrate solutions to these four types of Sequencing problems.

a. Sequencing of Operations: Two Machine Centers: Lima beans are received at a freezing plant.   There they are blanched and passed through a quality grader where the mature "sinkers" are separated out.   They are then passed through a size grader where the "floaters" are further separated into midget, tiny, small, medium, and large categories, and finally they are passed over sorting inspection belts, filled, and packaged.   The various size groups come out simultaneously and are held in different bins from which they are fed onto the sorting belt.   A lot of beans required the following time in hours to pass through sorting belt and filler:

	Time in hours		
Job or lot	Machine A sorting belt	Machine B filler	Sequence
1. Sinkers	1	5	1
2. Midgets	5	2	6
3. Tiny	7	6	3
4. Small	8	9	4
5. Medium	3	4	5
6. Large	2	3	2

We wish to run these lots in such a sequence that there will be a minmum of idle time for the equipment.   First, we examine the column for machine A and find the smallest value is for row 1.   Therefore sinkers are put in first place as the lot to be handled first.   We next examine column B and find the smallest value of 2 in row 2.   We therefore place midgets as the last (6th) lot to be handled.   Having disposed of rows 1 (sinkers) and 2

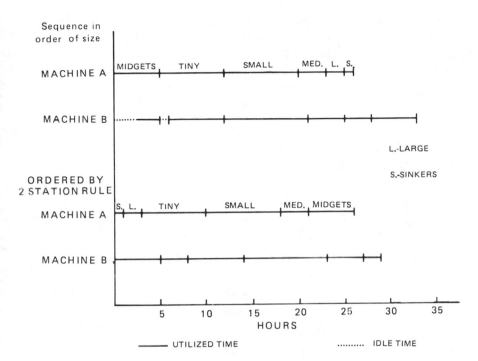

FIGURE 2 Gantt chart showing the reduction in idle machine time when lots are ordered by the two-machine rule.

(midgets), we reexamine column A, and find the next smallest value to be 2 for row 6. We therefore list the lot of large beans to be handled second, immediately after the sinkers. Reexamining column B for the remaining rows 3, 4, and 5, we select row 5 or the medium-size lot as next to last to be handled (fifth). Of the two remaining lots, the value for row 3 or tiny in column A is smaller than the value for row 4; hence, we select the lot of tiny beans as third in the sequence, with the lot of small beans as fourth.

This sequence will provide the minimal idle time for the filler as shown on the Gantt chart in Fig. 2, where this sequence is compared to one in which the different sizes are handled in order, beginning with the smallest.

b. Sequencing Operations: Three Machine Centers: We change the situation of the previous example somewhat. We now wish to blanch the lima beans only after they pass over the inspection tables, so that we must now consider time in the blancher as well as time on the sorting belt and in the filler. This problem of a number of lots going through three machine centers can be solved only if the longest time required for any one lot to pass through the middle machine is no greater than the shortest time required for any one lot to pass through either the first or the last machine. This condition is almost met as shown in the follow data:

Job or lot	Time in hours		
	Machine A sorting belt	Machine B blancher	Machine C filler
1. Sinkers	1	2	5
2. Midgets	5	0.5	2
3. Tiny	7	1	6
4. Small	8	2	9
5. Medium	3	1.5	4
6. Large	2	0.5	3

We eliminate middle column (B) by adding the values in this column to both the A and the C column as follows:

Job or lot	A + B	B + C	Sequence
1. Sinkers	3	7	2
2. Midgets	5.5	2.5	6
3. Tiny	8	7	3
4. Small	10	11	4
5. Medium	4.5	5.5	5
6. Large	2.5	3.5	1

We proceed to select the order of handling lots exactly as we did in the previous example. This time we select large as the first lot to be processed and the midgets as the last. Sinkers then become second in the sequence and medium second from last, with tiny third, and small fourth. These are shown in the form of a Gantt chart in Fig. 3.

c. Two Jobs on m Machines: Figure 4 illustrates a graphical approach used to solve the two-job, m-machine problem where the order of the machines used in one job may differ from the order used in the other job, but where the order in each job is fixed. Like the previously described sequencing problems, the object of this technique is to minimize idle machine time. Several feasible solutions are likely to be obtained from this technique, one of which may be optimal, though not guaranteed to be optimal. Assume baby lima beans and Fordhook lima beans are to be processed on four machines. The following shows the sequence to be used on the processing time for the two types of lima beans.

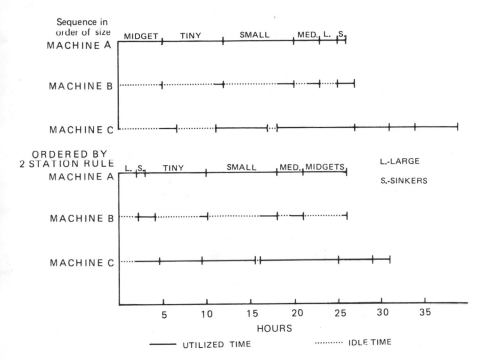

FIGURE 3   Gantt chart showing the reduction in idle time when three machines are involved in sequencing of lots.

Type	Sequence	Machine	Processing time (hr)
Baby	1	Sizer	2.0
	2	Quality grader	1.0
	3	Froth flotation washer	0.5
	4	Blancher	1.0
		Total	4.5
Fordhook	1	Quality grader	0.6
	2	Froth flotation washer	0.8
	3	Sizer	1.0
	4	Blancher	0.8
		Total	3.2

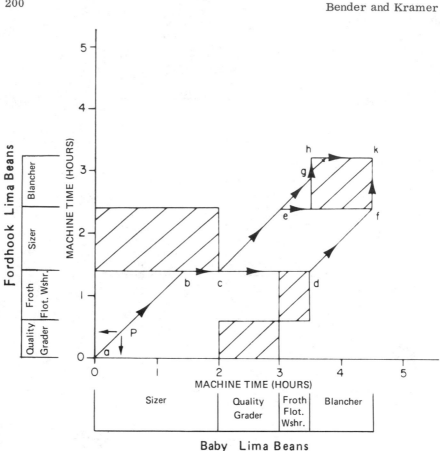

FIGURE 4   Graphical procedure for two-job m-machine problem.

The graph is prepared by using one axis for each job. In our case, one job is that of processing baby lima beans, and the x-axis depicts both the processing time and the machines which are used in this job. Beginning at the origin, the machines are placed in the required sequence with sufficient space allotted for the processing time. It can be seen that 2 hr were allotted for the sizer, followed by 1 hr for the quality grader, followed by 0.5 hr for the froth flotation washer, followed by 1 hr for the blancher. Using the appropriate figures, comparable information is shown on the y-axis for Fordhook lima beans. Choosing any point, for example, point P, the x-axis indicates the machine which is to be used for job 1, and the y-axis indicates the machine which, at the same time, is to be used for job 2. Point P, therefore, represents an instant when the sizer has been used for 0.4 hr in job 1, while at the same time, the quality grader has been used for 0.4 hr in job 2. The graphical procedure begins by blocking those areas (shown as crosshatched areas) which represent the intersection of the same machine. This eliminates the possibility of using the same machine in both jobs at the same time. Recognizing that the origin represents the start of both jobs, and point K represents the completion of both jobs, it will now be necessary to trace a path from the origin to point K which minimizes idle machine time. Three kinds of lines can be drawn:

1. Diagonal lines at an angle of 45°. These lines indicate that both jobs are proceeding simultaneously.

2. Vertical lines. These lines indicate idle machine time on job 1, which in our case is the processing of baby lima beans.

3. Horizontal lines. These lines indicate idle machine time in job 2, which in our case is the processing of Fordhook lima beans.

Obviously, it follows that the path with minimum idle time will be the one with a minimum of horizontal and vertical segments and a maximum of diagonal segments. Starting from the origin, every attempt is made to proceed along a diagonal path to point K. If a crosshatched area is encountered, it represents an obstacle which forces us to proceed vertically or horizontally.

Paths abcdfk, abcefk, and abceghk represent possible paths in processing from the origin to point K. The table below tabulates the idle machine time encountered along each of these three paths. The idle time for baby lima beans was obtained by adding vertical segments encountered along a given path, and the idle time for Fordhook lima beans was obtained by adding any horizontal segments encountered along a given path.

	Idle time (hr)		
	Baby	Fordhook	Total idle time
abcdfk	4.0	10.5	14.5
abcefk	4.0	10.5	14.5
abceghk	1.5	8.0	9.5 (minimum)

The results indicate that path abceghk is preferable to the other two paths because total idle time for this path is 9.5 hr as compared to 14.5 hr for the other two paths. Although this solution is probably optimal, there is no way of proving it.

d. n Jobs on m Machines: A heuristic technique for n jobs processed on m machines, but subject to serious limitations, was reported by Campbell, Dudek, and Smith (1970). The technique is applicable if the machines must be used in the same order for each product. Unlike the other techniques which have been described, however, we will minimize total elapsed time rather than idle time. The solution will not necessarily be optimal but will probably be close to optimal.

In applying the technique to m machines, the algorithm used for n jobs and two machines is applied to machine 1 and machine m. Next, the processing times for machines 1 and 2 are added together, and the processing times for machines m and m - 1 added together. Once again, the n-job, two-machine algorithm is applied, but this time the procedure calls for the summation of the processing time for machines 1, 2, and 3 and the summation of the processing times for machines m, m - 1, and m - 2. And again, the n-job, two-machine algorithm is applied.

This procedure continues until summation of the processing times for machines 1, 2, 3, ..., m - 2, m - 1, and summation of the processing times for machines m, m - 1, m - 2, ..., 3, 2 are obtained, and the algorithm applied. Each time the algorithm is applied, another sequence is obtained for a total of m - 1 sequences. The total elapsed time for each of these sequences is determined by use of a Gantt chart. The smallest of these times denotes the optimal or near-optimal sequence.

A vegetable canning company processes three varieties of lima beans: Thaxter, Bridgeton, and Fordhook. Four machines are used for each of three varieties. In order of use, they are a sizer, quality grader, froth flotation washer, and blancher. The following table shows the hours of processing time required for each variety on each machine.

	Sizer	Quality grader	Froth flotation washer	Flotation blancher
Machine order of use	1	2	3	4
Thaxter	1.8	0.9	0.6	0.9
Bridgeton	2.0	1.0	0.5	1.0
Fordhook	1.0	0.6	0.8	0.8

Applying the n-job, two-machine algorithm to the sizer and blancher, machines 1 and 4 in order of use, results in the following sequence: Bridgeton, Thaxter, Fordhook. Summing the processing times for the sizer and quality grader (machines 1 and 2 in order of use) and the processing times for the froth flotation washer and blancher (machines 3 and 4 in order of use) and applying the algorithm results in the following sequence: Fordhook, Bridgeton, Thaxter. Repeating this process once again by summing the processing times for the first three machines and the last three machines in order of use, and once again applying the algorithm results in the same sequence obtained earlier, namely, Bridgeton, Thaxter, and Fordhook. These steps are summarized in the following data:

Machine (identified by order of use)	(1)	(4)	(1) + (2)	(3) + (4)	(1) + (2) + (3)	(2) + (3) + (4)
Thaxter (T)	1.8	0.9	2.7	1.5	3.3	2.4
Bridgeton (B)	2.0	1.0	3.0	1.5	3.5	2.5
Fordhook (F)	1.0	0.8	1.6	1.6	2.4	2.2
Sequence	B-T-F			F-B-T		B-T-F

Figure 5 is a Gantt chart of the two sequences obtained by time application of this technique. Examination of the chart quickly reveals that the best sequence is Bridgeton, Thaxter, and Fordhook because the total elapsed time for this sequence is 7.0 hr as compared to 7.2 hr for the Fordhook-Bridgeton-Thaxter sequence. There is no way of proving whether the technique used which defined the 7.0 hr obtained with the Bridgeton, Thaxter, Fordhook sequence yields an optimal solution. However, since there are only six possible sequences in our small problem, a Gantt chart can be drawn for each of the possible sequences to reveal the answer. The following table shows the results of just such an effort.

FIGURE 5   Gantt charts illustrating processing times.

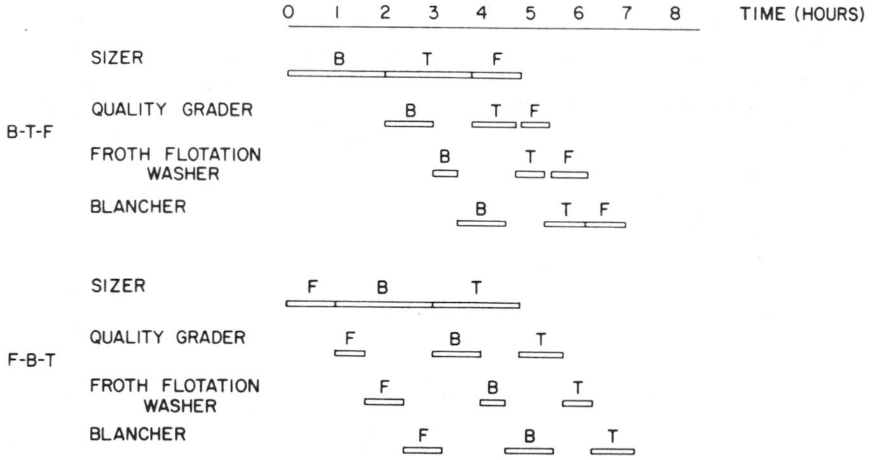

Sequence	Total elapsed time (hr)
B-T-F	7.0
F-B-T	7.2
B-F-T	7.2
F-T-B	7.3
T-B-F	7.1
T-F-B	7.3

C. Replacement or Repair of Equipment

1. Parts Replacement or Repair

As equipment is used continuously, the point is reached where parts need to be adjusted, sharpened, repaired, or replaced. This is particularly true of parts such as knives, screens, belts, gaskets, and the like. If the process is permitted to proceed until action is indicated, this may mean expensive "down" time. If this down time is very expensive, it may be preferable to wait until such a time as the entire operation comes to a stop, such as at the end of a shift; however, if this is done, then the end product may fail to meet specifications of one kind or another. It may be possible to avoid such costly breakdowns by a study of the equipment and a knowledge of the probability of the need for replacement or repair.

Consider the case of a packer of whole-kernel corn who has 50 machines for cutting the kernels from the cobs. Each machine has two cutter heads, which consist of sets of knives requiring adjustment and sharpening. If a machine must be stopped during production and the head replaced, the cost of this replacement, including the cost of delaying production, is estimated at $6.00 per head. The cost of replacing a head when the machine is not in operation, on the other hand, is only $2.50 per head. A study is made to determine how long cutter heads last before they need replacement. It is found that about a fourth of all the heads can be expected to fail at the end of one double shift of 10 hr, about half can be expected to fail after 2 days, and the remaining fourth will fail at the end of 3 days, or 60 hr of operation. Thus the probability that heads need replacement is as follows:

1 Days	2 Probability
1	0.25
2	0.50
3	0.25

The average number of cutter heads that will require replacement daily can be determined by dividing the sum of products of column 1 and 2 above, into the total number of cutter heads, as follows:

$1 \times 0.25 = 0.25$

$2 \times 0.50 = 1.00$

$3 \times 0.25 = \underline{0.75}$

Sum   2.00

and 100 cutter heads divided by 2.0 = 50 cutter heads that will require replacement daily. Since the cost of replacing a cutter head as needed is $6.00, then the cost of replacements as needed is

$50 \times \$6.00 = \$300.00$ daily

An alternative procedure would be to replace all cutter heads at the end of each day, during the 4-hr period when the machines are not operating, whether they need replacement or not. Since such replacements cost only $2.50 per head and there are 100 heads to be replaced, the total daily cost of such a policy would be as follows: 100 X $2.50 = $250.00, which is the more economical policy for this situation.

2. Machinery Replacement

Machinery may be replaced for any one of three reasons: (1) increasing operating costs more than compensate for reduced capital costs; (2) new machinery performs better than the old; (3) new equipment is needed for prestige or morale purposes. Only the first two reasons will be considered here.

a. When to Replace a Machine with an Identical New Unit: A machine has an initial cost of $6000. Costs of maintaining the machine in operating condition during the first year are $1200. These costs include all repair, maintenance, and parts replacement costs, as well as any down-time costs. If the machine is to be replaced after only 1 year of operation, and if it has no resale value, then total cost per year of operating is $6000 + $1200 = $7200. However, if the machine is to be kept for 2 years, then the capital costs per year are reduced by half:

$\$6000 \div 2 = \$3000$

During the second year of operation, however, operating costs begin to increase. Thus, whereas the first year's operating expenses are only $1200, the second year's operating expenses are $1600, or an average annual

TABLE 7   Machine Cost Per Year as Affected by Duration

A Machine cost $6000; operating cost $1200, increasing annually by $400

1	2	3	4	5	6
		Operating cost			Total Ave.
Replace	Capital Cost per Year	Current Year	Accumulative	Average per Year	Cost per Year
Every year	6000	1200	1200	1200	7200
After 2 years	3000	1600	2800	1400	4400
3 years	2000	2000	4800	1600	3600
4 years	1500	2400	7200	1800	3300
5 years	1200	2800	10000	2000	3200[a]
6 years	1000	3200	13200	2200	3200[a]
7 years	857	3600	16800	2400	3257
8 years	750	4000	20800	2600	3350

B—Machine cost $10,000; operating cost $300, increasing annually by $200

Every year	10000	300	300	300	10300
After 2 years	5000	500	800	400	5400
3 years	3333	700	1500	500	3833
4 years	2500	900	2400	600	3100
5 years	2000	1100	3500	700	2700
6 years	1667	1300	4800	800	2467
7 years	1429	1500	6300	900	2329
8 years	1250	1700	8000	1000	2250
9 years	1111	1900	9900	1100	2211
10 years	1000	2100	12000	1200	2200[a]
11 years	909	2300	14300	1300	2209
12 years	833	2500	16800	1400	2233

[a]Lowest cost indicating replacement time.

operatings cost of $1400 per year. Utilizing the machine for 2 years, therefore, involves an annual cost of

$3000 + 1400 = $4400

Such calculations for an 8-year period are shown in Table 7A. The total average costs per year are also plotted in Fig. 6. The indicated time for replacement is obviously when the total average cost per year is at the minimum, which in this instance is 5 or 6 years. Note that the total average cost drops sharply at first, reaches a minimum value of $3200 per year, and then proceeds to rise again very gradually. This is typical of such data. From a practical standpoint, this means that an error in the direction of keeping machinery for an additional period beyond the optimal time is probably not a serious error and less costly than replacing with new machinery too soon. Thus if such operating costs can be estimated precisely, then replacement time can be predicted in advance; however, it is almost as satisfactory to maintain a record of costs, as shown in Table 7, and replace only after average costs per year level off, or even begin to rise again.

b. When to Replace a Machine with an Improved Unit: A new machine is now available which will accomplish the same job as the machine in the previous example. Its cost is $10,000, substantially higher than the machine now owned; however, operating costs are much lower and are expected to increase more gradually with use. The estimated average annual costs for this new machine are shown in Table 7, part B with the minimal average cost being reached with a 10-year use period. Since this value of $2200 is far less than the average cost of $3200 for the old machine, and everything else being equal, then obviously the $10,000 machine should replace the $6000 machine.

If, however, the old machine has been purchased, say, just 1 year ago, should it be replaced immediately? If we assume that the old machine has no resale value, we must figure that if a new machine is purchased, then the old machine has no capital value, only operating costs. The new machine should therefore be purchased only after operating costs with the old machine exceed total average annual costs for the new machine. Since this

FIGURE 6 Total average costs per year for two machines.

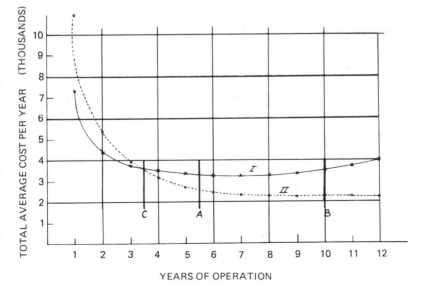

does not occur until after 3 years use of the old machine, the purchase of
the new machine is not justified until after the old machine has been in use
for 3 years.

This problem has been simplified for the sake of brevity. In actual
practice, cost of money, differences in capacity, and quality of performance must all enter into such an evaluation. For more detailed calculation
of similar situations refer to Terborgh (1958a, b; 1964), Lyle (1957), Henry
and Haynes (1978) and Brigham and Pappas (1976).

D. The Computer as an Aid in Decision Making

The previous example illustrates a common situation for managerial decision making. That is, the principle involved is relatively simple. Namely,
we should keep a machine until the average cost (fixed plus variable) is minimized and then replace it. Although the principle is straightforward, the
computational burden can be rather large. If the cost of borrowed money is
included, many individuals are inclined to feel that the time and effort involved in doing the calculations may not be worth the return in terms of improved decisions. In this regard, computers can be an important aid.

Table 8 lists the instructions for a simple program written in BASIC.
BASIC is an easy-to-learn computer language which is available on many
small home and office computers as well as nearly all large computers.
The program in Table 8 is designed to perform the calculations needed and
print the results for Table 7.

The computer program in Table 8 can be divided into the following components:

1. Instruction 100. Clears register A which will be used to accumulate operating costs.

2. Instructions 110 to 180. Request the user to enter the purchase
   price of the machine (B), the annual operating cost (C), the annual
   increase in operating costs (D), and the maximum life of the machine (E).

3. Instructions 190 to 360. Print the table headings. Of course, these
   instructions are not essential. They merely provide us with an attractive, easy-to-read answer in table form.

4. Instructions 370 to 440 are the part of the program which performs
   all of the needed calculations:

   a. Instruction 370 says that each year (N) will be calculated start-
      ing with 1 and going to the maximum considered (E).

   b. Instruction 380 calculates the capital cost per year.

   c. Instruction 390 calculates operating cost for the current year.

TABLE 8   Computer Program in Basic to Calculate Machinery Replacement
Tables

```
100 A = 0
110 PRINT 'INPUT MACHINE COST';
120 INPUT B
130 PRINT 'INPUT OPERATING COST';
140 INPUT C
150 PRINT 'INPUT ANNUAL INCREASE';
160 INPUT D
170 PRINT 'FOR HOW MANY YEARS';
180 INPUT E
190 PRINT
200 PRINT
210 PRINT
220 PRINT TAB(5); '1'; TAB(16); '2'; TAB(29); '3'; TAB(42); '4';
230 PRINT TAB(54); '5'; TAB(68); '6'
240 FOR J=1 to 75
250 PRINT ' ';
260 NEXT J
270 PRINT
280 PRINT TAB(36); 'OPERATING COST'
290 PRINT TAB(26);'---------------------';TAB(63);'TOTAL AVE.'
300 PRINT TAB(10);'CAPITAL COST    CURRENT'; TAB(52);'AVERAGE COST PER'
310 PRINT   '   REPLACE';TAB(13);'PER YEAR';TAB(27);'YEAR'
320 PRINT TAB(37);'ACCUMULATIVE      PER YEAR';TAB(66);'YEAR'
330 FOR J=1 TO75
340 PRINT '-';
350 NEXT J
360 PRINT
370 FOR N=1 TO E
380 F=INT(B/N)
390 G=C+(D*(N-1) )
400 A=A+G
410 H=INT(A/N)
420 I=F+H
430 PRINT TAB(5);N;TAB(14);F;TAB(27);G;TAB(40);A;TAB(53);H;TAB(66);I
440 NEXT N
450 END
```

   d. Instruction 400 accumulates the annual operating costs.

   e. Instruction 410 calculates the average annual operating cost.

   f. Instruction 420 adds the capital cost per year and the average
       operating cost per year.

   g. Instruction 430 prints all of the values just calculated.

   h. Instruction 440 states that N is to be incremented by 1 and the
       program returned to statement 370 in order to continue the cal-
       culations.

5. Instruction 450 is a nonexecutable statement necessary at the end of
    every BASIC program.

From this brief presentation, it can be seen that a relatively simple
program (only 14 statements are essential, the remainder are window dres-
sing) can do a great deal of work. With this program a large number of machines may be evaluated without incurring an unbearable computational burden.

A final word of caution. The computer is not capable of making judgments. It only performs the calculations that are explicitly given to it. It
cannot tell us what to do if a machine breaks down. It cannot predict whether a better machine is available. However, if we know what calculations
need to be made, we can program a computer to perform them for us quickly, easily and accurately, thereby freeing up valuable managerial time to
make the decisions that only management can make.

III. PERT

PERT and CPM are techniques used as aids in planning and scheduling the
individual activities which comprise a complete project. Both techniques
have proven to be particularly helpful to planners responsible for converting new food product ideas into grocery store realities. However modest
the development effort, these techniques save time and money in the long
run by the simple expedient of providing management with an organized,
clear and concise picture of the overall project, as well as specific details
concerning those activities demanding immediate and concentrated attention.
Virtually every research department and food technology laboratory, regardless of its size, can benefit from these valuable tools.

A. Basic Concepts

Critical path method (CPM) and program evaluation and review technique
(PERT) were developed independently, though at about the same time. In
1957, Morgan R. Walker of Dupont and James E. Kelley, then of Remington
Rand, were attempting to devise a better method for solving scheduling

problems. Their work resulted in CPM, and with it came one of the first
truly innovative solutions to this class of problems since the development of
the bar graph or Gantt chart. The distinguishing feature of the critical path
method is the arrow diagram. Figure 7, illustrating the development of a
new food product, reveals that an arrow diagram is nothing more than a
configuration of interconnecting arrows, where each arrow represents a job
within a project and where all of the arrows together represent the entire
project. The tail of each arrow represents the beginning of a particular job,
and the head of each arrow represents the end. In other words, each arrow
represents a duration of time. The numbers on either side of every arrow
merely serve to label each job. For example, when we discuss the first job
we may refer specifically to job 1, 2; when we discuss the last job, it is job
6, 9. The term "node" is applied to each of the numbered points, i.e.,
where the heads and tails of arrows meet. In fact, these are points in time
where some jobs end and others begin. Arrows which follow each other
represent a sequence of jobs in which each job must be completed before
commencement of another. Arrows which run parallel to each other represent groups of jobs which can be done at the same time. In our example,
jobs 1, 2, 0, 0, and 0, 4 must be completed in sequence; however, jobs 2, 3
and 2, 7 can be completed in parallel or at the same time.

Dashed arrows, called "dummies," merely serve to maintain the work
sequence. In our example, jobs 4, 5; 4, 6; and 7, 8 precede job 6, 9 and
must be completed before job 6, 9 can even begin. At this point, it should
be quite evident that in using CPM the planner must first clearly specify
which jobs are to comprise the entire project, which of the various jobs
must be completed before others can begin, and which groups of jobs can be
performed at the same time as others in the project. Only at this point will
the planner be in a position to draw an arrow diagram, which is, of course,
the backbone of the CPM technique, and as we shall presently see, the backbone of PERT as well.

Before going into any more detail on CPM, we turn our attention to
PERT because the differences between the two techniques, thus far, are essentially only differences in nomenclature. PERT was developed in 1958 by
the U.S. Navy as a management planning and control system for the Polaris
missile program. The Navy has stated that PERT cut 2 years off of the construction time and launching of the first Polaris missile. In PERT, an arrow diagram becomes a network, a node becomes an event, and a job becomes an activity.

B. A PERT Illustration

A typical PERT network is shown in Fig. 8. Here, the same project shown
in Fig. 7, the development of a new food product, is used to illustrate the
technique. Concepts which applied in Fig. 7 now apply to Fig. 8. Arrows
again represent durations of time and circles (they can just as well be ellip-
ses or squares) now represent points in time. The circle with the small

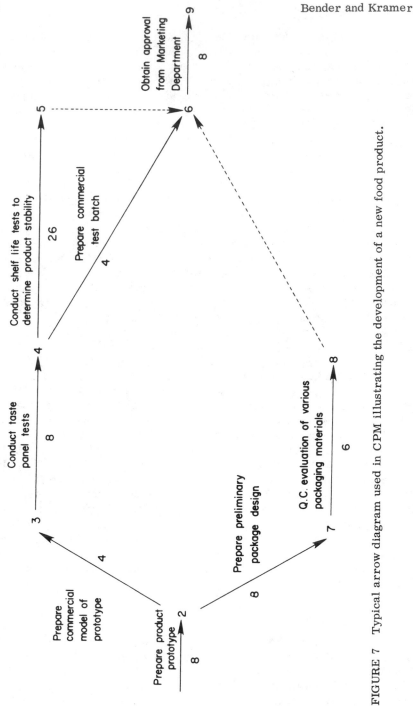

FIGURE 7   Typical arrow diagram used in CPM illustrating the development of a new food product.

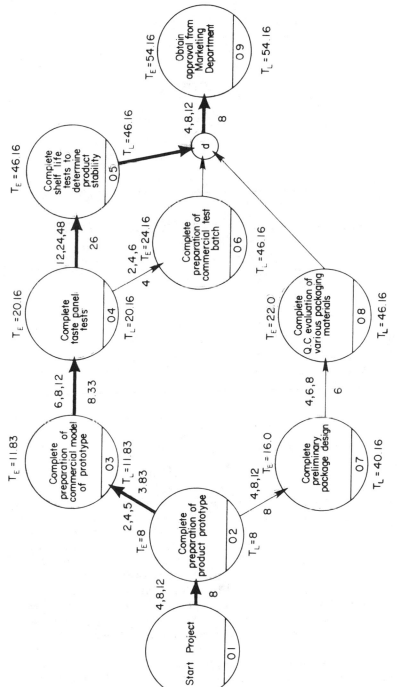

FIGURE 8   Typical network used in PERT illustrating the development of a new food product.

TABLE 9   CPM and PERT Notations

CPM	PERT
Arrow diagram	Network
Node: a point in time represented by a number	Event: a point in time represented by an ellipse, circle or square
Job: a duration of time represented by an arrow	Activity: a duration of time represented by an arrow
Float time	Slack time

"d" represents a dummy event which indicates that activities 05, 06, and 08 must be completed before activity 09 can even begin. And once again, jobs shown in sequence must be performed in sequence, and those shown in parallel can be performed at the same time. A summary of CPM and PERT notation is given in Table 9.

1. The Forward Pass

The main idea of both CPM and PERT is to identify those jobs or activities which ultimately dictate the completion time of the entire project. In the project shown in Fig. 8, the activities may be divided into three groups where each group has a different completion time. These three groups are as follows:

    Group I: Activities 01, 02, 03, 04, 05, and 09

    Group II: Activities 01, 02, 03, 04, 06, and 09

    Group III: Activities 01, 02, 07, 08, and 09

The object, in this instance, is to determine which of these three groups will take the longest time to complete. The group which takes the longest time is said to lie on the critical path. Any delay in any activity which lies on the critical path will delay the entire project. The differences in time required to complete a group of activities on the critical path and a group of activities which does not lie on the critical path is called the "float time" in CPM notation and the "slack time" in PERT notation. Both of these terms appropriately convey the idea that a delay in a group of activities which does not lie on the critical path can be tolerated so long as the delay does not cause the completion time to exceed the completion time for the group on the critical path. Since the methodology involved in identifying the jobs or activities which lie on the critical path is essentially the same for both CPM and PERT, only PERT notation will be used in the description

which follows, except where differences are encountered which would war-
rant a separate and detailed discussion.

Estimating the time required to complete each activity is the first step
in identifying the critical path. Generally, time estimates are given in
7-day calendar weeks where each week is assumed to be a 5-day, 40-hr
work week. Furthermore, in PERT, three estimates of time are made for
each activity: an optimistic estimate (a), a most likely estimate (m), and a
pessimistic estimate (b). An "optimistic estimate" is the minimum completion time for an activity or the time required to complete the activity if all
goes well and the best of luck is encountered. Statistically, an optimistic
time estimate represents a time which is so early that the planner feels
that the odds of finishing the activity on or before this time are 1 in 100. A
"most likely estimate" is one which, in the opinion of the planner, constitutes the most probable completion time for the activity in question. A "pessimistic estimate" is the maximum completion time for an activity or the
time required to complete the activity if bad luck is encountered. Statistically, a pessimistic time estimate represents a time which is so late that
the planner feels that the odds of finishing the activity on or beyond this time
are 1 in 100. As can be seen in Fig. 8, each of these three estimates is
placed over the arrow denoting the appropriate activity. For example, activity 03 has been assigned an optimistic estimate of 2 weeks, a most likely
estimate of 4 weeks, and a pessimistic estimate of 5 weeks. The next step
consists of utilizing a formula to calculate the mean or expected time.

The mean or expected time t_e represents the midpoint of possible completion times and is calculated by the formula

$$t_E = \frac{a + 4m + b}{6} \tag{1}$$

where

a = optimistic estimate
m = most likely estimate
b = pessimistic estimate

Using the time estimates for activity 03, the expected time would be calculated as follows:

$$t_E = \frac{2 + 4(4) + 5}{6} = 3.83 \text{ weeks}$$

Formula (1) is merely the formula used to calculate the mean of a set of
data having a beta distribution.

In PERT, the possible completion times are assumed to have a beta
distribution, and so formula (1) enables us to calculate the midpoint of that
distribution. It follows that the probability of obtaining a completion time

greater than t_E is 50% and that obtaining one lower is also 50%. Calculated values of t_E are shown below each arrow in Fig. 8.

Finally, we are in a position to perform those calculations which lead directly to the critical path. First, the cumulative expected time T_e is calculated for each event. As the name implies, the cumulative expected time T_E for any event, say, event 04 in Fig. 8, is the sum of the expected times t_e of each event leading directly to event 04. Therefore, the cumulative expected time for event 04 is the sum of the expected times for events 02, 03, and 04 or $8 + 3.83 + 8.33 = 20.16$, which is shown above event 04 in Fig. 8.

Since three paths lead into event 09, with events 05, 06, and 08 directly preceding event 09, and have T_e values of 46.16, 24.16, and 22.0, respectively, the largest T_E (46.16), the T_E for event 05, is added to 8, the t_e value for event 09, to obtain 54.16, the T_e for event 09. In other words, the cumulative expected time for event 09 is that obtained by proceeding along the longest path. It should now be clear that each T_E value represents the earliest completion time for an event.

2.   The Backward Pass

Having completed the "forward pass" through the network by calculating T_e values for each event, we can now begin the "backward pass" by calculating T_L values, where T_L represents the latest allowable time for an event. We begin at event 09, the last event in the network, where we set $T_L = 54.16$, which is the calculated value of T_E for event 09. We then calculate the latest allowable time for an event, say, event 05, by substracting the expected time t_e from the T_L value of event 09. In other words, the T_L value for event 05 is $54.16 - 8.0 = 46.16$; or the T_L value for event 09 less the t_e value for event 09. The T_L value for event 06 is calculated in the exact same manner. The T_L value for event 04 is calculated by subtracting the t_e value for event 05 from the T_L value for event 05. And so performing the calculation, we obtain $46.16 - 26 = 20.16$. Note that the difference between the T_L value for event 06 less the t_e value for event 06, or $46.16 - 4 = 42.16$, is not the T_L value for event 04 because the T_L value of 20.16 is smaller than 42.16, and where there is a choice, the smallest value is always chosen. In general, when situations arise where there is a choice of alternative T_E and T_L values, the largest T_E value and the smallest T_L value are always chosen.

Now, to determine which events lie on the critical path, it is only necessary to examine the network and connect those events together which have equal T_E and T_L values. In our network, those events which lie on the critical path are connected by an extra heavy arrow.

3.   Slacks in the Network

The value defined by the quantity $T_L - T_E$ is termed the "positive slack" of the event. In Fig. 8, it can be seen that the positive slack of event 07 is $24.16(= 40.16 - 16.0)$. It should be noted that the positive slack of every event along the critical path is zero. The term "negative slack" is applied to those situations where a target date T_T is assigned to the project, which is earlier than the cumulative expected time T_E of the last event on the critical path. In other words, negative slack is merely $T_T - T_E$, where $T_E > T_T$. In our example, we began our calculation of latest allowable times by setting the latest allowable time for event 09 equal to the cumulative expected time for event 09. There is no rule that says it must be done this way. If, for example, a target date for completion of the project has been set prior to construction of a PERT network, this target date can be used as the latest allowable time for the last event in the network. Let us suppose that we had established a target date of 60 weeks. Allowing T_L values for all of the other events in out network, and obtain the results shown in Fig. 9. The critical path is now identified as the sequence of events, which, taken individually, have the same slack as the last event. In other words, as may be seen from an examination of Fig. 9, the slack of event 09 is 5.84 weeks, and this is precisely the slack of every event on the critical path of the network shown in Fig. 9.

4. Actions Which May Be Repeated

One of the characteristics of all product development work, indeed, of almost every form of research, is the need to adjust the product or formula after it has been evaluated by a taste panel or analytical laboratory. This gives rise to a loop in a PERT network which the planner may be inclined to depict as shown in Fig. 10. This approach is incorrect. Figure 11 illustrates the proper way to handle this problem. Note that zero has been used as the optimistic estimate for reformulation of the product and taste tests of the reformulated product. This indicates that these two activities may not have to occur, thus eliminating the need for looping.

C. Summary

PERT is a deceptively simple procedure. The discipline that it imposes upon management is considerable. Just drawing a PERT chart and defining the activities and events can be a challenging task. Fortunately, computer programs are available to perform the necessary calculations and to identify the critical path.

Aside from differences in nomenclature and pictorial representation, by far the greatest and most important difference between PERT and CPM is the method for handling time. In PERT, as we have seen, three estimates are made for each activity; in CPM, only one estimate is made. If the CPM user wants to account for unforeseen circumstances, he must overestimate his or her time for the job in question. This approach works quite well in the construction industry where the time required to complete various jobs is known with considerable accuracy because of the remendous backlog of experience which exists. If a project being managed by CPM is delayed for some reason, jobs on the critical path are simply put on a crash basis.

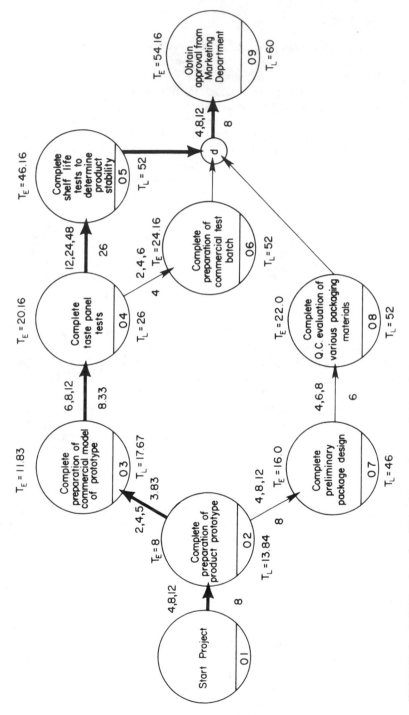

FIGURE 9 PERT network with a 60-week completion date.

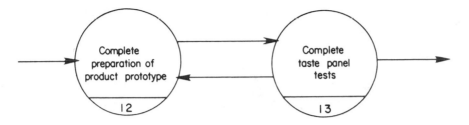

FIGURE 10   Part of a PERT network which incorrectly illustrates how to handle looping.

One of the truly interesting as well as significant facts to emerge from CPM and PERT is that usually only about 10% of the total number of activities of a project are found to lie on the critical path. Also, jobs thought to be trivial have frequently turned out to be critical.

IV. TRANSPORTATION PROBLEMS

There are many facets to a firm's transportation problems. Here we will limit our concern to the scheduling of shipments and to problems of waiting lines.

A. Scheduling of Shipments

Chapter 7 discussed the concept of linear programming and some of its applications. One of the first important applications of linear programming was in solving transportation problems. A simple example can illustrate how this can be done.

A typical transportation problem is shown in Fig. 12. We have two plants manufacturing mayonnaise. Plant 1 has three truckloads ready for shipment. Plant 2 has six truckloads ready for shipment. In addition, it is shown that warehouse 1 needs three truckloads; warehouse 2 needs two truckloads, and warehouse 3 needs four truckloads.

Figure 12 shows that to ship a truckload of mayonnaise from plant 1 to warehouse 1 costs $17.00. To ship from plant 1 to warehouse 2 costs $11.00 per truckload. To ship from plant 1 to warehouse 3 costs $15.00 per truckload. Shipping from plant 2 to warehouses, 1, 2, and 3 costs $13.00, $14.00, and $12.00, respectively.

We have a total of nine truckloads available at our points of origin. We need a total of nine truckloads delivered to our destinations. Our problem is to decide how much to ship from each plant to each warehouse in order to meet our needs (i.e., the requirements of the warehouses) without exceeding our resources (i.e., the quantities available at our plants). However,

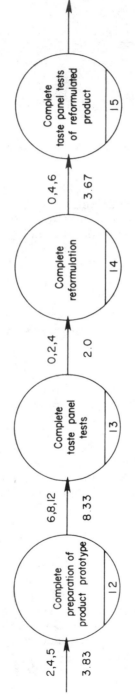

FIGURE 11   Part of a PERT network which correctly illustrates how to handle looping.

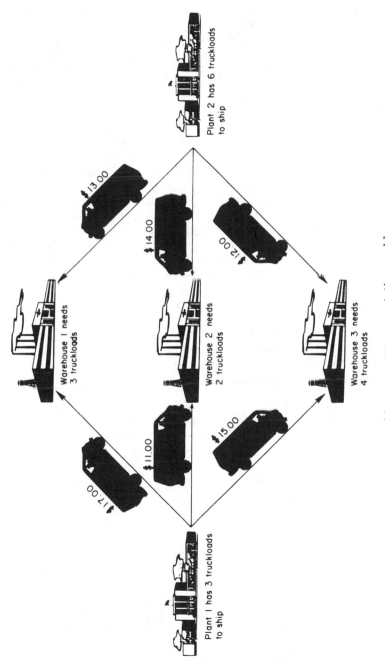

FIGURE 12   Schematic presentation of least-cost transportation problem.

it is not enough just to satisfy these physical needs. We want a plan that achieves these goals at least cost.

We can visualize this as a linear programming problem. The possibility of each origin–destination combination is an activity. The level of the activities that ship from plant 1 will have a constraint to reflect the fact that only three truckloads are available to plant 1. Similarly, shipments from plant 2 cannot exceed six truckloads.

A constraint must be added to ensure that at least three truckloads go to warehouse 1. A constraint must exist in order to ensure that at least two truckloads go to warehouse 2. Finally, a constraint must exist to ensure that at least four truckloads go to warehouse 3. Unless these constraints are specifically incorporated into the problem, we cannot be sure that they will be met.

Table 10 shows the initial tableau for our least-cost transportation problem. The columns represent shipping activities. It is important to note that every possible way of shipping from plants 1 and 2 to warehouses 1, 2, and 3 is included. Even high-cost routes that appear unlikely are included in the formulation of the problem. This is done because if they are included but not used in the final plan, the cost of the extra information is small. On the other hand, if a shipping possibility is omitted, it will never be considered. As a general rule, it pays to include alternatives that appear unlikely rather than to exclude them.

Table 11 shows the initial tableau of our problem in standard simplex form with appropriate disposal and artificial activities. The least-cost solution to our problem is

Plant 1 ships

   Two truckloads to warehouse 2 at $11.00 per truckload = $22.00

   One truckload to warehouse 3 at $15.00 per truckload = $15.00

Plant 2 ships

   Three truckloads to warehouse 1 at $13.00 per truckload = $39.00

   Three truckloads to warehouse 3 at $12.00 per truckload = $36.00

Total cost = $112.00

There is no other shipping pattern which will meet the warehouse demands without exceeding the plant availabilities and cost less than $112.00.

One of the features of the problem that we just solved was that the sum of available supplies (i.e., nine truckloads at two plants) exactly equaled the sum of requirements (i.e., nine truckloads at three warehouses). If our supplies had exceeded our demands, some would have been left at one or both plants. If our demands had exceeded our supplies, the problem would have been infeasible (i.e., could not have been solved).

TABLE 10 Initial Tableau for Least-Cost Transportation Problem

C_j	Shipping activities						Right-hand side
	-17 P1-W1	-11 P1-W2	-15 P1-W3	-13 P2-W1	-14 P2-W2	-12 P2-W3	
Plant 1	1	1	1	0	0	0	$\leqslant 3$
Plant 2	0	0	0	1	1	1	$\leqslant 6$
Warehouse 1	1	0	0	1	0	0	$\geqslant 3$
Warehouse 2	0	1	0	0	1	0	$\geqslant 2$
Warehouse 3	0	0	1	0	0	1	$\geqslant 4$

TABLE 11 Initial Tableau in Simplex Form for a Least-Cost Transportation Problem

$C_j \rightarrow$		0	0	0	0	0	-17	-11	-15	-13	-14	-12	$-m$	$-m$	$-m$
			Disposal Activities					Real Activities						Artificial Activities	
		Plant 1	Plant 2	Whse 1	Whse 2	Whse 3	P1-W1	P1-W2	P1-W3	P2-W1	P2-W2	P2-W3	Q_1	Q_2	Q_3
	B	x_1	x_2	x_3	x_4	x_5	x_6	x_7	x_8	x_9	x_{10}	x_{11}	x_{12}	x_{13}	x_{14}
0 Plant 1	3	1	0	0	0	0	1	1	1	0	0	0	0	0	0
0 Plant 2	6	0	1	0	0	0	0	0	0	1	1	1	0	0	0
$-m$ Q_1 (Whse 1)	3	0	0	-1	0	0	1	0	0	1	0	0	1	0	0
$-m$ Q_2 (Whse 2)	2	0	0	0	-1	0	0	1	0	0	1	0	0	1	0
$-m$ Q_3 (Whse 3)	4	0	0	0	0	-1	0	0	1	0	0	1	0	0	1
Z_j	$-9m$	0	0	m	m	m	$-m$	$-m$	$-m$	$-m$	$-m$	$-m$	$-m$	$-m$	$-m$
$Z_j - C_j$	$-9m$	0	0	m	m	m	+17	+11	+15	+13	+14	+12	0	0	0

Since the sum of supplies just equals the sum of requirements, it is not necessary to set up our problem with a series of inequalities. That is, since we know that in order to meet the warehouse requirements everything will be shipped from the two plants, we do not have to account for leftover truckloads. In other words, there is no need for disposal activities.

In general, transportation problems are composed of a tableau with only zeros and ones. In addition, if we ensure that the sum of the supplies available at origins equals the sum of requirements needed at destinations, a new algorithm can be used which will solve the problem far more efficiently. This alternative algorithm can result in a substantial reduction in computations. However, its use is restricted to problems where the initial tableau is composed of zeros and ones.

In practice, if a computer is available with a linear programming (LP) algorithm, it is easier and cheaper to solve transportation problems within the framework of an ordinary LP. However, if very large transportation problems (e.g., in excess of 400 rows and 1000 columns) are to be solved repeatedly, it would be advantageous to utilize a transportation linear programming algorithm that has been written for a computer. Such commercial codes are readily available.

With very large problems, just defining and coding the problem for submission to a computer can be a major undertaking. One means of reducing this task is to use a computer program that takes data in a form that is easy for management to generate and transforms it into a coded form suitable for the computer. Using the simple transportation problem just discussed, the following discussion illustrates a computer program written in FORTRAN which generates the matrix needed to define the initial problem. Such matrix generators can become very sophisticated. This very simple example merely illustrates the concept.

The basic idea of matrix generation is shown in Fig. 13. A FORTRAN program deck reads a data set and generates on magnetic tape the LP problem in proper format. The LP algorithm reads the problem from the magnetic tape and solves it. Table 12 shows the complete card set to define and solve the problem stated in Table 10. The first part of the FORTRAN program reads and stores the input data. In this case, the following:

1. Number of plants

2. Number of warehouses

3. Shipping costs from the various plants to warehouses

4. Inventory of the plants

5. Requirements of the warehouses

Once this is accomplished, the creation of the matrix follows. The matrix tableau can be divided into three sections: namely, row ID, activities, and the RHS (right-hand side).

```
'  FOR,SIX   DECK,DECK
        DIMENSION COST (99,99) ,PLTINV(99),WHREQ(99)
        READ (5,10) NPLNTS
        WRITE(6,10) NPLNTS
        READ (5,10) NWHSE
        WRITE(6,10) NWHSE
    10 FORMAT(10I5)
        DO 100 I=1,NPLNTS
        READ (5,110) (COST(I,J),J=1,NWHSE)
        WRITE(6,110) (COST(I,J),J=1,NWHSE)
   110 FORMAT(8F10.2)
   100 CONTINUE
        DO 200 I=1,NPLNTS
        READ (5,210) PLTINV(I)
        WRITE(6,210) PLTINV(I)
   200 CONTINUE
   210 FORMAT(1F10.0)
        DO 300 I=1,NWHSE
        READ (5,310) WHREQ(I)
        WRITE(6,310) WHREQ(I)
   300 CONTINUE
   310 FORMAT(1F10.0)
   150 FORMAT(1H0)
  3040 FORMAT(1H1)
        N9 = 6
        N9 = 10
        WRITE(6,3040)
C
C      ROW ID CREATION
C
        WRITE(N9,350)
   350 FORMAT(4HNAME,10X,8HPROBNAME)
        WRITE(N9,400)
   400 FORMAT(4HROWS)
        WRITE(N9,401)
   401 FORMAT(8H   N COST)
        DO 420 I=1,NPLNTS
        WRITE(N9,425) I
   425 FORMAT(2X,1HL,1X,1HP,I2)
   420 CONTINUE
        DO 440 I=1,NWHSE
        WRITE(N9,445) I
   445 FORMAT(2X,1HG,1X,1HW,I2)
   440 CONTINUE
C
C      MATRIX CREATION
C
        REALP = 1.00
        REALN =-1.00
        WRITE(N9,600)
   600 FORMAT(7HCOLUMNS)
        DO 605 I=1,NPLNTS
        DO 610 J=1,NWHSE
        COST(I,J) = COST(I,J) * REALN
        WRITE(N9,620) I,J,COST(I,J)
```

226

TABLE 12 (Continued)

```
  620 FORMAT(4X,1HP,I2,1HW,I2,4X,8HCOST        ,2X,F12.2)
      WRITE(N9,640) I,J,I,REALP
  640 FORMAT(4X,1HP,I2,1HW,I2,4X,1HP,I2,5X,2X,F12.2)
      WRITE(N9,660) I,J,J,REALP
  660 FORMAT(4X,1HP,I2,1HW,I2,4X,1HW,I2,7X,F12.2)
  610 CONTINUE
  605 CONTINUE
C
C     RHS CREATION
C
      WRITE(N9,700)
  700 FORMAT(3HRHS)
      DO 710 I=1,NPLNTS
      WRITE(N9,715) I,PLTINV(I)
  715 FORMAT(4X,2HB1,8X,1HP,I2,7X,F12.2)
  710 CONTINUE
      DO 720 I = 1,NWHSE
      WRITE(N9,725) I,WHREQ(I)
  725 FORMAT(4X,2HB1,8X,1HW,I2,7X,F12.2)
  720 CONTINUE
      WRITE(N9,800)
  800 FORMAT(6HENDATA)
      WRITE(N9,810)
  810 FORMAT(4HNAME,10X,8HZZZZZZZZ)
      WRITE(N9,800)
      IF(N9-10)900,901,900
  901 REWIND N9
  900 STOP
      END
'XQT
      2
      3
      1700        1100        1500
      1300        1400        1200
          3
          6
          3
          2
          4
'XQT UOM*FMPS.FMPS
      TITLE   TEST PROBLEM
      CALL ENTER(LP)
      ASSIGN 400 TO KINV
      CALL ATTACH('MYFILE',10,CARD,OLD)
      ADATA = 'PROBNAME'
      ARHS = 'B1           '
      AOBJ = 'COST'
      FOBJWT = -1.
      CALL INPUT(FILE,'MYFILE')
      CALL OUTPUT
      CALL OPTIMIZE
      CALL SOLUTION
      STOP
  400 CALL INVERT
      RETURN
      END
```

Card 1 tells the computer to compile a FORTRAN program.

Card 2 dimensions the three arrays COST, PLTINV (plant inventory), and WHREQ (warehouse requirements).

Cards 3 to 7 read and write NPLNTS (number of plants) and NWHSE (number of warehouses). These are the first and second cards on the data deck.

FIGURE 13   Flow diagram illustrating matrix generation for linear programming problems.

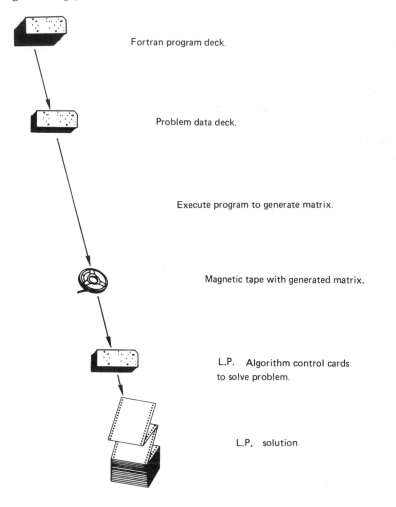

Fortran program deck.

Problem data deck.

Execute program to generate matrix.

Magnetic tape with generated matrix.

L.P.   Algorithm control cards
to solve problem.

L.P.   solution

Cards 8 to 12 are a do-loop which reads and writes the shipping costs from the plants to the warehouses. The cost array stores these values with plants as rows and warehouses as columns. For example, the cost of shipping for plant 2 to warehouse 3 would be stored in the cost array with the subscripts COST (2, 3).

Cards 13 to 21 read and write PLTINV(I) (plant inventory) and WHREQ(I) (warehouse requirements).

Cards 22 to 27 contain format statements. The card N9 = 6 is a "key" card. A write (6, xxx) means write on paper and write (10, xxx) means write on magnetic tape. From here forward, all write statements will be in the form write (N9, xxx) (i.e., write on tape). If we reversed the order of the cards and N9 were assigned a value of 6, all write statements in the (N9, xxx) form would write on paper. This could be done to verify the data format of the FORTRAN program.

Card 31 writes format 350. This statement is the first card required for the FMPS LP package.

Card 32 writes format 400. The program is now duplicating the coding required for the FMPS system.

The remaining statements in the FORTRAN program generate the problem in standard LP format, ending by rewinding the tape in preparation for the LP algorithm.

The last FORTRAN instruction is END, a nonexecutable statement which simply indicates that the FORTRAN program is complete.

The card XQT tells the computer to execute the FORTRAN program. The next nine cards are the data set for this problem. It is at this point that we can change our problem by changing the number of plants or warehouses, the cost of shipping from a specific plant to each warehouse, etc.

The remaining cards starting with XQT UOM*FMPS.FMPS tell the computer to read the LP problem previously written on logical unit 10, solve it and print the solution.

Obviously once such a matrix generator has been developed, it is easy to use it repeatedly to solve new distribution problems as availability, requirements, and shipping-costs change.

One of the disadvantages of using linear programming to solve transportation problems is that the LP model assumes that everything is infinitely divisible. With very large problems this may not be important, but in many cases the LP assumption that one-half of a truck costs half as much and ships half as much as a whole truck can cause serious problems. To overcome this, IBM developed a vehicle scheduling program (VSP-360). Nearly all large computer companies have a similar package. The initial logic is identical to that of a transportation linear programming model. That is, every origin and destination must be identified, the cost of shipping from each origin to each destination must be determined, etc. The difference between a transportation LP and a vehicle scheduling program is that the scheduling program can include such features as a mix of trucks of different sizes, the fact that certain streets are one-way part of the day and two-way at other times, driver A can work with drivers B and C but not drivers D and E, certain trucks can not use specific streets, etc. In other words, the problems faced by the ordinary dispatcher have been incorporated into the algorithm. Although these vehicle scheduling programs do not guarantee an optimal solution, experience has shown 10 to 35% reductions in transportation costs through the use of vehicle scheduling programs.

B. Waiting Lines

Having determined where to ship and how much to ship, we now consider what arrangements to make for receiving the shipment. Thus, for example, having decided to send X number of truck loads from origin Y to destination Z, we now must consider what our receiving facilities at the destination are. We may have one unloading platform(K = 1), so that only one truck can be unloaded at one time, or we may have several unloading platforms (K > 1). We may have fewer or more personnel available for unloading the trucks. This will have an effect on the time (u) each truck will be at a dock. In studying a particular situation, we may recommend to extend or reduce platform space, to increase or decrease size of unloading crews, or to change facilities or manner of handling the unloading operations. These waiting-line problems can be solved only if the number of units that can be unloaded in a given time period is greater than the number of arrivals.

Our objective is to minimize costs by providing an adequate number of receiving facilities and an unloading crew of sufficient size to ensure maximum efficiency at minimum cost. Our problem is illustrated graphically in Fig. 14. As we increase the number of receiving platforms, the cost of unutilized platform space increases because of the increased expense required for maintaining unused platforms. On the other hand, each truck is waiting for shorter and shorter periods of time so that cost of waiting is continuously decreasing. The minimum total cost is a combination of these two costs as shown in Fig. 14.

Such problems involve "waiting-line" or "queueing" theory (Saaty, 1957) and may be solved by the use of the equations derived from Erland distributions, or in complex situations, by the use of "simulation" or "Monte-Carlo" techniques (Meyer, 1956). We shall consider only three situations, beginning with the simplest kind, and ending with one involving the choice of different kinds of unloading facilities.

1. Determining Size of Unloading Crew, Lot Size Constant

A tomato packing plant receives truckloads of tomatoes during the 10-hr working day. Although an attempt is made to schedule the arrival of these

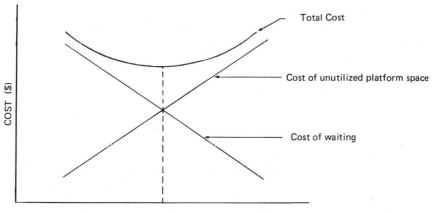

COST ($)

NUMBER OF RECEIVING PLATFORMS

FIGURE 14   An illustration of the trade-off between the cost of waiting and the cost of idle service facilities.

loads at one per hour, they actually appear to arrive in a random fashion (apparently in accordance with a Poisson distribution), averaging one truck arrival per hour. Unloading labor cost per person day is $15.00 and idle time per truck hour is $6.00. The current operation consists of a crew of two, who unload a truck in exactly 45 min. The packer suspects that if he were to add more men to the unloading crew, he might more than make up for the added labor cost, in saving in truck waiting time. He finds that three people will unload a truck in exactly 30 min and four in 20 min.

In such a situation, where unloading time (u) is fixed, we may find the average waiting or idle time (E_w) by the formula

$$E_w = \frac{a}{2u(u - a)}$$

where

a = arrivals per hour = 1

u = unloadings per hour = $\frac{60}{45}$ = 1.33

Thus,

$$E_w = \frac{a}{2u(u - a)} = \frac{1}{2.67(.33)} = 1.135 \text{ hr}$$

waiting time per truck or total waiting time for the 10 trucks for the day, of 11.35 hr, which at $6.00/hr represents a cost of $68.10/day.

Now if the packer were to add a third person to the unloading crew, he would reduce unloading time from 45 min per truck to 30 min, so that unloadings per hour (u) would now be 60/30 = 2, and average waiting time per truck would be reduced as follows:

$$E_w = \frac{a}{2u(u - a)} = \frac{1}{4(2 - 1)} = 0.25 \text{ hr per truck}$$

times 10 trucks = 2.5 waiting hours per day, times $6.00/hr = $15.00. Thus by increasing the number of people on the unloading crew from 2 to 3, the cost of idle, or waiting, time of trucks has been reduced from $68.10 to $15.00 or by $53.10, at a cost of $15.00 for the added labor, a net savings of $38.10.

Adding a fourth person to the crew would reduce unloading time to 20 min, so that u = 60/20 = 3. Hence,

$$E_w = \frac{a}{2u(u - a)} = \frac{1}{6(3 - 1)} = \frac{1}{12} \text{ hr}$$

Multiplied by the 10 trucks for the day, we now have total truck waiting time of 10/12 hr, which multiplied by $6.00 is $5.00. Compared to the three-crew situation, this represents a further savings of $10.00/day in reduced truck idle time; however, since the labor cost of the fourth person at $15.00 is greater than this saving of $10.00, we are not justified in adding the fourth person to the unloading crew.

We may, if we wish, obtain additional information on the characteristics of such a waiting line, where unloading time is constant, by the use of the following formulas; our example is the three-person crew with an unloading time (u) of 30 min, or 2 per hour:

Average number of waiting trucks:

$$E_m = \frac{a^2}{2u(u - a)} = \frac{1^2}{4(2 - 1)} = \frac{1}{4} \text{ truck}$$

Average number of trucks in yard:

$$E_n = \frac{a^2}{2u(u - a)} + \frac{a}{u} = \frac{1}{4(2 - 1)} + \frac{1}{2} = \frac{3}{4} \text{ truck}$$

Average time of truck in yard:

$$E_v = \frac{a}{2u(u - a)} + \frac{1}{u} = \frac{1}{4(2 - 1)} + \frac{1}{2} = \frac{3}{4} \text{ hr}$$

2. Determining Size of Unloading Crew, Lots Varying in Size

We now consider a somewhat different situation, where truck size and shape may vary substantially, so that unloading time by a crew of a certain size is no longer constant but may vary exponentially. Then our equations differ somewhat as follows:

Average waiting time per truck:

$$E_w = \frac{a}{u(u - a)} = \frac{1}{2(2 - 1)} = \frac{1}{2} hr$$

Average number of waiting trucks:

$$E_m = \frac{a^2}{u(u - a)} = \frac{1}{2(2 - 1)} = \frac{1}{2} truck$$

Average number of trucks in yard.

$$E_n = \frac{a}{u - a} = \frac{1}{2 - 1} = 1 truck$$

Average time of trucks in yard:

$$E_v = \frac{1}{u - a} = \frac{1}{1} = 1 hr$$

To determine the optimal crew size, we proceed as in the previous example, but use the somewhat different equations shown above for E_w.

In plants of any size, a single-station unloading platform may very quickly become a bottleneck, so that a multistation arrangement is required. In such an event, in addition to other considerations, the packer should know whether he or she has sufficient, or an excessive, number of unloading stations for most efficient operation.

3. Determining Number of Unloading Stations

A tomato packer has trucks arriving at the platform at the average rate of 10 per hour ($a = 10$). Cost of idle truck time is $6.00/hour. There are three unloading stations ($K = 3$). The crew at each station is capable of unloading four trucks per hour ($u = 4$). Total cost of maintaining and operating an unloading station, including all labor costs, is $10.00/hr. The packer could, of course, reduce the number of unloading stations. He or she could also find the platform space to increase the number of unloading stations to four or even five. First determine whether the use of three stations is the most economical, or whether the number should be increased to four, or possibly reduced to two. The packer is also considering the possibility of changing the facilities at the unloading platform from conveyors handling individual boxes to palletized crates.

We first consider the present system and determine the average waiting per truck (E_w). With more than one station involved (K is greater than 1) our equation becomes considerably more complex:

$$\text{Average waiting time per truck} = E_w = \frac{u(a/u)^K}{(K-1)!(Ku-a)^2}(P_0)$$

where

     u = unloading per hour at each platform = 4 trucks

     a = arrivals per hour = 10 trucks

     K = number of stations for unloading = 3 stations

     P_0 = probability that no trucks will arrive

We first need to find P_0. This can be done by the use of the following formula:

$$P_0 = \left\{ \left[\sum_{n=0}^{K-1} \frac{1}{n!}\left(\frac{a}{u}\right)^n \right] + \frac{1}{K!}\left(\frac{a}{u}\right)^K \frac{Ku}{Ku-a} \right\}^{-1}$$

where

$$\sum_{n=0}^{K-1} (1/n!)(a/u)^n \text{ means the sum of the values of } (1/n!)(a/u)^n,$$

beginning with n = 0 plus the same term when n = 1, and so on, until n = K - 1. Since in our example K = 3, we sum the terms of $(1/n!)(a/u)^n$ when n = 0, n = 1, and n = 2 as follows:

When n = 0,

$$\frac{1}{0!}\left(\frac{a}{u}\right)^0 = 1 \quad (\text{since } 0! = 1)$$

When n = 1,

$$\frac{1}{1}\left(\frac{10}{4}\right)^1 = 2.5$$

When n = 2,

$$\frac{1}{2(1)}\left(\frac{10}{4}\right)^2 = 3.125 \qquad \text{(since 2! = 2 X 1)}$$

Since K = 3, then

$$\frac{1}{K!} = \frac{1}{3 \text{ X } 2 \text{ X } 1} = \frac{1}{6}$$

$$\left(\frac{a}{u}\right)^K = \left(\frac{10}{4}\right)^3 = 15.625$$

$$\frac{Ku}{Ku - a} = \frac{3(4)}{3(4) - 10} = \frac{12}{2} = 6$$

Collecting all the terms in the equation,

$$P_0 = \left[\sum_{n=0}^{K-1}\frac{1}{n!}\left(\frac{a}{u}\right)^n + \frac{1}{K!}\left(\frac{a}{u}\right)^K \frac{Ku}{Ku - a}\right]$$

$$= \frac{1}{1 + 2.5 + 3.125 + \dfrac{1}{6}(15.625)(6)} = \frac{1}{22.25}$$

Having found that the probability of having no trucks arriving per hour was $P_0 = 1/22.25$, we may now solve for the average waiting time per truck as follows:

$$E_w = \frac{u(a/u)^K}{(K - 1)!(Ku - a)^2}P_0 = \frac{4(10/4)^3}{(3 - 1)![3(4) - 10]^2}\left(\frac{1}{22.25}\right)$$

$$= \frac{62.5}{8}\frac{1}{22.25} = 0.35 \text{ hr}$$

Thus with the present unloading facilities and with an average of 10 trucks per hour, the idle time of trucks waiting to unload is 0.35 hr times 10, or 3.5 hr, which at a cost of $6.00/hr represents a total truck waiting-time cost of $21.00/hr. Having previously established the cost of maintaining and operating an unloading station at $10.00/hr, it is logical to consider the advisability of adding a fourth unloading station.

Again we must first solve for P_0. Unloadings per hour per station remain the same at u = 4. Arrivals per hour also remain the same at a = 10. Number of stations, however, is now changed from K = 3 to K = 4.

$$P_0 = \frac{1}{1 + 2.5 + 3.125 + 2.609 + 1/24(10/4)^4 \, [16/(16 - 10)]} = \frac{1}{13.569}$$

$$E_w = \frac{4(10/4)^4}{(4 - 1)! [4(4) - 10]^2} \frac{1}{13.569} = \frac{156.25}{2930.90} = 0.053 \text{ hr}$$

Thus idle time of the trucks waiting to be unloaded can be reduced to an average of 0.53 hr per truck, or 0.53 truck hours for the 10 arrivals per hour. At a cost of $6.00 per idle truck hours, this amounts to $3.18/hr, as compared to $21.00/hr when only three unloading platforms are available, or a cost reduction of $21.00 - $3.18 = $17.82. Subtracting $10.00 hourly cost of maintaining and operating a fourth station, we have a net savings of $7.82/hr by operating the fourth station.

Since setting up and operating a fifth unloading station will add $10.00 to the cost and the maximum cost reduction by reducing truck waiting time is now only $3.18, then clearly, net costs cannot be reduced further by adding a fifth unloading station. Our answer to this problem is that the most economical method of unloading these trucks is by the use of four stations, where the average waiting time per truck is 0.053 hr, or a little more than 3 min.

We may determine the average number of waiting trucks by the following formula (example is for the four-station problem):

$$E_m = \frac{au(a/u)^K}{(K - 1)! (Ku - a)^2} (P_0) = \frac{40(10/4)^4}{3! (16 - 10)^2} \frac{1}{13.569}$$

$$= 0.533 \text{ waiting truck}$$

Average number of trucks in the yard may be determined as follows:

$$E_n = \frac{au(a/u)^K}{(K - 1)! (Ku - a)^2} (P_0) + \frac{a}{u} = \frac{40(10/4)^4}{3! (16 - 10)^2} \frac{1}{13.569} + \frac{10}{4}$$

$$= 0.533 + 2.5 = 3.033 \text{ trucks in yard}$$

The average time of trucks in yard is

$$E_v = \frac{u(a/u)^K}{(K - 1)! (Ku - a)^2} P_0 + \frac{1}{u} = \frac{4(10/4)^4}{3! (16 - 6)^2} \frac{1}{13.569} + \frac{1}{4} = 0.783 \text{ hr}$$

If a palletizing arrangement is considered, data should be obtained for determining new values for number of trucks unloaded per hour at each station (u) and the cost of constructing, maintaining, and operating the new stations. Number of arrivals (a) remains the same. With such data, and using the procedure described above, calculations may be made to determine what the optimum number of pallet-handling stations should be and whether a change should be made from handling individual boxes to pallets.

4. A Simple Computer Program

As waiting-line problems become more complex, the computational burden quickly becomes excessive. Using a small pocket calculator one can greatly reduce this burden. A simple program in BASIC can make the burden of calculations insignificant. Table 13 shows a program in BASIC which performs the calculations needed to determine P_0 and E_w for the case with multiple unloading stations.

Instructions 100 to 150 request the user to enter U (unloadings per hour at each station), A (arrivals per hour), and K (number of stations for unloading).

Instructions 160 to 230 perform a loop necessary to calculate the factorials in the equations.

Instructions 280 to 320 calculate P_0.

Instructions 340 to 360 calculate E_w.

Instruction 380 prints the values of P_0 and E_w.

TABLE 13   Computer Program in BASIC to Calculate
Waiting Time for Multiple Unloading Stations

```
100 PRINT 'INPUT U ';          220   S=S+(X*Y)
110 INPUT U                    230 NEXT J
120 PRINT 'INPUT A ';          280 W1=W*K
130 INPUT A                    290 X1=(1/W1)
140 PRINT 'INPUT K ';          300 Y1=A/U)**K
150 INPUT K                    310 Z1=(K*U)/(K*U-A)
160 S=0                        320 P=(1/(S+X1*Y1*Z1)
170 W=1                        340 X2=U*Y1
180 FOR J = 0 TO K-1           350 Y2=W*((K*U-A)**2)
185    IF J = 0 THEN 200       360 E=(X2/Y2)*P
190    W=W*J                    380 PRINT 'P= ';P, 'E= ';E
200    Y=(1/W)                  390 END
210    X=(A/U)**J
```

Such a program enables a decision maker to consider a variety of alternatives quickly and easily.

V. SUMMARY

This chapter examined a number of management tools. In each case, the basic concept was relatively simple. At the same time, the computational burden was or could easily become overwhelming. It is in situations such as these that a computer can become a significant aid to decision makers. However, with or without a computer, the first step must be a clear statement of the problem including what the decision maker seeks to achieve.

ACKNOWLEDGEMENTS

The authors wish to express their appreciation to AVI Publishing Co., Westport, Conn. for permission to incorporate into this chapter all figures and tables (except for Tables 8 and 13), which were previously published in Bender, Kramer, and Kahn (1976; pp. 195-197, 207, 211-213, 216, 219, 220, 290-292, 314, 315, 318, 337, 347, 348, 354, 356, 357, 361, 364, 365). In addition, the authors wish to acknowledge the assistance of Susan Blatchley Epstein in the development of the examples used, Richard Dillon for writing the FORTRAN program, and Frederick Bender for writing the programs in BASIC.

REFERENCES

Agrawal, R. C. and Heady, E. O. (1972). Operations Research Methods for Agricultural Decisions, The Iowa State University Press, Ames, Iowa.

Akers, S. B., and Freidman, J. (1955). A non-numerical approach to production scheduling problems, J. Operations Res. Soc. Amer. 3: 429-442.

Bender, F. E., Kramer, A., and Kahan, G. (1976). Systems Analysis for the Food Industry, Avi, Westport, Conn.

Brigham, E. F. and Pappas, J. L. (1976). Managerial Economics, 2d ed. Dryden Press, Hinsdale, Ill.

Campbell, H. G., Dudek, R. A., and Smith, M. L. (1970). A heuristic algorithm for the n job, M machine sequencing problem, Management Sci. 16(10):B630-B637.

Churchman, C. W., Ackoff, R. L., and Arnoff, E. L. (1957). Introduction to Operations Research, Wiley, New York.

Fabrycky, W. J., Ghare, P. M., and Torgersen (1972). Industrial Operations Research, Prentice-Hall, Englewood Cliffs, N.J.

Henry, W. R. and Haynes, W. W. (1978). Managerial Economics: Analysis and Cases, 4th Ed. Business Pulblications Inc., Dallas, Tex.
Horowitz, J. (1967). Critical Path Scheduling. Ronald Press, New York.
Levin, R. I., and Kirkpatrick, D. A. (1966). Planning and Control with PERT/CPM. McGraw-Hill, New York.
Lyle, P. (1957). Regression Analysis of Production Costs and Factory Operations, Oliver and Boyd, London.
Mauchly, J. W. (1962). Critical-Path Scheduling. Chem Eng. 69:139-154.
Maynard, H. B. (1967). Handbook of Business Administration. McGraw-Hill, New York.
Meyer, H. A. (1956). Symposium on Monte Carlo Methods, University of Florida, March, 1954. Wiley, New York.
Moder, J. J. and Phillips, C. R. (1964). Project Management with CPM and PERT. Van Nostrand Reinhold, New York.
Saaty, T. L. (1957). Resume of useful formulas in queueing theory. Operations Res. 5(2):161-200.
Sasieni, M., Yaspan, A., and Friedman, L. (1959). Operations Research Methods and Problems. Wiley, New York.
Siemens, N., Marting, C. H., and Greenwood, F. (1973). Operations Research. Free Press, New York.
Starr, M. K. (1971). Systems Management of Operations. Prentice-Hall, Englewood Cliffs, N.J.
Steir, H. L. (1953). The basic concepts of management control—men, machines, and money. Canning Trade 75(30):50-56.
Terborgh, G. (1958a). Dynamic Equipment Policy. Machinery and Allied Products Institute, Washington, D.C.
Terborgh, G. (1958b). Business Investment Policy. Machinery and Allied Products Institute, Washington, D.C.
Terborgh, G. (1964). Investment Policy—A Challenge to American Management. Machinery and Allied Products Institute, Washington, D.C.

9

Application of Computers in Food Rheology Studies

MICHA PELEG   University of Massachusetts, Amherst, Massachusetts

I. INTRODUCTION

Rheological studies often require considerable amounts of mathematical calculations. Naturally, therefore, the computer is an indispensible tool in dealing, with rheological problems. As in other fields, the computer can be used in different ways. In food rheology there are two main applications. In the first, the computer is used to facilitate and expedite routine calculations that can also be carried out by other (although considerably more cumbersome) methods. Notable examples are statistical analyses of experimental data and curve fitting when the rheological model is already established or assumed.

The second type of application involves problems that are especially formulated for a computer and would be otherwise extremely difficult, if not practically impossible, to solve by other means, e.g., the solution of certain types of differential equations and the calculation of internal stress distributions in bodies with complex geometry. It is obvious that the distinction between the two types is not at all clear, and there are many cases in which the amount of necessary calculations by itself transforms an otherwise routine problem into one that requires a special computer program.

In this chapter the discussion is focused on the second type of application with emphasis on how the computer can and has been employed to solve specific problems in food rheology. The theoretical and practical implications of these rheological problems has only briefly been mentioned. Their detailed discussions can be found in numerous publications and textbooks, e.g., Sherman (1970, 1979), Mohsenin (1970), and deMan et al. (1976).

Also described in the chapter is a method of treating shear analysis data for food powders flowability evaluation based on Jenike's (1961) methodology. Although the mechanics of particulated solids is sometimes treated as a separate branch, it certainly belongs to the realm of rheology as past and recent publications indicate.

II. MACROPHENOMENOLOGICAL RHEOLOGICAL MODELS

A. Liquids

1. Newtonian Liquids

Most light beverages (including milk but not all fruit juices) can be considered as newtonian liquids which can be described by

$$\tau = \eta_N \dot{\gamma} \tag{1}$$

where τ is the shear stress, $\dot{\gamma}$ is the shear rate, and η_N is the newtonian viscosity. Since in most cases the dependency of the viscosity on temperature can satisfactorily be described by an Arrhenius type of equation, the mathematical treatment of newtonian behavior is rather straightforward (Rao, 1977). In certain cases, however, especially where the effect of solute concentration is the purpose of the analysis, the relationship is more complex and ought to be treated by detailed physical and chemical analysis. When the latter, for practical considerations, cannot be carried out, an empirical relationship such as polynomial equation (Fernandez Martin, 1972) or response surface can be calculated through routine library programs (Myers, 1971).

2. Nonnewtonian Liquids

Most nonnewtonian liquid foods can be described by the equation

$$\tau = \tau_0 + k_1 \dot{\gamma}^n \tag{2}$$

where τ_0 is a constant representing the yield stress, k_1 is a constant, largely referred to as consistency, and n a power (usual $n < 1$), referred to as a flow behavior index.

It ought to be remembered that since most liquid foods show shear thinning over a large range of shear rates, their flow curve will generally appear in the shape schematically depicted in Fig. 1. This may tempt the experimenter to fit the data by a nonlinear regression program (e.g., SPSS) in order to calculate the constants. The risk here is that the constants so calculated do not necessarily represent the physical characteristics of the system. This is especially true with regard to the yield stress (τ_0), i.e., that neither its existence nor its magnitude can be inferred through mathematical manipulation of the flow curve as shown in Fig. 1. Since the existence of a yield stress may bear technological significance, its determination ought to be carried out independently by physical methods. This is true for other types of models (e.g., the Casson's equation) that contain a yield constant. It may be added that factors such as time-related effects should also be considered before the flow pattern is established (Mizrahi, 1979).

D. Solids

1. Linear Viscoelastic Models

The definition of "solid" is not as straightforward as most people may think. Let us therefore consider as solids those food materials that

FIGURE 1 Schematic representation of the flow curve of pseudoplastic liquid with and without yield stress (explanation in the text).

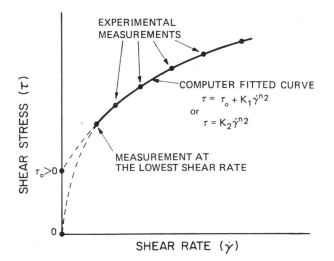

EXPERIMENTAL MEASUREMENTS

COMPUTER FITTED CURVE
$$\tau = \tau_o + K_1 \dot\gamma^{n2}$$
or
$$\tau = K_2 \dot\gamma^{n2}$$

MEASUREMENT AT THE LOWEST SHEAR RATE

SHEAR STRESS (τ)

$\tau_o > 0$

SHEAR RATE ($\dot\gamma$)

are referred to as solids in everyday speech (e.g., fruits, nuts, frank-furters, candy bars). Their common mechanical feature is that they can support appreciable levels of shear stresses and therefore do not flow under their own weight or small external loads. This mechanical strength is apparently the crucial psychological criterion of solidity de-spite the fact that it is not always or necessarily identical to rheolog-ical criteria.

The simplest behavior of solids is described by Hooke's law, which is only applicable to small elastic deformations. Most, if not all, food materials are viscoelastic materials, and therefore their mechanical re-sponse to loading depends on the strain and the strain rate history of the specimen.

The viscoelastic behavior of materials is mainly characterized through mathematical models that describe their stress-strain, stress-relaxation, and creep curves. In the case of linear viscoelasticity the models are frequently constructed as mechanical analogs that help visu-alize the behavior pattern. The most common elements of the models are the following (σ is the stress, and ϵ is the strain):

Basic

Elastic (spring) $\qquad \sigma = K\epsilon$ $\hfill$ (3)

Viscous (dashpot) $\qquad \sigma = \eta \dot{\epsilon}$ $\hfill$ (4)

Combinations

Maxwell $\qquad\qquad \dot{\epsilon} = \dfrac{\dot{\sigma}}{K} + \dfrac{\sigma}{\eta}$ $\hfill$ (5)

Kelvin-Voigt $\qquad\quad \sigma = K\epsilon + \eta \dot{\epsilon}$ $\hfill$ (6)

It has been shown that for even a rough approximation of the behavior of foods, models containing at least three basic elements are necessary. The analytical solution of the resulting equation may be quite elaborate, and the aid of a computer program becomes inescapable.

a. Examples: In the cases of stress-strain relationships at constant strain rate or constant deformation rate at small deformations, the con-stant speed (deformation rate) in which most testing machines operate in-deed produces constant strain rate. In such cases ($\dot{\epsilon} = $ constant) the analyt-ical solution of the phenomenological equation is relatively simple. As an example, a Maxwell element relation will be

$$\sigma(\epsilon) = \eta \dot{\epsilon} \left[1 - \exp\left(-\frac{K\epsilon}{\eta \dot{\epsilon}}\right) \right] \hfill (7)$$

If we have an array of Maxwell m elements in parallel, the solution will be simply the sum of the stresses, or

$$\sigma(\epsilon) = \dot{\epsilon} \sum_{i=1}^{m} \eta_i \left[1 - \exp\left(-\frac{K_i \epsilon}{\eta_i \dot{\epsilon}}\right) \right] \tag{8}$$

In the case of large deformations, which are of particular interest in texture studies, the strain rate can by no means be considered as constant because in compression

$$\dot{\epsilon} = \frac{V}{H_0 - Vt} \tag{9}$$

and in tension.

$$\dot{\epsilon} = \frac{V}{H_0 + Vt} \tag{10}$$

where H_0 is the initial specimen length, $H(t)$ the length after time (t), and V the constant deformation rate (V = dH/dt). Similarly, the true strain deviates significantly from the nominal (or engineering strain) and should be calculated from

$$\epsilon = \ln \frac{Ho}{H_0 \quad Vt} \tag{11}$$

in compression and

$$\epsilon = \ln \frac{H_0}{H_0 + Vt} \tag{12}$$

in tension.
The solution for a single Maxwell element for compression now becomes (Peleg, 1977)

$$\sigma(t) = K \exp[a(b - t)] \left\{ \log_e \frac{b}{b - t} + \sum_{n=1}^{\infty} \frac{(-a)^a \, [b^a - (b - t)^n]}{n(n!)} \right\} \tag{13}$$

where 1/a is the relaxation time of the element and 1/b is the initial strain rate.
For any discrete maxwellian model with m elements the solution becomes

$$\sigma(t) = K_o \log_e \frac{b}{b-t} + \sum_{i=1}^{m} \left[K_i \exp[a_i(b-t)] \right.$$

$$\left. \left\{ \log_e \frac{b}{b-t} + \sum_{n=1}^{\infty} \frac{(-a_i)^n [b^n - (b-t)^n]}{n(n!)} \right\} \right] \tag{14}$$

It can be shown that for small values of a and b the series converges fairly rapidly, and therefore the calculation of $\sigma(t)$ can be made by a straight-forward computer program.

2. Nonlinear Models

Most food materials exhibit nonlinear viscoelasticity and rheological memory. In special cases there are also specific modes of behavior that are characterized by discontinuities in the stress-strain relationships (e.g., a bio-yield point in certain fruits), stress generation (e.g., in muscle tissues as a result of rigor mortis), and strong time effects (e.g., due to enzymatic activity).

Representation of the true behavior of solid foods therefore can rarely be done by linear models whose applicability, even as first approximation, is limited to very small strains. To account for nonlinearity, the models must be modified, and there are three different ways by which this was done in foods.

1. By introducing nonlinear components into the model, e.g., transforming a linear Kelvin-Voigt element [Eq. (6)] into forms such as

$$\sigma = C_1 \epsilon + C_2 \dot{\epsilon} + C_3 \epsilon \dot{\epsilon} \tag{15}$$

or

$$\sigma = C_1 + C_2 \dot{\epsilon} + C_3 (\dot{\epsilon})^2 + C_4 (\dot{\epsilon})^3 + \cdots \quad \text{(Green-Rivlin model)} \tag{16}$$

and/or introducing a memory function into the model equation (Clark, Fox, and Welch, 1971). The resulting models are usually of considerable mathematical complexity and are difficult to solve analytically. With the aid of computers, however, they can be treated with relative ease, as demonstrated by Clark, Fox, and Welch (1971) in the case of cottonseeds (see below).

2. By modifying the geometry of the array of otherwise linear elements. (Nonlinear elements will only increase the mathematical complexity.) This approach has been suggested for meat by Segars and Kapsalis (1976). Their published model array had a hexagonal shape, and it included 10 elastic and viscous elements. (The principle of introducing nonlinearity through geometrical means is schematically demonstrated in Fig. 2).

NONLINEAR
VISCOELASTIC MODEL

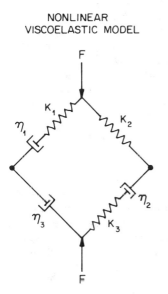

FIGURE 2   A nonlinear rheological model
produced from linear elements.

Treatment of the resulting set of differential equations, which also include
trigonometric relationships, requires an elaborate computer program especially if the model contains more than a few basic elements.

    3. By introducing additional types of elements into the model array.
A list of such elements and their mathematical properties is given below.
(ϵ and σ are strain and stress, respectively, and the subscript c represents
preassigned critical value.)

Element	Descriptive function	Mathematical representation
St. Venant (frictional)	Plastic deformation	For $\sigma < \sigma_c$, $\epsilon = 0$ else $\epsilon = \epsilon$
Fracture (shear pin)	Irreversible elimination of elements, discontinuity	For $\sigma < \sigma_c$, $\epsilon = 0$ else $\sigma = 0$
Contact	Activation of elements, strain hardening	For $\epsilon < \epsilon_c$, $\sigma = 0$ else $\sigma = \sigma$
Contractile	Stress generation, rigor mortis in muscle	$\Delta\sigma = \Delta\sigma(t)$

It should be mentioned that the incorporation of such elements does allow for effects that cannot be accounted for by the two other methods. Notable cases are discontinuities in the stress–strain relationships and reversal of the slope sign in stress-relaxation curves. Obviously, one can try to combine the methods, e.g., by including nonlinear and discontinous elements in the same model. In many cases, however, such an approach would have little advantage over purely empirical relationships that may have simpler mathematical forms.

III. SIMULATION OF RHEOLOGICAL BEHAVIOR

A. Analog Computers

One of the main reasons for using mechanical analogs as phenomenological models is that they help to visualize relationships that are otherwise described by differential equations. In actual simulation of rheological behavior, however, the mechanical analog is rarely a practical tool, for obvious technical reasons. These technical difficulties can easily be overcome if the model is expressed in terms of an analogous electrical circuit. The direct (force–voltage) analogy is given by the following (Duirling, 1969):

Mechanical component	Electrical equivalent
Force (F)	Voltage (e)
Velocity $V = \dfrac{dx}{dt}$	Current $i = \dfrac{dq}{dt}$
Displacement (x)	Charge (q)
Mass (M)	Inductance (L)
viscous damping (η)	Resistance (R)
Compliance $\dfrac{1}{K}$	Capacitance (C)

There is, however, another analogy, known as the mobility analogy, in which the mechanical force is represented by current and not by voltage.
    In such a system the analogy is given by the following:

Mechanical component	Electrical equivalent
Force (F)	Current $i = \dfrac{dq}{dt}$
Velocity $V = \dfrac{dx}{dt}$	Voltage (e)

Mechanical component	Electrical equivalent
Displacement (x)	Flux linkage (ϕ)
Mass (M)	Capacitance (C)
Viscous damping (η)	Conductance $G = \dfrac{1}{R}$
Compliance $\dfrac{1}{K}$	Inductance (L)

Since electrical circuit components can easily be added and replaced, different models can easily be constructed. Furthermore, the forcing function $\epsilon(t)$ can also be allowed to vary. Constant (DC source), sinusoidal (AC source), step (switch), and ramp voltage functions can easily be obtained, thus facilitating simulations of different test conditions.

Despite their numerous methodological advantages, the analog computer which is based on electrical circuits has had limited application if food rheology research. The main reason is the rise of digital computers and the rapid development of techniques and programs by which differential equations can be dealt with. A notable example of an analog computer array for testing the applicability of different rheological models in food materials is demonstrated in Fig. 3.

D. Digital Computers

Digital computer simulations have mainly been employed in two different ways. In the first, the computer was programmed to plot or calculate the behavior of a preconceived rheological model, and the results were compared with the observed behavior of experimental specimens under similar physical conditions. The comparison could also yield numerical values for the model constants, thus providing a tool of quantifying the rheological characteristics of the system under study. This approach has been demonstrated in a variety of food materials, and in each the program it was made to fit the specific problems that are associated with the selected type of mathematical model. Recent examples are the simulation of rigor mortis in muscle by Stoner et al. (1974) (included in the model is the contractile element), deformation and relaxation of apple tissue (DeBaerdemaeker and Segerlind, 1976), dynamics of cushioning (Kennish and Henderson, 1978), and deformation and relaxation of American cheese (Chen and Rosenberg, 1977) (included in the model is a St. Venant element).

The second approach is that not only the test conditions are allowed to vary but also the model itself. The conceptual basis for it is that the accurate behavior of foods is characterized by nonlinearity irreversibility

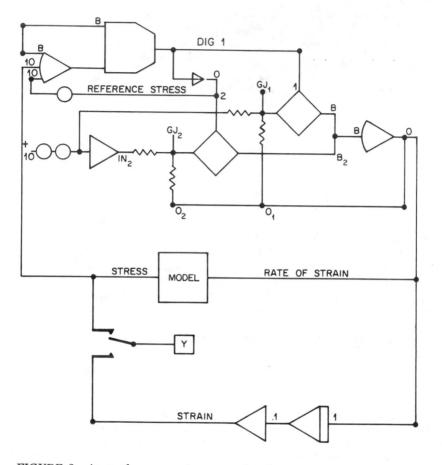

FIGURE 3   An analog-computer array for determination of the rheological behavior mode of foods. (From Clark, Fox, and Welch, 1971.)

and memory. Therefore, any model with a fixed number of elements (or fixed characteristic constants) cannot have general applicability. This notion has led Drake (1979) to develop a general program in which the model characteristics can vary during the simulation by the activation and inactivation of elements. An output of a program based on similar principles but with much smaller combinations permitted is demonstrated in Fig. 4.

IV. CONTACT STRESSES AND THERMAL STRESSES

One of the major goals of rheological studies in foods is the establishment of failure criteria and conditions under which failure occurs. The

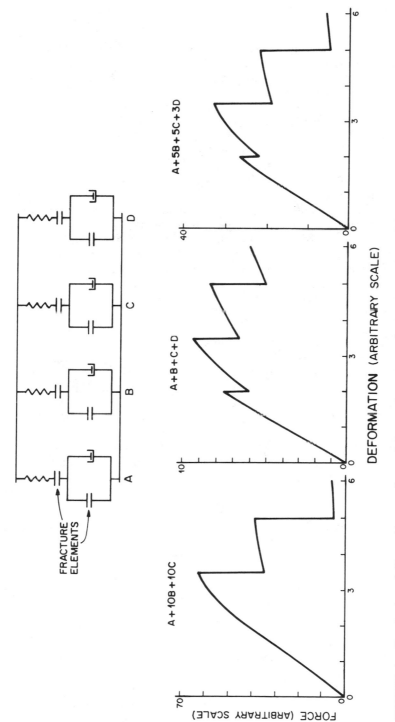

FIGURE 4 Simulation of discontinuous force deformation curves by incorporation of fracture elements into the model array. (From Cox and Peleg, 1976.)

practical implications of this problem are obvious. Mechanical damage
(e.g., in fruits, vegetables, grains) is often a result of stresses that are
produced through contact between units or between a food unit and another
object such as a container wall.

Schematically the contact stress problem is demonstrated in Fig. 5.
The consequence of force operating on bodies in contact is the development
of internal stresses (both normal and shear) whose distribution depends on
the geometry of the system and the rheological properties of the bodies involved. For hard elastic materials having well-defined shapes (e.g., disks,
spheres), the stress distribution can be calculated analytically, and the results are summarized in the mechanical literature.

In more complex geometries, but mainly for bodies made of a uniform
material, solutions can be obtained through photoelastic analysis and similar methods. In most foods these methods are of limited applicability since
both the geometries and the rheological properties of the materials are far
from being defined. Approximation of the internal-stresses situation can
be achieved through the method of "finite element analysis" (Zienkiewicz,
1971). According to this method the body in question is divided by a grid
as shown in Fig. 6. The number of nodes and elements can be selected at
will according to the needed degree of accuracy. For each element a set
of equations that represent the mechanical state is written. Their simultaneous solution, by a special computer program, is then plotted in a form
that reflects the internal-stress distribution. An example of the application
of this methodology is shown in Figs. 7 and 8 (Rumsey and Fridley, 1977).
It should be mentioned that the result of the analysis may change quite
dramatically according to the mechanical properties that are assumed when
writing the elements equations matrix. The program, however, can be rerun with different constants, thus providing an indication of the sensitivity

FIGURE 5   Different types of contact stress problems.

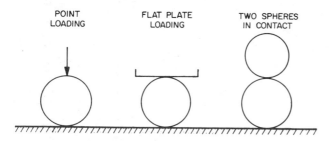

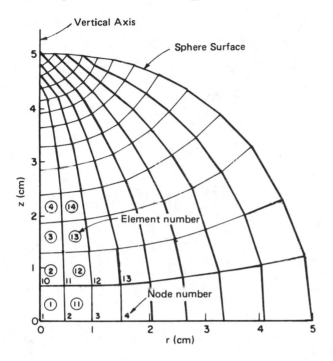

FIGURE 6   Elements and nodes in the solution of contact stress problem in apple through finite element analysis.   (From Rumsey and Fridley, 1977.)

of the stress distribution not only to the external geometry but also to the material properties (Sherif and Segerlind, 1976).

The mechanical properties of foods are known to be greatly affected by both temperature and moisture. In certain cases (e.g., drying cracks in grain, heat cracks in frankfurters) the mechanical damage is induced by what may be called "thermal" and "moisture" stresses. In such systems, however, the problem is not static since there usually exists a gradient with respect to both. In such cases the finite element analysis has to be carried out with the inclusion of the temperature and/or moisture transient profile in the elements equations. Results of such analysis in the study of the effect of temperature on the internal stresses in a corn kernel is shown in Figs. 9 to 11.

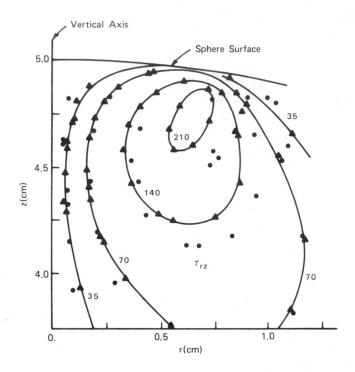

FIGURE 7 Distribution of shear stresses in flat-plate loading: Solid lines, 140-element grid; triangles, 116-element grid; and dots, 92-element grid. (From Rumsey and Fridley, 1977.)

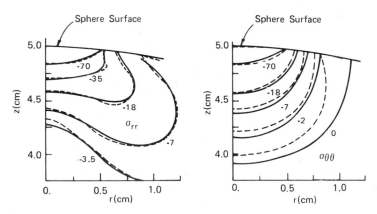

FIGURE 8 Distribution of stresses in a simulated apple in contact with another apple. (σ_{rr} are radial normal stresses, $\sigma_{\theta\theta}$ are tangential normal stresses, σ_{Z7} are vertical normal stresses, and τ_{rZ} are shear stresses). (From Rumsey and Fridley, 1977.)

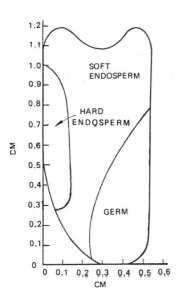

FIGURE 9   Schematic presentation of corn kernel. (From Gustafson, Thompson, and Sokhansanj, 1979.)

FIGURE 8   (Continued)

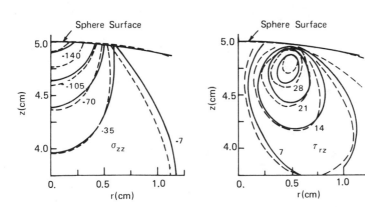

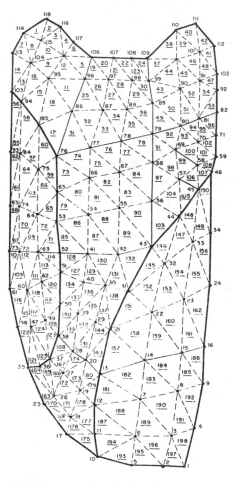

FIGURE 10   The finite element grid for a cross section of a corn kernel.
(From Gustafson, Thompson, and Sokhansanj, 1979.)

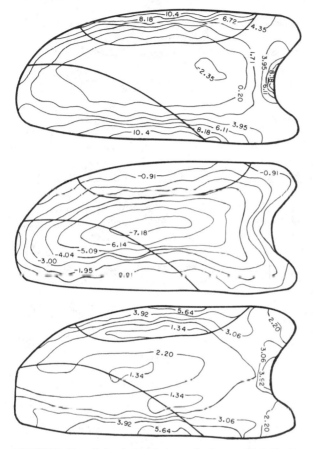

FIGURE 11   Calculated internal stress distribution in corn kernel due to cooling: maximum normal principal stresses (top), minimum principal normal stresses (middle), and maximum shear stresses (bottom). (From Gustafson, Thompson, and Sokhansanj, 1979.)

V. DETERMINATION OF THE FLOW FUNCTION OF FOOD POWDERS

One of the effective means of quantifying a powder's flowability is the flow function (Jenike, 1961). The flow function (FF) is the relationship between the unconfined yield stress -F (a measure of the powder surface capability to resist collapse) and the major consolidation stress -V. The values of the FF are derived from shear analysis in which the yield stress in shear is measured at various consolidation levels and under various normal loads. For a single test (Fig. 12) a typical curve called the yield loci curve is obtained and is used to calculate a single pair of F and V values as shown schematically in Fig. 12. The yield curves of many nonfood powders could be fitted by the Warren spring equation (Ashton, Farley, and Valentine, 1965)

$$\left(\frac{\tau}{C}\right)^n = \frac{\sigma + T}{T} \qquad (17)$$

where τ is the shear stress, σ the normal stress, C the powder cohesion, and T is tensile strength.

When the validity of the equation can be assumed, a curve-fitting method combined with geometrical considerations can serve as the mathematical basis for the flow function calculation by a computer (Stainforth, Ashley, and Monely, 1970/1; Eelkman-Rooda, 1975). In a variety of food powders, it has been shown that the pertinent part of the yield loci curve is well approximated by a straight line (Peleg and Mannheim, 1973; Peleg, Mannheim, and Passy, 1973). In such a case (see Fig. 13) the calculation i is facilitated since the solution of the following equations will yield (the unknowns τ_1, σ_1, and M)

$$(\sigma_1 - M)^2 + \tau_1^2 = (\sigma_0 - M)^2 + \tau_0^2 \qquad (18)$$

$$a\sigma_1 + b = \tau_1 \qquad (19)$$

FIGURE 12   The yield loci curve of compacted cohesive powder.

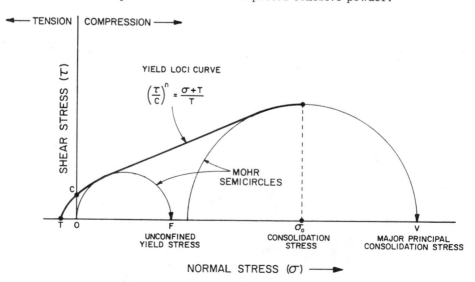

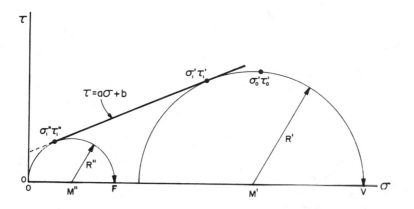

FIGURE 13   The geometry for computerized calculation of the unconfined yield stress (F) and the major consolidation stress (V) in the case of linear yield loci curve.

$$a = -\frac{\sigma_1 - M}{\tau_1} \tag{20}$$

(for F calculation, σ_0 and $\tau_0 = 0$), which in turn (after selection of the proper solutions of the quadratic equation) will yield

$$R^2 = (\sigma_0 - M)^2 + \tau_0^2 \tag{21}$$

$$V = M' + R \tag{22}$$

$$F = 2M'' \tag{23}$$

A computer program for these calculations, including special corrections for abnormal experimental results, has been written by Saguy (1970).

ACKNOWLEDGMENT

The author expresses his thanks to Mr. R. J. Grant for his graphic assistance and to Mrs. Roberta Zidik for typing the manuscript.

REFERENCES

Ashton, M. D., Farley, R., and Valentine, F. H. H. (1965). Some investigations into the strength of and flow of powders, Rheol. Acta 4: 206.

Chen, Y. and Rosenberg, J. (1977). Non-linear viscoelastic model containing a yield element for modelling a food material, J. Texture Studies 8:477.

Clark, R. L., Fox, W. R., and Burns Welch, G. (1971). Representation of mechanical properties of non-linear viscoelastic materials by constitutive equations, Trans. ASAE 14:511.

Cox, D. E. and Peleg, M. (1976). Simulation of discontinuous force deformation curves by a computer. Unpublished.

DeBaerdemaeker, J. G. and Segerlind, L. J. (1976). Determination of the viscoelastic properties of apple flesh, Trans. ASAE 19:346.

deMan, J. M., Voisey, P. W., Rasper, V. F., and Stanley, D. W. (1976). Rheology and Texture in Food Quality. Avi Publ. Co., Westport, Conn.

Drake, B. (1979). A FORTRAN Program FRAMOD for simulation of large fractoviscoelastic models, J. Texture Studies 10:165.

Duirling, A. E. (1969). An Introduction to Electrical Engineering. Macmillan, New York.

Eelkman-Rooda, J. (1975). A numerical method for the calculation of the powder flow properties obtained with the Jenike Flow Factor Tester, Powder Technol. 12:97.

Fernandez Martin, F. (1972). Influence of temperature and composition on some physical properties of milk and milk concentrates. J. Dairy Res. 39:75.

Gustafson, R. J., Thompson, D. R., and Sokhansanj, S. (1979). Temperature and stress analysis of corn kernel—finite element analysis, Trans. ASAE 22:955.

Kennish, W. J. and Henderson, J. M. (1978). Formulation of models for cushion materials for impact applications, Trans. ASAE 21:793.

Mizrahi, S. (1979). A review of the physiochemical approach to the analysis of structural viscosity of fluid fruit products, J. Texture Studies 10:67.

Mohsenin, N. N. (1970). Physical Properties of Plant and Animal Materials, Vol. 1. Gordon and Breach, New York.

Myers, R. H. (1971). Response Surface Methodology. Allyn and Bacon, Boston.

Peleg, M. (1977). Operational conditions and the stress-strain relationship of solid foods—theoretical evaluation, J. Texture Studies 8:283.

Peleg, M. and Mannheim, C. H. (1973). Effect of conditioners on the flow properties of powdered sucrose, Powder Technol. 7:45.

Peleg, M., Mannheim, C. H., and Passy, N. (1973). Flow properties of some food powders, J. Food Sci. 38:959.

Rao, M. A. (1977). Rheology of liquid food: A review, J. Texture Studies 8:135.

Rumsey, T. R. and Fridley, R. B. (1977). A method for determining the shear relaxation function of agricultural materials, Trans. ASAE 20: 386.

Saguy, I. (1970). A program for Jenike's flow function calculation. Unpublished.

Segars, R. A. and Kapsalis, J. G. (1976). Contributions of the U.S. Army Natick Research and Development Center to the objective measurement of the textural quality of meat, J. Texture Studies 7:129.

Sherif, S. M. and Segerlind, L. J. (1976). Contact stresses in a spherical body composed of a nearly incompressible material. ASAE paper 76-3594.

Sherman, P. (1970). Industrial Rheology. Academic, London.

Sherman, P. (ed.) (1979). Food Texture and Rheology. Academic, New York.

Stoner, D. L., Haugh, C. G., Forrest, J. C., and Sweat, V. E. (1974). J. Texture Studies 4:483.

Stainforth, P. T., Ashley, R. C., and Morely, J. N. B. (1970/1). Computer analysis of powder flow analysis, Powder Technol. 4:250.

Zienkiewicz, O. C. (1971). The Finite Element Method in Engineering Science. McGraw-Hill, New York.

10

Optimization Methods and Applications

ISRAEL SAGUY The Volcani Center, Agricultural Research Organization, Bet Dagan, Israel

*Present affiliation:
Massachusetts Institute of Technology, Cambridge, Massachusetts

I. INTRODUCTION

The application and implementation of optimization theories and techniques are essential in any competitive field today. Optimization is used to design or improve more efficient processes. There have been tremendous advances in optimization theory and techniques in the last two decades, along with comparable progress in computers, applied mathematics, numerical analysis, and engineering.

Food engineering has lagged behind some of the other engineering disciplines in implementing optimization techniques in processing and research. The primary reasons for this are twofold (Saguy and Karel, 1982): (1) the physicochemical characteristics of foods are extremely complex, thus causing difficulties in the mathematical modeling and simulation of their behavior, and (2) food engineers and scientists rarely have training in optimization techniques beyond classical calculus and may fear entering the field due to preconceived views of its complexity. Therefore, the main scope of this chapter is to give the food practitioner the necessary mathematical and optimization theory background to choose and implement a specific software capable of solving an optimization problem. Like any tool, the more one understands its structure and capabilities (or limitations), the more valuable it becomes. Thus, as a consumer user of optimization procedures, one must develop the knowledge to characterize the process, construct the optimization problem, and implement techniques which already exist.

In this chapter we could not include all the optimization theories, methods, and software. The emphasis here is to simplify the material for the food engineer or scientist who is merely seeking an efficient, effective software and does not wish to become involved with rigorous theoretical derivations. Consider this chapter on optimization as a user's guide rather than a theoretical treatment.

The main emphasis of the chapter will be (1) to summarize certain areas of activity where optimization is fundamental and assist the reader in getting some current endeavors in optimization, (2) to present a summary and interpretation of some optimization methods, and (3) to highlight the software and computer codes available.

Our approach to optimization was based on two generally occurring problems. The first involving the choice of the values for parameters and/or independent variables to optimize (minimize or maximize) the function defined as the "objective function." This family of problems is known as "static optimization" and is to be discussed herewith. The second type of problem is the optimization of an "objective function," which depends upon time or the path of the independent variables. This "dynamic optimization" is the subject of Chap. 11.

The following chapters are intended to tear down some of the barriers preventing application of optimization in the food industry by presenting several techniques involved with the available software. Furthermore, a complete set of typical food-oriented optimization problems and software will be evolved.

A. Objective Function

The function J to be optimized (maximized or minimized) is most frequently known as the "objective function" or "performance index." This function is also called "merit," "effectiveness," or "criterion." The objective function may be obtained from either a mathematical model or by fitting an equation through experimental points. It may be continuous or in some cases discrete valued.

The objective function for a single-independent-variable case is expressed

$$J = f(x) \tag{1}$$

where x is the independent variable. Using a simplified vector notation, an independent multivariable $\underline{x}$ will be a point or a vector in n-dimensional space. Thus,

$$\underline{x} = (x_1, x_2, \cdots, x_n) \tag{2}$$

and

$$J = f(\underline{x}) = f(x_1, x_2, \cdots, x_n) \tag{3}$$

Cases where the objective function is expressed by a differential equation will be treated in Chap. 11.

One of the main responsibilities of the optimizer is to construct a suitable objective function to be minimized. This function may be quite simple and relatively easy to calculate or quite complicated (e.g., maximizing nutrient retention during sterilization of food in cans). Hence, special attention is required before any optimization attempts are to be made to avoid illusive, conflicting, and misleading objective functions.

It is sufficient to seek the minimum of an objective function, since maximization can always be transformed into a minimization problem by seeking the minimum of $-f(x)$. Thus, our approach throughout this chapter is guided by minimization only.

B. Function Behavior

We start our description by defining terms that are useful in understanding function behavior, and will provide basic knowledge concerning terminology.

1. Monotonic Increasing

A function is monotonic increasing when $f(x_2) > f(x_1)$ for any $x_2 > x_1$.

2. Monotonic Decreasing

A function is monotonic decreasing when $f(x_2) < f(x_1)$ for any $x_2 > x_1$.

In situations in which $f(x_2) \geq f(x_1)$ or $f(x_2) \leq f(x_1)$ for any $x_2 > x_1$, the functions are monotonic nonincreasing and monotonic nondecreasing, respectively. Hence, the function $f(x) = x^2$ is monotonic increasing for $x > 0$, whereas the function is monotonic decreasing for $x < 0$.

3. Unimodal Function

If changing the independent variable (x) in a given direction results in increasing (or decreasing) of the function value up to a maximum (or minimum), followed by a decrease (or increase), the function is termed "unimodal." Unimodal functions may include not only continuous concave or convex behavior (see below) but also discontinuities or discrete-valued functions, as depicted in Fig. 1. It follows from the unimodal definitions for the one-dimensional case that the minimum will be (for $x_1 < x_2 < x^* < x_4$)

$$f(x_1) > f(x_2) > f(x^*) < f(x_4)$$

In an n-dimensional problem (where $n = 2, 3, \cdots$), normally only concave or convex forms of unimodal functions are considered.

4. Concave and Convex Functions

Concavity and convexity are special cases of unimodal functions. A concave one-variable function is defined by

$$f[(1 - \lambda)x_1 + \lambda x_2] \geq (1 - \lambda)f(x_1) + \lambda f(x_2) \tag{4}$$

where $0 \leq \lambda \leq 1$. It is strictly concave if $\geq$ can be replaced by $>$ in Eq. (4).

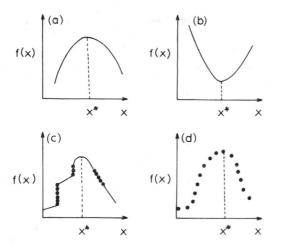

FIGURE 1   Typical unimodal functions: a = concave; b = convex; c = discontinuous; d = discrete valued.

A "convex" function has a similar definition, but with the inequality
sign reversed. The latter definitions may be illustrated in a one-dimensional typical situation by a surface f(x) and a line segment (dashed, Fig. 2)
connecting any two points on this surface. If this segment lies on or below
the surface the function is concave; on or above the surface means convex.

A more convenient test for convexity-concavity involves testing the
second derivative. If a function f(x) is continuous and the second derivative
exists over the tested range, than f(x) is convex over this region if and
only if

$$\frac{d^2 f}{dx^2} \geq 0 \tag{5}$$

FIGURE 2   A typical (a) convex and (b) concave function.

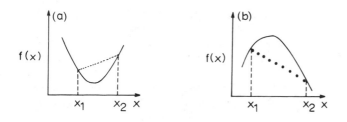

If, however, the above inequality sign is reversed, the function is concave.

The definition of convexity-concavity functions has a vast importance
in any optimization procedure. For instance, any relative minimum of a
convex function is also the global minimum. In a similar fashion, any
local maximum of a concave function is also the global maximum.

C. Contour Representation of Surfaces

To avoid difficulties when dealing with a multidimensional problem, we now
focus on the level contours of a graph. Geometrically, the objective function constitutes an n-dimensional surface in the R^{n+1} space which represents the n independent variables x_1, x_2, $\cdots$, x_n ($\underline{x}$) and the dependent
variable $J = f(\underline{x})$.

Describing such a surface on a two-dimensional page is very difficult.
It is more convenient to resort to a contour plot or map of the objective
function where the values of J (at a given $\underline{x}$) are projected onto the page
plane.

These lines of equal objective function values (known also as level
lines) projected onto the horizontal plane are called contour lines. Examples
of contour usage include topographic weather maps and heat flow diagrams.
For instance, we may observe that the objective function $J = x_1 x_2$ can be
interpreted in a R^3 surface. The contour map is a suitable substitute for
the perspective sketch of this graph. In other words, we say that the contour lines can represent a relation in an R^3 surface by picturing its domain
in R^2. In the derived plot the value of J changes from one contour to another.

Figure 3, an illustration of a typical contour map, indicates several
essential definitions: point A is a local minimum, point B is a saddle point,
point C is the global minimum, and the sharp rise from B to C is called a
"ridge."

II. CLASSICAL OPTIMIZATION

We now explain "classic" optimization by differentiation and Lagrange
multipliers after defining the basic terms and giving several applications of
the calculus needed in basic optimization (min-max) problems.

Classical calculus allows us to locate the extreme of an unconstrained
continuous and differentiable function. However, its applicability in many
engineering situations is severely limited by the complexity, dimensions,
and nonlinear nature of most objective functions. Since the optimum can
lie in the interior of a given region at a stationary point, or at a disconti-
uity either within or on the boundary of the region, only certain cases can
be treated using analytical methods. The use of analytical methods is restricted to interior points for which the objective function and its derivatives
exist and are continuous.

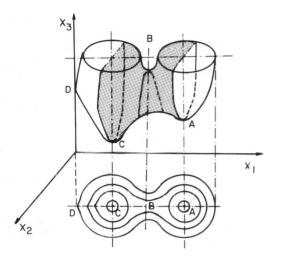

FIGURE 3   A typical contour map (A = local minimum; B – saddle point, C = global minimum; D = sharp ridge).

We shall start our treatment with a single-variable problem, followed by the multivariable case.

A. Single Variable Problem

For cases where the objective function is continuous and depends only on a single independent variable, and certain of its derivatives are also continuous, we are able to optimize analytically. A single-variable optimization problem is the search for the value of x, which on the closed interval [A ≤ x ≤ B] causes f(x) to be minimized. At the extreme x^* the first derivative vanishes. Hence,

$$f'(x^*) = 0 \tag{6}$$

Equation (6) is also known as the "necessary condition for extremum." Observing the second derivative, if it exists, one may conclude that

If $f''(x^*) > 0$	f has a minimum at x^*	(7)
If $f''(x^*) < 0$	f has a maximum at x^*	(8)
If $f''(x^*) = 0$	no conclusion may be drawn	(9)

In general, the point at which $f'(x^*) = 0$ is called a "stationary point." Sufficient conditions for an extremum are given by Eqs. (6) to (8). A typical representation of the above conditions is given in Fig. 4.

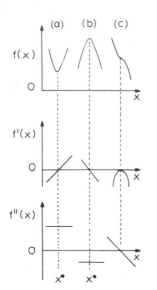

FIGURE 4   Typical functions with a = minimum, b = maximum, c = non-extreme (saddle point), and their first and second derivatives.

In the search for the extremum, we must first determine the location of the stationary point by using $f'(x^*) = 0$. In some cases the determination of this point will probably require numerical root-finding methods for the solution (see Chap. 2). By checking the nonvanishing, even derivative, we can discover whether a maximum, minimum, or an inflection point was derived.

B. Multivariable Problem

Many optimization problems may be reduced by an appropriate series of direct eliminations to a problem which deals only with a single independent variable. However, practical situations and experience in food engineering and other related fields show that maximization or minimization of a multivariable objective function (i.e., the objective function includes more than one independent variable) may be required. Moreover, elimination of variables in most practical situations is either complex or not possible.

Multivariable problems have a structure entirely different from that of a single-variable problem. In most cases, methods derived for single variable are not applicable in solving multivariable cases. This problem of dimensionality makes the unimodal assumption less plausible. An excel-

lent introduction to these various methods, along with their shortcomings
(e.g., ridges) may be found in the book by Wilde and Beightler (1967).
 The multivariable optimization problem involves the determination
of values of n independent variables, x_1, x_2, $\cdots$, x_n (i.e., the vector $\underline{x}$),
which minimizes the objective function $J = f(\underline{x})$. A function $f(\underline{x})$ takes its
global minimum at a point $\underline{x}^*$ if $f(\underline{x}) > f(\underline{x}^*)$ for all values of $\underline{x}$ over which
$f(\underline{x})$ is defined. This is based on the assumption that the value of $\underline{x}^*$ at
which $f(\underline{x}^*)$ reaches its minimum is actually in the region over which $\underline{x}$ is
defined.
 Using the Taylor's series expansion procedure (see Chap. 2) to expand
the objective function about the point $\underline{x}^*$ results in some pertinent conclusions. Note that the necessary condition for optimality is

$$\left(\frac{\partial f}{\partial x_i}\right)_{\underline{x}^*} = 0 \qquad i = 1, 2, \ldots, n \tag{10}$$

or

$$\nabla f(\underline{x}^*) = 0$$

where $\underline{x}^*$ is a stationary point.
 Secondly, for all nontrivial points $\underline{x}^*$, if

$$\nabla^2 f(\underline{x}^*) < 0 \qquad f(\underline{x}^*) \text{ is maximum at } \underline{x}^* \tag{11}$$

$$\nabla^2 f(\underline{x}^*) > 0 \qquad f(\underline{x}^*) \text{ is minimum at } \underline{x}^* \tag{12}$$

where $\nabla^2 f$ is given as

$$\sum_{i=1}^{n} \sum_{k=1}^{n} \frac{\partial^2 f(\underline{x}^*)}{\partial x_i \, \partial x_k} \, \Delta x_i \, \Delta x_k \tag{13}$$

where

$$\Delta x_i = x_i - x_i^*$$
$$\Delta x_k = x_k - x_k^*$$

 The above conditions can be confusing; let us summarize them in a
different way. Let $f = f(x_1, x_2, \ldots, x_n)$, and define the determinant D_i
(also known as the "Hessian," denoted by H):

$$
D_i = \begin{vmatrix} \dfrac{\partial^2 f}{\partial x_1^2} & \cdots & \dfrac{\partial^2 f}{\partial x_1 \, \partial x_i} \\[2ex] \dfrac{\partial^2 f}{\partial x_2 \, x_1} & \cdots & \dfrac{\partial^2 f}{\partial x_2 \, \partial x_i} \\[2ex] \dfrac{\partial^2 f}{\partial x_i \, \partial x_1} & \cdots & \dfrac{\partial^2 f}{\partial x_i^2} \end{vmatrix} \tag{14}
$$

Considering the above equation, the following two statements apply (Edel-baum, 1962):

1. Sufficient conditions for a minimum to exist at a stationary point $\underline{x}^*$ are that all the determinants must be positive definite at $\underline{x}^*$:

$$
D_i > 0 \qquad \text{for all } i = 1, 2, \ldots, n \tag{15}
$$

2. Sufficient conditions for a maximum to exist at a stationary point $\underline{x}^*$ are that

$$
D_i < 0 \qquad \text{for all odd } i \ (\text{i.e., } i = 1, 3, 5, \ldots) \tag{16a}
$$
$$
D_i > 0 \qquad \text{for all even } i \ (\text{i.e., } i = 2, 4, 6, \ldots) \tag{16b}
$$

Example 1   Maxima and minima (Korn and Korn, 1968):   Find the maxima and minima of the function

$$
f(\underline{x}) = 3x_1^3 - x_1 + x_2^3 + 3x_2^2 - 1
$$

The necessary conditions are [following Eq. (10)]

$$
\frac{\partial f}{\partial x_1^*} = 9x_1^2 - 1 = 0 \quad (\text{e.g., } x_1^* = \pm \frac{1}{3})
$$

$$
\frac{\partial f}{\partial x_2^*} = 3x_2^2 - 6x_2 = 0 \quad (\text{e.g., } x_2^* = 0, \, 2)
$$

Sufficient conditions are derived from the sign of the determinant D_i [Eqs. (15) and (16)]:

$$
D_1 = \frac{\partial^2 f}{\partial x_i^2}
$$

$$D_2 = \begin{vmatrix} \dfrac{\partial^2 f}{\partial x_1^2} & \dfrac{\partial^2 f}{\partial x_1 \, \partial x_2} \\[20pt] \dfrac{\partial^2 f}{\partial x_2 \, \partial x_1} & \dfrac{\partial^2 f}{\partial x_2^2} \end{vmatrix}$$

As

$$\frac{\partial^2 f}{\partial x_1^2} = 18x_1 \qquad \frac{\partial^2 f}{\partial x_2^2} = 6x_2 - 6 \qquad \frac{\partial^2 f}{\partial x_1 \, \partial x_2} = 0$$

it follows

$$D_1 = 18x_1$$

$$D_2 = \begin{vmatrix} 18x_1 & 0 \\ 0 & 6x_2 - 6 \end{vmatrix} = (18x_1)(6x_2 - 6)$$

When testing for minimum according to Eq. (15), all determinants should be positive definite; thus,

$$18x_1 > 0 \qquad (18x_1)(6x_2 - 6) > 0$$

yielding

$$x_1 > 0 \qquad x_2 > 1$$

At the stationary points we have

$$x_1^* = \frac{1}{3} \qquad x_2^* = 2 \qquad f(\underline{x}^*) = -\frac{47}{9} \quad \text{minimum}$$

Testing for a maximum according to Eq. (16), we find

$$D_1 < 0 \qquad D_2 > 0$$

Hence,

$$18x_1 < 0 \qquad (18x_1)(6x_2 - 6) > 0$$

yielding

$$x_1 < 0 \qquad x_2 < 1$$

At the stationary point we have

$$x_1^* = -\frac{1}{3} \qquad x_2^* = 0 \qquad f(\underline{x}^*) = -\frac{7}{9} \text{ maximum}$$

Unfortunately, real cases are more complicated than Example 1 implies. First, the foregoing sufficient conditions are of little practical importance; most problems in the sciences, engineering, or economics require considerable numerical computation to determine whether the Hessian matrix is positive definitive. More information about the character of a stationary point can usually be obtained from physical consideration (Cooper and Steinberg, 1970).

C. Lagrange Multipliers

Frequently, the problem of optimization has to be performed in accordance with constraints imposed on the system. These constraints are equality and inequality. In most practical systems, engineering and economic constraints are natural and occur repeatedly.

The optimization problem for which "Lagrange multipliers" are applicable could be summarized as follows: minimize the objective function $J = f(\underline{x})$ subject m **equality** constraints (m < n) expressed by

$$e_k(x_1, x_2, \cdots, x_n) = 0 \qquad \text{for } k = 1, \cdots, m \tag{17}$$

In principle, the m constraint equations can be solved for m of the variables x_i. These variables are then eliminated from the original objective function J. However, in practice this can be a tedious and time-consuming process. To overcome these difficulties, we can apply the "Lagrange multipliers" (named for Joseph Louis Lagrange).

The method consists of multiplying the equality constraint equations by unknown constants, λ_k (k = 1, $\cdots$, m; where m = number of equality constraints), and adding this derived set of equations to the original objective function, forming a new objective function, L:

$$L = f(\underline{x}) + \sum_{k=1}^{m} \lambda_k e_k \tag{18}$$

where λ_k is the Lagrange multiplier.

The optimum is determined by solving the set of n + m equations derived from the necessary conditions for extremum:

$$\frac{\partial L}{\partial x_i} = 0 \qquad \frac{\partial L}{\partial \lambda_k} = 0 \tag{19}$$

To elucidate the above procedure, we will illustrate an application of Lagrange multipliers in a minimization problem where only one equality constraint exists (m = 1). According to Eq. (19), we have

$$dL = \frac{\partial f}{\partial x_1} \, dx_1 + \frac{\partial f}{\partial x_2} \, dx_2 + \cdots + \frac{\partial f}{\partial x_n} \, dx_n = 0 \tag{20}$$

and from Eq. (17), it follows that

$$de = \frac{\partial e}{\partial x_1} \, dx_1 + \frac{\partial e}{\partial x_2} \, dx_2 + \cdots + \frac{\partial e}{\partial x_n} \, dx_n = 0 \tag{21}$$

Multiplying Eq. (21) by λ and adding the resultant to Eq. (20) yields

$$\left(\frac{\partial f}{\partial x_1} + \lambda \frac{\partial e}{\partial x_1} \right) dx_1 + \cdots + \left(\frac{\partial f}{\partial x_n} + \lambda \frac{\partial e}{\partial x_n} \right) dx_n = 0 \tag{21a}$$

If each of the above parentheses vanishes, it results in the n equations

$$\frac{\partial f}{\partial x_i} + \lambda \frac{\partial e}{\partial x_i} = 0 \qquad \text{for } i = 1, 2, \ldots, n \tag{22}$$

Including the original equality constraint Eq. (17) yields a set of n + 1 equations of n + 1 variables (i.e., $x_1, x_2, \ldots, x_n$, and λ). Once solved, these equations define the optimal solution. Thus, instead of the original objective function J, we have minimized the new devised function L, which satisfies the constraint.

Based on the above example, we may conclude that the method of Lagrange multipliers is carried out by solving the system of equations derived from the sufficient extremum conditions for the unknown values of x_i^* and λ as an additional variable.

Example 2 Lagrange multiplier: Let us consider finding the maximum of a rectangular volume that fits inside a sphere of unit radius. The volume (objective function) to be maximized is $J = x_1 x_2 x_3$, where J is one-eighth of the total volume (i.e., $V = 8x_1 x_2 x_3$). The equality constraint imposed on the variables is given as

$$e(x_1, x_2, x_3) = x_1^2 + x_2^2 + x_3^2 - 1 = 0$$

Therefore, we consider the maximization of the new devised function

$$L \equiv J + \lambda e = x_1 x_2 x_3 + \lambda (x_1^2 + x_2^2 + x_3^2 - 1)$$

The sufficient conditions for the extreme will yield the following set:

$$\frac{\partial L}{\partial x_1} = x_2 x_3 + \lambda 2 x_1 = 0$$

$$\frac{\partial L}{\partial x_2} = x_1 x_3 + \lambda 2 x_2 = 0$$

$$\frac{\partial L}{\partial x_3} = x_1 x_2 + \lambda 2 x_3 = 0$$

and the original equality constraint

$$x_1^2 + x_2^2 + x_3^2 - 1 = 0$$

This set consists of four equations with four unknowns: x_1, x_2, x_3, and λ. Once determined, the optimal solution may be established, and solving the set yields

$$x_1^* = x_2^* = x_3^* = \frac{1}{3} \qquad J^* = \frac{1}{3} \qquad V^* = \frac{8}{3}$$

To generalize the Lagrange multiplier method for n variables and m equality constraints, we merely use m Lagrange multipliers λ_k (k = 1, 2, ..., m)

Beveridge and Schechter (1970) illustrate the application of this method in an optimal design of a heat exchanger, and Ray and Szekeley (1973) have used it to optimize an extraction process. Numerous other examples may be found in the literature.

We would like to point out that Lagrange multipliers transform a constrained optimization problem into an unconstrained problem. However, this could be difficult to solve due to the complex nature and nonlinearity of the derived system. This inconvenience illustrates the need for a numerical scheme (such as Newton–Raphson under certain circumstances) for solving the set.

A system subjected to inequality constraints may be converted to an equality-constraint system by adding one or more slack variables (see Chap. 7) and again applying the Lagrange multipliers. The essential conditions for the extremum in a system subjected to inequality constraints were established by Kuhn and Tucker (1951). The conditions, derived by the Kuhn and Tucker method, exploit the Lagrange multipliers and may be depended upon to yield the global minimum if both the objective function and constraints are convex. For further details, see Beveridge and Schechter (1970).

III. SEARCH TECHNIQUES

There are a number of drawbacks in the implementation of classical optimization techniques; thus their use is restricted in practical engineering systems. Generally, the classical approach requires either continuity or differentiability of $f(\underline{x})$, or both. These conditions are not always met in practical problems. In some cases even if these conditions are satisfied, the solution of the derived set of nonlinear equations presents a considerable task.

For these complicated cases, search methods known as "direct search" methods may be most practical.

A. Univariate Searches

1. Direct Search

Direct search methods evaluate and compare the objective function at a sequence of trial points, $\underline{x}^{(1)}$, $\underline{x}^{(2)}$, . . ., $\underline{x}^{(k)}$ (where 1, 2, . . ., k denote the trial number, and $\underline{x}$ is the independent variable vector), in order to reach the optimal point $\underline{x}^*$. The only requirement that the objective function must satisfy is that it be computable. The various search methods differ in their approach to selecting the trial points.

Direct methods are generally applied in the following circumstances (Walsh, 1976):

1. The objective function $f(\underline{x})$ is not differentiable or is subject to random error.

2. The derivatives $\partial f / \partial x_i$ are discontinuous, or their evaluations require much more effort than obtaining $f(\underline{x})$ itself.

3. Computer facilities are inadequate to sustain more sophisticated algorithms.

4. An approximate solution may be required at any time during the course of the calculations.

For these reasons, direct search methods may not be very efficient when applied alone. Hence, many direct search methods are used in conjunction with gradient methods (see Sec. IV).

We classify the search methods to be discussed in the following sections as follows:

Univariate search: the simplest class of minimization with just one variable

Multivariable search: the successor of univariate methods capable of handling more than one independent variable

We describe only a few existing methods selected for their practicability and simplicity. For a complete discussion, consult more elaborate texts, e.g., Avriel (1976); Beveridge and Schechter (1970); Box, Davies, and Swann (1969); Gill, Murry, and Wright (1981); Flecther (1981); Walsh (1976).

2. Univariate Search

"Univariate search" methods may be defined as methods of approaching the value of x^* in a given interval ($a < x < b$) by an iterative procedure which successively decreases this interval. The improved estimate of x^* for each iteration will lie within the region $x_\ell < x^* < x_u$, where x_ℓ and x_u denote the lower and upper values of the uncertainty interval (the latter region is also defined as the resolution expressed by α).

Another way of defining univariate search methods is to find the position of the extremum by an approximating point, rather than defining the interval of the extremum as was previously stated.

3. Grid Search

The grid search method is the simplest in concept, but the most inefficient; therefore, it is known as the "exhaustive" method. This search is presented here mainly for illustrative purposes. When using the algorithm, the interval to be searched is subdivided into $\Delta x/2$ equally spaced sections (Fig. 5), and the objective function is evaluated at each of the points $k\,\Delta x/2$ [k = 1, 2, ..., $2/(\Delta x) - 1$]. The least value of $f(x)$ is then selected and the minimum is derived.

The number of function evaluations is specified by $2/\Delta x - 1$. After the search has been completed, the interval of uncertainty, α, is reduced to Δx (Fig. 5 indicates graphically the features of this method; in this case, $x^* = 0.3$ and $\alpha = 0.2$).

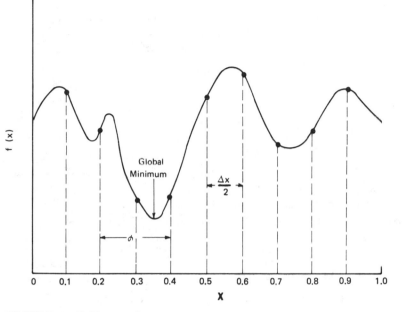

FIGURE 5   Grid search.

The major disadvantages of this scheme are

1. It can be used only when the evaluation of the objective function is very simple and the optimal point is within a small region; otherwise, a large number of evaluations are required.

2. The choice of Δx may inadvertently miss the global minimum.

3. Information acquired during the evaluations of f(x) is not used to speed up the search.

Random search is a slightly different method in that is evaluates the objective function at random points within the interval. In this scheme, a random number R in the unit interval is generated, and the objective function is evaluated at this point.   This procedure is repeated n times to derive f(x*).

4. Fibonacci Search

This method uses the sequence of positive integers known as Fibonacci numbers, defined by

TABLE 1   Fibonacci Numbers

n	0	1	2	3	4	5	6	7	8	9	10	11	12	13	14	...
F_n	1	1	2	3	5	8	13	21	34	55	89	144	233	377	610	...

$$F_0 = F_1 = 1 \tag{23}$$

$$F_n = F_{n-1} + F_{n-2} \qquad n > 2 \tag{24}$$

Kiefer (1953), the first to use these series, applied them to an optimum sequential elimination. This method is known as "Fibonacci search." The sequence of the first Fibonacci numbers are given in Table 1. The role of Fibonacci numbers in the sequential search may be demonstrated by the following:

Consider the initial interval $[a^{(1)}, b^{(1)}]$, designated as $L^{(1)}$, where the objective function $f(x)$ is supposed to be unimodal over this range. The first two trial points $x_3^{(1)}$ and $x_4^{(1)}$ (where the superscript denotes the iteration number) are placed at a distance $\ell^{(1)}$ from each boundary.

Let $\ell^{(1)}$ be calculated as

$$\ell^{(1)} = L^{(1)} \frac{F_{n-2}}{F_n} \tag{25}$$

where F_n is the appropriate Fibonacci number and n is the total number of experiments finally required. This must be specified in advance. Using the above relation, we have

$$x_3^{(1)} = a^{(1)} + \ell^{(1)} \tag{26a}$$

$$x_4^{(1)} = b^{(1)} - \ell^{(1)} \tag{26b}$$

The objective function is evaluated at these trial points, and the interval is reduced by discarding the region not containing the minimum according to the following test (note that for maxima the opposite criteria should be applied):

$$\text{If } f[x_3^{(1)}] < f[x_4^{(1)}] \qquad \text{then } a^{(1)} < x^* < x_4^{(1)} \tag{27a}$$

discard interval $[x_4^{(1)}, b^{(1)}]$, and

If $f[x_3^{(1)}] > f[x_4^{(1)}]$ then $x_3^{(1)} < x^* < b^{(1)}$ (27b)

discard interval $[a^{(1)}, x_3^{(1)}]$, where x^* is the location of the minimum. The first iteration is completed with these two function evaluations (calculations of $f[a^{(1)}]$ and $f[b^{(1)}]$ are not included).

The new interval designated $L^{(2)}$ will be given by

$$L^{(2)} = L^{(1)} - \ell^{(1)} = L^{(1)} - L^{(1)} \frac{F_{n-2}}{F_n} = L^{(1)} \frac{F_{n-1}}{F_n} \qquad (28)$$

The boundaries of this interval are determined by the decision taken following the test given in Eq. (27). The new derived boundaries are designated $a^{(2)}$ and $b^{(2)}$

The new trial points will be placed at $x_3^{(2)}$ and $x_4^{(2)}$ so that they are at distance $\ell^{(2)}$ from each end of the interval $L^{(2)}$. Using the relation cited previously, Eq. (25), we have

$$\ell^{(2)} = L^{(2)} \frac{F_{n-3}}{F_{n-1}} \qquad (29)$$

and following the same symmetric rule, we get

$$x_3^{(2)} = a^{(2)} + \ell^{(2)}$$

$$x_4^{(2)} = b^{(2)} - \ell^{(2)}$$

In this case, only one functional evaluation is required since the others correspond to values retained from the previous cycle. Again, these points are tested, and the subinterval not containing the minimum is discarded.

Continuing the process according to the above scheme, we may conclude that the general sequence $x_3^{(i)}$ and $x_4^{(i)}$ is

$$x_3^{(i)} = \frac{F_{n-1-i}}{F_{n+1-i}} [b^{(i)} - a^{(i)}] + a^{(i)} \qquad (30a)$$

$$x_4^{(i)} = \frac{F_{n-i}}{F_{n+1-i}} [b^{(i)} - a^{(i)}] + a^{(i)} \qquad (30b)$$

$$\ell^{(i)} = L^{(i)} \frac{F_{n-i-1}}{F_{n-i+1}} \quad \text{and} \quad L^{(i)} = L^{(1)} \frac{F_{n-i+1}}{F_n} \tag{30c}$$

where i is the iteration number and n is the total number of iterations to be performed.

We can show that the ith iteration reduces the interval containing the minimum by a factor of F_{n-i}/F_{n+1-i}; after n iterations (and n function evaluations) the uncertainty region α is

$$\alpha = \frac{1}{F_n} \tag{31}$$

This relation allows us to determine the number of experiments required to obtain the desired uncertainty region for locating the minimum. Remember that the number n of iterations must be known before the search is initiated.

After n - 1 function evaluations and subinterval discarding at each step, the remaining point is in the exact center of the remaining interval (since $F_0 = F_1 = 1$). Thus, the next evaluation should also have been at this point. To determine in which half of the range the minimum actually lies, one of the final test points is displaced arbitrarily by a small value ϵ. However, if we set $\epsilon = 0$, then only n - 1 evaluations are required. The following example elucidates this case.

Example 3 Fibonacci search: Minimize the function, $f(x) = x^2 - 6x$ in the range $0 < x < 10$ to 3% of x.

In this case, we have $\alpha \leq 0.03$, and from Eq. (31), $1/F_n < 0.03$ or $F_n > 33.3$. Looking at Table 1 of Fibonacci numbers we see that this is satisfied for n = 8.

Search results [using subroutine ZXFIB (IMSL, 1977)] are given in Table 2. The region determined by the search indicates that x^* lies within the interval (2.94, 3.23) and $f(x^*)$ is approximately -8.996. The exact solution is $x^* = 3.0$, $f(x^*) = -9.0$. Thus, we have indeed determined the location of x to less than 3% of the initial region.

It may be concluded that Fibonacci search is the most efficient univariate procedure since no other method can guarantee an interval reduction as large as F_n in only n function evaluations (Box, Davies, and Swann, 1969). However, it does require initially the number of interations to be performed.

In the food industry practical experience shows that some variables are discrete valued and may not be treated in the same manner (e.g., the choice of a pump size or a pipe diameter where only few values are available). In this situation, the discrete-valued Fibonacci search method should be applied (Beveridge and Schechter, 1970).

5. Golden Section Method

The major disadvantage of Fibonacci search is that in many practical situations, specifying the exact number of experiments to be performed to meet the desired accuracy is not possible. Thus, it is more desirable to have a method that allows the alteration of the accuracy as the search progresses.

It can be shown that the Fibonacci series can be approximated by (Kiefer, 1953)

$$F_n = \frac{1}{\sqrt{5}} \left[\left(\frac{1 + \sqrt{5}}{2} \right)^{n+1} - \left(\frac{1 - \sqrt{5}}{2} \right)^{n+1} \right] \qquad n = 0, 1, 2, \ldots \tag{32}$$

From this formula and for large values of n, we find

$$\tau \equiv \frac{F_n}{F_{n-1}} \sim \frac{1 + \sqrt{5}}{2} = 1.6180 \qquad n > 4 \tag{33}$$

This relation suggests that the successive interval reduction factors achieved by the Fibonacci search approaches r (= 1.6180). Historically, it was known that the golden mean (section) divides a line segment into two unequal parts so that the ratio of the larger to the original segment equals the ratio of the smaller to the larger. The present elimination technique was thus called "search by golden section" (Wilde, 1964). In this search scheme, we note that

$$\frac{F_{n-2}}{F_n} = \frac{1}{\tau^2} = 0.3820 \qquad n > 4 \tag{34}$$

and thus using the relation cited previously, Eq. (30c),

$$\ell^{(i)} = 0.3820 L^{(i)} \tag{35}$$

where ℓ is the distance from each boundary and L is the search interval. For large values of i (i.e., i > 4), the above relation is a reasonable representation of the Fibonacci search.

The accuracy obtained by k experiments will be

$$\alpha = \frac{L^{(k)}}{L^{(1)}} = \frac{L^{(k)}}{L^{(k-1)}} \frac{L^{(k-1)}}{L^{(k-2)}} \cdots \frac{L^{(2)}}{L^{(1)}} = (0.6180)^{k-1} \tag{36}$$

TABLE 2 Fibonacci Search $f(x) = x^2 - 6x$ ($\alpha = 0.03$; $n = 8$)

Iteration (i)	$a^{(i)}$	$b^{(i)}$	$L^{(i)}$	$\ell^{(i)}$	$x_3^{(i)}$	$x_4^{(i)}$	$f[x_3^{(i)}]$	$f[x_4^{(i)}]$	Interval discarded
1	0	10	10	3.82	3.82	6.18	-8.33	+1.12	[6.18, 10.0]
2	0	6.18	6.18	2.35	2.35	3.82	-8.58	-8.33	[3.82, 6.18]
3	0	3.82	3.82	1.47	1.47	2.35	-6.66	-8.58	[0, 1.47]
4	1.47	3.82	2.35	0.881	2.35	2.94	-6.66	-8.99	[1.47, 2.35]
5	2.35	3.82	1.47	0.588	2.94	3.23	-8.99	-8.95	[3.23, 3.82]
6	2.35	3.23	0.88	0.294	2.64	2.94	-8.87	-8.99	[2.35, 2.64]
7	2.64	3.23	0.59	0.294	2.94	2.94	-8.99	-8.99	[2.64, 2.94]

since

$$L^{(k)} = L^{(k-1)} - \ell^{(k-1)} = (1 - 0.3820)L^{(k-1)} \qquad \frac{L^{(k)}}{L^{(k-1)}} = 0.6180$$

Figure 6 illustrates a flow diagram of the logical decisions needed during computation. Table 3 presents the results obtained from the minimization of $f(x) = x^2 - 6x$ (Example 3) by subroutine ZXGSP (IMSL, 1977) using the golden-section search method.

The Fibonacci search reduces the interval of uncertainty by a factor of $1/F_n$, while the golden-section procedure reduces the same interval by a factor of $1/\tau^{n-1}$. Hence, Fibonacci search achieves a 17% greater interval reduction (Box, Davies, and Swann, 1969).

The efficiency of Fibonacci and golden-section searches are compared in Table 4 for various values of n iterations. Each entry represents the reduction ratio $L^{(n)}/L^{(1)}$ (i.e., the ratio of the length of the final interval of uncertainty to that of the starting interval). For high values of n, the golden-section interval exceeds the Fibonacci interval by about 17% as anticipated.

6. Quadratic Interpolation Search Method

While Fibonacci and golden-section techniques are very efficient in guaranteeing a desired accuracy, other methods have been found in practice to be more efficient. This efficiency may be expressed by substantially greater accuracy in the same number of function evaluations (Coggins, 1964).

These methods are based on fitting a low-order polynomial (quadratic; for cubic, see Davidon, 1959) through a number of points in the neighborhood of the minimum. The derived polynomial is differentiated and equated to zero. Solving this equation provides an estimate of the extremum. This procedure is the basis of an algorithm developed in the multivariate search methods of Davis, Swann, and Campey (Swann, 1964) and Powell (1964).

The Davies-Swann-Campey univariate search starts with an initial point (e.g., initial guess), then a known step size is taken along the search direction. At this point the objective function is evaluated. The step size is doubled for the next function evaluation until a function value exceeds the preceding value. A typical situation is illustrated in Fig. 7; at point 6 the minimum has been overshot. In this case, the step length is halved, and another step is taken from the last successful point.

The procedure yields four points equally spaced along the search axis for which the function values have already been calculated. The point furthest from the one corresponding to the smallest function value is discarded (point 6). A quadratic equation is fitted through the three retained "best" points (4, 5, and 7; denoted for simplicity x_1, x_2, and x_3).

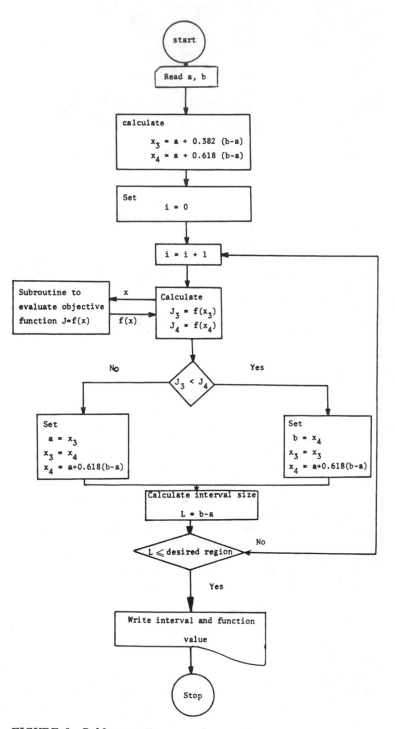

FIGURE 6   Golden-section search procedure.

286

TABLE 3 Search by Golden-Section Technique $[f(x) = x^2 - 6x]$

Iteration (i)	$a^{(i)}$	$b^{(i)}$	$L^{(i)}$	$\ell^{(i)}$	$x_3^{(i)}$	$x_4^{(i)}$	$f[x_3^{(i)}]$	$f[x_4^{(i)}]$	Interval discarded
1	0	10	10	3.82	3.82	6.18	-8.33	+1.12	[6.19, 10.0]
2	0	6.18	6.18	2.36	2.36	3.82	-8.59	-8.33	[3.82, 6.18]
3	0	3.82	3.82	1.46	1.46	2.36	-6.63	-8.59	[0, 1.46]
4	1.46	3.82	2.36	0.902	2.36	2.92	-8.59	-8.99	[1.46, 2.36]
5	2.36	3.82	1.46	0.558	2.92	3.26	-8.99	-8.93	[3.26, 3.82]
6	2.36	3.26	0.902	0.344	2.71	2.92	-8.91	-8.99	[2.36, 2.71]
7	2.71	3.26	0.557	0.212	2.92	3.05	-8.99	-8.99	[2.71, 2.92]
8	2.92	3.26	0.344	0.131	3.05	3.13	-8.99	-8.99	[3.13, 3.26]

TABLE 4   Reduction Ratios $[L^{(n)}/L^{(1)}]$ for
One-Variable Search Strategies

n	Fibonacci	Golden section
2	0.5	0.618
3	0.333	0.382
4	0.200	0.236
5	0.125	0.146
6	0.077	0.090
8	0.029	0.034
10	0.011	0.013
14	1.6E-3	1.9E-3
20	9.1E-5	1.1E-4

It can be shown that the minimum of the fitted quadratic curve is at

$$x^* = x_2 + S_m \tag{37a}$$

and

$$S_m = \frac{f(x_1) - f(x_3)}{2[f(x_1) - 2f(x_2) + f(x_3)]} \tag{37b}$$

We use the derived value of S_m, Eq. (37b), to obtain an estimate of the minimum location x^*. A new search stage with a reduced step length may then be initiated, using as the initial base point either x_2 or $x_2 + S_m$, whichever corresponds to a smaller function value.

Powell's algorithm (Powell, 1964), based on the above guidelines, locates the position of x^* from the values of $f(x_1)$, $f(x_2)$, and $f(x_3)$ calculated at the best three points retained. The minimum x^* is then estimated:

$$x^* = \frac{1(x_2^2 - x_3^2)f(x_1) + (x_3^2 - x_1^2)f(x_2) + (x_1^2 - x_2^2)f(x_3)}{2(x_2 - x_3)f(x_1) + (x_3 - x_1)f(x_2) + (x_1 - x_2)f(x_3)} \tag{38}$$

It can be shown that this formula diminishes to that used previously, according to the Davies-Swann-Campey algorithm, Eq. (37), when the minimum is located between x_1 and x_3 and all three points x_1, x_2, and x_3 are equidistant.

The objective function at x^* is then compared with the best previous point, subject to a convergence limit:

$$f(x^*) - f[x^* = best)] < \epsilon \qquad\qquad (39)$$

If the above criterion is satisfied, the procedure ends. Otherwise, the worst point (the greatest function value) is replaced by x^*, a new quadratic surface is fitted, and a better x^* is derived. This procedure is repeated until the convergence criterion is satisfied (see also Sec. III.C).

B. Multivariable Search Methods

Previously, we outlined some methods often used in single-variable optimization problems. As the number of independent variables increases, the search techniques tend to be more complicated and have to be capable of

FIGURE 7 A typical scheme of Davies-Swann-Campey algorithm (numbers refer to sequence in which the points were introduced).

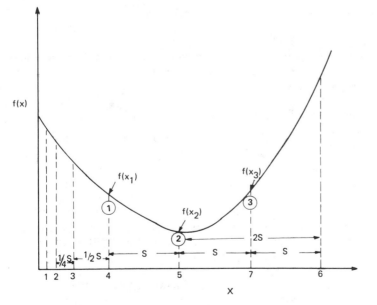

handling multidimensional searches. Nevertheless, many methods are based on one-dimensional search techniques applied at different levels of sophistication.

We discuss some of the most well known methods, but we would like to emphasize that there is no general technique available that can be applied in all cases. Although particular methods are favored in specific applications, most practical problems can be solved in several ways.

1. Alternating Variable Search

The simplest progression of test points toward the extremum is one in which only a single independent variable x_i is searched at a time while the other $n - 1$ variables (i.e., $x_1, x_2, \ldots, x_{i-1}, x_{i+1}, \ldots, x_n$) are held constant. The technique is based on moving parallel to the x_i axis. This means that the optimization problem is reduced to a one-dimensional search, where any univariate search technique or analytical method can be applied.

The univariate search procedure is carried out until a minimum is located in one direction. This value is retained as the new base point, and the search is continued in a new direction parallel to a different variable (axis) until all of the n directions have been searched sequentially. Any systematic order may be used to choose the variables.

Searching all the n variables completes the first stage; the entire procedure is then repeated until no further function improvement is observed (see Sec. III.C for details). Figure 8 illustrates this technique in a two-dimensional case.

This "one-at-a-time" procedure is also known as "sectioning." Its effectiveness in solving optimization problems is restricted to cases where no interactions of the independent variables exist. Otherwise, the procedure tends to be inefficient in locating the extremum, especially when dealing with sharp ridges that cause severe decreases in progression toward the extremum and oscillation.

2. Pattern Search

To overcome some drawbacks of the alternating variable search methods, an improved procedure was suggested by Hooke and Jeeves (1961). The technique is known as pattern search, or the Hooke and Jeeves' method.

The method is based on local exploration about a selected (arbitrary) original base point, B_1. To visualize the procedure, follow each step in Fig. 9, a two-dimensional application of the algorithm. Local exploration consists of changing a single independent variable x_i by a given step size d_i, which can be individually chosen to match the magnitude of different variables (Fig. 9 shows an initial value of $d_1 = 3/2d_2$).

Exploration is based on evaluating the objective function at the base point and at $\pm d_i$ perturbations until the first "success," i.e., the point where a reduced function value is observed. The point is denoted the new

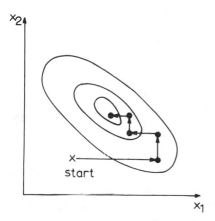

FIGURE 8   A typical "alternating variable search."

"temporary base point" (or "temporary head"), and x_i takes a new value (on Fig. 9 this is expressed by an arrow). The search is repeated for each variable until all variables have been explored. If a greater function value is obtained, the point is rejected and called "failure." At most, $2n + 1$ function evaluations are required. The final temporary head is then designated as the second base point B_2.

After establishing the second base point B_2, we can create the first pattern ("leap frog") move to locate the first pattern point P_1. The latter

FIGURE 9   Hooke and Jeeves' method for a typical two-dimensional case.

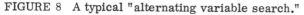

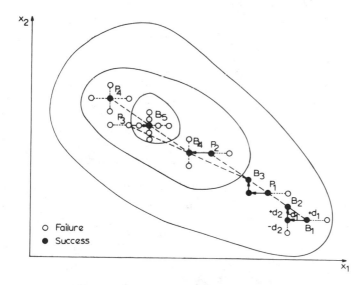

is defined as the terminal point of a directed line passing through B_1 and B_2 and located twice the distance between B_1 and B_2. In other words, P_1 is the reflection of B_1 through B_2. Hence,

$$P_1 = 2B_2 - B_1 \qquad\qquad\qquad (40)$$

The reason for doubling the distance is based on the assumption that similar results would be expected if a search was carried out about B_2. Moreover, it provides an acceleration associated with increasing the choice of the distance between the exploratory base points.

At this derived pattern point P_1, a new exploratory search starts. The Hooke and Jeeves' method avoids testing the pattern points for success or failure. The second stage is completed by establishing a third base point, B_3 (and so on points P_2, B_4, P_3, B_5, and P_4).

If the exploratory moves yield higher function values (failures), the pattern is abandoned (as at point P_4), and a new sequence of exploratory moves starts about the last base point (point B_5).

The procedure outlined above is repeated until no further improvement is observed. This could mean that the minimum has been located to the accuracy limit or the search has descended into a steep skew. In both cases, a new procedure is required. This may lead to reducing the step size by half (i.e., $d_1^{(new)} = d_1/2$ and $d_2^{(new)} = d_2/2$) and repeating the whole procedure. If the step size falls below a preassigned value (or any other criterion), the search is terminated (point B_5).

The acceleration obtained in this method ensures a rapid convergence even if the initial step size is small. Moreover, Hooke and Jeeves found empirically that computation time of their method increased only linearly with the number of variables, while with classical optimization methods, computation time grows with the cube of dimensionality. The latter is very important when dealing with a high number of independent variables.

3. Powell's Method

A review of univariate search methods revealed that quadratic curve-fitting procedures were very valuable. Similarly, for multivariable systems Powell's (1964) algorithm is extremely efficient and probably the most effective direct search method, with an impressive convergence in the region of the extremum (Box, Davies, and Swann, 1969). For simplicity, we limit our explanation to a two-dimensional case; however, the same rules apply to a more general multivariable situation.

The simplified Powell's algorithm is

1. A starting base point B_1 and initial search directions (M_1 and M_2) parallel to the original axes are selected.

2. From the original base point B_1, search along M_1 and then along M_2 using Davies, Swann, and Campey's quadratic algorithm to locate the improved base point B_2. Retain $f(B_1)$ and $f(B_2)$.

3. Determine the direction M_3 as

$$M_3 = B_2 - B_1 \tag{41}$$

Along the M_3 direction, calculate and retain the objective function at the expanded point, defined as

$$X_t = 2B_2 - B_1 \tag{42}$$

The expanded point is located at the same distance from B_2 as the original base point B_1.

4. Check if the following inequalities hold at the expanded point

$$f(X_t) > f(B_1) \tag{43}$$

and/or

$$[f(B_1) - 2f(B_2) + f(X_t)][f(B_1) - f(B_2) - \Delta]^2 > \frac{\Delta[f(B_1) - f(X_t)]^2}{2} \tag{44}$$

where Δ is the greatest change in the objective function achieved during either of the two linear searches along M_1 and M_2.

The first inequality Eq. (43), once satisfied, means that the M_3 direction is probably not worth searching since the objective function starts to rise soon after B_2. The second inequality Eq. (44) checks whether the M_3 direction points across a deep valley.

If either of the above inequalities are satisfied, the old search directions are retained, and a new sequence of single-variable searches is started from the last derived base point B_2, according to the guidelines given above.

However, if neither inequality is satisfied, a single search is performed along the improved direction M_3 until the minimum is found. This point is now designated B_1, and new search directions are chosen $[M_1^{(new)} = M_2, M_2^{(new)} = M_3]$, allowing the initiation of the new searches.

5. Convergence is assumed when the values of the base points between successive iterations are less than a preassigned limit.

The method outlined above forms part of Powell's more general method for finding the minimum value of an objective function $f(\underline{x})$ without calculating derivatives. Nevertheless, it may be used together with any optimization technique that requires a one-dimensional search.

4. Evolutionary Operation and Response Surface Analysis

Evolutionary operation (EVOP) is an optimization technique introduced by

Box (1957). The method was designed to quantify the response of a system due to changes of the operation conditions without disturbing the process dynamics or substantially endangering the plant profits.

Evolutionary operations are particularly useful when the measurements of the objective function are subject to random errors, as may be the case in industrial food production and quality control measurements, or whenever no suitable mathematical model exists (Box, 1975).

Searching for an improved product or process by EVOP is accomplished by a systematic slight variation of each of the variables under study from its normal standard-plant operating level. In its simplest form, EVOP is based on factorial or factorial-factorial designs whereby the objective function is evaluated at the vertices (see Sec. III.B.5) and the center of a hypercube in the space created by the independent variables. The strategy is based on repetitive construction of a new design about the best vertex that corresponds to the smallest function value [as distinct from the center point (Box, Davies, and Swann, 1969)]. Searching for an improved product or process by EVOP requires basic statistical knowledge beyond the scope of this chapter. However, the reader should consult several excellent sources that discuss the subject in detail (Box, 1957; Bender, Kramer, and Kahan, 1976; Kramer and Twigg, 1966).

The EVOP search results in locating a region containing the optimal solution. However, in some cases we would further consider analyzing this region in more detail to locate the true optimal point. Response surface analysis (RSA) provides the mathematical ground for locating the extremum.

The RSA method is based on the assumption that for n independent variables, the response will be a function of the levels at which these variables are combined. Thus, for instance, a second-order equation has the form

$$J = a_0 + a_1 x_1 + a_2 x_2 + a_3 x_1^2 + a_4 x_2^2 + a_5 x_1 x_2 \qquad (45)$$

where J is the response function, x_1 and x_2 represent the independent variables, and a_0 through a_5 represent parameters. Other equations may be applied.

After fitting a surface to this entire region, we can approach the optimal point by applying an appropriate search technique. The RSA method and its numerous potential applications in the food industry and food science are discussed in many publications (Aguilera and Kosikowski, 1976; Bender, Kramer, and Kahan, 1976; Herika, 1972; Liener, 1972; Maddox and Richert, 1977; Min and Thomas, 1980; Sood and Kosikowski, 1979).

The RSA method furnishes a valuable technique for the optimal formulation of food products (Kissell, 1967; Roozen and Pilnik, 1976; Smith and Ross, 1963) and may be regarded as a supplement to linear programming.

One common problem facing a food practitioner is the development of an optimal product. This development may involve the "eclipse" approach (Moskowitz, Stanley, and Chandler, 1977), which is a statistical approach similar to the RSA methods. The eclipse method provides the food technologist with a rapid method for converging on potentially optimal formulation through a consumer-generated ideal sensory profile.

Another class of response surface designs, termed "rotatable designs," was introduced by Mullen and Ennis (1979) for food product development. These designs permit the researcher to look at the response surface of a particular characteristic as a function of several experimental inputs with a minimum number of experimental trials. The minimum number of experiments required for obtaining an optimum has a crucial importance in reducing and minimizing the efforts, labor, and time invested in any real problem. This may be accomplished through the simplex method.

5. Simplex Method

The EVOP approach inspired Spendley, Hext, and Himsworth (1962) to devise a method based on the geometric design known as a "regular simplex." The objective function is evaluated at $n + 1$ mutually equidistant points in the space created by the n independent variables. In our application we are concerned only with the $n + 1$ points, which are the vertices of the simplex. The simplex is regular if its vertices are equally spaced, e.g., an equilateral triangle in a two-dimensional problem or a regular tetrahedron in a three-dimensional problem.

The simplex procedure requires that the objective function be evaluated at each of the vertices and a projection be made from the point that yields the highest objective function value through the center of the remaining points. The basic procedure (Walsh, 1976) is

1. Estimate the initial guess of the optimal point x^* and let this guess be one vertex of the initial simplex. After scaling the variables (if-necessary), the remaining vertices of the initial simplex are chosen.

2. Evaluate the objective function at each vertex.

3. Replace the vertex with the highest function by its reflection through the centroid of the remaining n vertices, thus creating a new simplex. Evaluate the objective function at the new vertex.

4. Repeat step (3) for each simplex.

5. If the highest function value occurs at the new vertex, reflect the vertex with the second highest function value. This step provides a precaution rule to avoid oscillations.

6. If one vertex of the simplex remains unchanged for M consecutive
    iterations, reduce the size of the simplex. This step is called
    "contraction." Spendley, Hext, and Himsworth (1962) noted that M
    depends on the number n of independent variables. They suggested
    M to be the smallest integer exceeding the following relation:

$$M = 1.65n + 0.05n^2 \tag{46}$$

where M is the number of consecutive iterations and n is the number of variables.

7. After a prescribed number of contractions have been carried out,
    the procedure is terminated

A simplex search for a two-dimensional case is illustrated in Fig.
10, which indicates that the triangle reflection through the side opposite the
highest-valued vertex is equivalent to movement down the steepest of the
three slopes at right angles to each face. It also shows that when the simplex is near the minimum, the contraction step is of major importance in
locating the true minimum by reducing the simplex.

FIGURE 10   Typical simplex search.

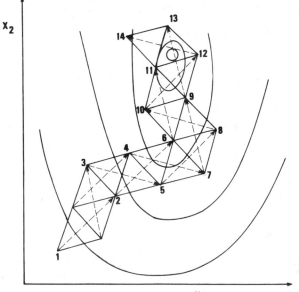

The search by a regular simplex presents certain drawbacks, mainly the lack of acceleration throughout the process and difficulties encountered when searching sharp ridges and curved valleys.

Modifications of the simplex method, i.e., "modified simplex" (MS), were proposed by Campey and Nickols (1961) and Nelder and Mead (1965). The MS technique suggests a procedure which overcomes the drawbacks of the regular simplex method. The Nelder and Mead method permits alteration of the simplex shape; thus, it is also known as the "flexible polyhedron." It too is based on n + 1 vertices. Furthermore, it provides an expansion step for the acceleration of the search procedure. Three constant parameters are used: the reflection coefficient α, the contraction coefficient β ($\beta < 1$), and the expansion coefficient γ ($\gamma > 1$). Nelder and Mead recommended the values of $\alpha = 1$, $\beta = 0.05$, and $\gamma = 2$, whereas other values for these parameters could be used (Himmblaue, 1972).

Recently, Morgan and Deming (1974) and Routh, Swartz, and Denton (1977) further improved the simplex method and introduced the modified simplex (MS) and the super-modified simplex (SMS), respectively. The SMS method has been successfully used to determine the optimum set of operating systems and processes in analytical chemistry. The methods were reported to be capable of dealing with systems where random error exists and of handling multifactor effects, which include variable interaction and stiff ridges, thus projecting a plausible avenue of optimizing food processes and research. All these modifications allow expansion and contraction of the simplex, and some can even handle constraints. The simplex method, as described here for unconstrained multivariable minimization, should not be confused with the simplex method in linear programming (Dantzig, 1963) described in Chap. 7.

C. Convergence Criteria

The decision to terminate an iterative optimization algorithm is very difficult, yet it is very important to the accuracy and precision of the solution. The main problem is the possibility of prematurely stopping at a remote value, yielding misleading results.

Usually the criteria for stopping an algorithm are based on two combined tests: the relative changes in both the objective function value and the independent variables values during the k + 1 iteration, as compared to previous values. These criteria should be below a given small positive number and may be expressed as follows:

$$\frac{f(\underline{x}^{(k+1)}) - f(\underline{x}^{(k)})}{f(\underline{x}^{(k)})} \leq \epsilon_1 \tag{47a}$$

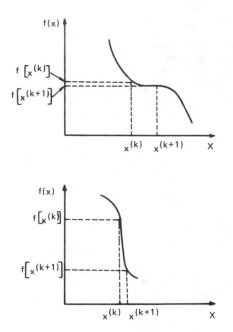

FIGURE 11   Situation avoided by joint criteria for convergence.

$$\frac{\underline{x}^{(k+1)} - \underline{x}^{(k)}}{\underline{x}^{(k)}} \leq \epsilon_2 \tag{47b}$$

The above criteria ensure that the algorithm is not prematurely terminated as a result of reaching a flat plateau or a very steep slope of the objective function, as depicted in Fig. 11.

The problem of convergence to a local minimum remains. To overcome this, the optimization procedure is restarted at several different,
widely distributed, initial points, and if all or most lead to a single solution,
then that point is conjectured to be the global optimum. This strategy
should be adopted whenever no other information on the optimum neighborhood is available. There is no strategy that guarantees the derivation of
the global optimum in all cases.

IV.   GRADIENT METHODS

By evaluating the objective function derivatives, the optimum may be found
in fewer steps. Gradient methods are those which select the direction of
search by using values of the partial derivatives of the objective function
with respect to the independent variables, as well as the values of $f(\underline{x})$.

Usually only the first-order derivatives are used; however, this is not always true.

The required derivatives are evaluated when possible by an explicit analytical formula obtained by differentiation of the objective function or numerically by using some finite differences scheme.

The gradient g is a vector directed from a given point p in n-dimensional space which expresses the greatest rate of improvement in the objective function. The direction of the gradient usually varies from point to point as depicted in Fig. 12 (for further explanation of this figure, see Sec. V).

The gradient methods are generally divided into two main categories. The first group uses the derivatives at the current test point to provide a search direction. The second class uses the gradient to guide the search (Beveridge and Schechter, 1970); however, the move is not necessarily in the direction of the steepest descent.

Gradient methods generally furnish a faster and more rapid convergence procedure as compared with direct search methods. This superior efficiency occurs only if satisfactory analytical expressions are available for evaluating the gradient and if the initial estimate of the extremum is not too remote from its real value. When numerical schemes are used to estimate the derivatives, this efficiency is severely impaired, and the decision on which method to be used has to be considered individually (Box, Davies, and Swann, 1969).

FIGURE 12   Gradient direction, feasible region, and active constraint.

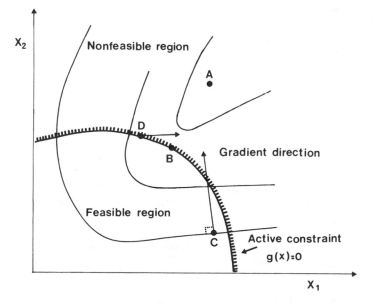

A. Steepest Descent Method

The method of the steepest descent was first derived by Cauchy in 1847, and since then, many versions of the method have been suggested. The unit vector in the direction of the negative gradient g of the objective function points in the direction of the most rapid decrease in $f(\underline{x})$ with respect to each of the components of $\underline{x}$, and it is orthogonal to the contour of the objective function at a given point $\underline{x}_p^{(k)}$.

Iterative techniques utilize the search direction g for the location of an improved estimate of the minimum and may be described by

$$\underline{x}_p^{(k+1)} = \underline{x}_p^{(k)} + h\underline{g} \qquad (48)$$

where k is the iteration number, g is a n-dimensional vector, and h is a distance (scalar) moved along the direction vector. The direction vector g is defined as

$$\underline{g} = - \frac{\nabla f(\underline{x}_p)}{f(\underline{x}_p)} \qquad (49)$$

Where the coefficients of the vector g are

$$g_i = \pm \frac{(\partial f / \partial x_i)_p}{\sqrt{\sum_{i=1}^{n} \left(\partial f / \partial x_i \right)_p^2}} \qquad i = 1, \ldots, n \qquad (50)$$

and g is normalized:

$$\sum_{i=1}^{n} g_i^2 = 1 \qquad (51)$$

The positive sign in Eq. (50) denotes the steepest ascent, and the negative sign denotes the steepest descent; p denotes the base point at which the gradient is evaluated.

Two general methods of selecting the step size h are normally applied. In the simplest case, h can chosen as a fixed constant during the whole iterative procedure. In this case the choice of the step length should be carefully selected to avoid unexpected overshooting (causing oscillation) when h is too large, or on the other hand, a small value of h may require an excessive number of steps to approach the optimum. Hence, a compromise between accuracy and efficiency should be adopted.

The basic search scheme of the steepest descent method using a constant preselected step length h may be summarized as follows:

1. The normalized gradient vector of the steepest descent direction is evaluated at a current point p [using Eqs. (50) and (51)].

2. An improved point is located by applying the general iterative scheme Eq. (48) using the specified step length h.

3. Steps (1) and (2) are repeated until no further reduction in the objective function values is observed. Otherwise, the step length is decreased, and the search is restarted from the last best point.

4. The procedure is terminated when convergence criteria are met or when the maximum number of step reductions has been reached.

Other methods use direct search techniques along the gradient instead of taking a fixed step in the direction of the steepest descent. Minimization along the gradient may benefit from linear accelerated search techniques such as Powell's method or any other direct search procedure described previously.

In cases when the contours form a sharp ridge, the steepest descent tends to zigzag and oscillate. These drawbacks may be overcome by several methods of scaling the variables to yield contours as spherical as possible. Nevertheless, the steepest descent method is not recommended for general use due to its slow convergence near the optimum.

B. Newton's Method

As we have already seen, the search direction in the steepest descent method is orthogonal to the contours of constant function values. Since most contours are not circular or spherical, the direction of the search does not always pass through the real extremum.

Newton's method originated from the quadratic approximation of $f(\underline{x})$. It makes use of the second-order information (i.e., second partial derivative with respect to $\underline{x}$ evaluated at $\underline{x}_p^{(k)}$) and provides a better estimation of the direction of the global minima. However, its efficiency is usually greatly reduced if the initial estimate is not in the neighborhood of the extremum.

Suppose that a function $f(\underline{x})$ is continuous and twice differentiable. Thus, in addition to the gradient vector $\nabla f(\underline{x})$, the hessian matrix H [e.g., $\nabla^2 f(\underline{x})$] exists for every point $\underline{x}$. Taylor second order expansion of the objective function $f(\underline{x})$ about the minimum $\underline{x}^{(k)}$, yields:

$$f(\underline{x}^{(k+1)}) = f(\underline{x}^{9(k)}) + \nabla f(\underline{x}^{(k)}) \, \Delta\underline{x}^{(k)} + \frac{1}{2} \, \nabla^2 f(\underline{x}^{(k)}) \, \Delta\underline{x}^{(k)} \, \Delta\underline{x}^{(k)} \qquad (52)$$

302   Saguy

where

$$\Delta \underline{x}^{(k)} = \underline{x}^{(k+1)} - \underline{x}^{(k)} \tag{53}$$

The minimum of $f(\underline{x})$ in the direction of $\Delta \underline{x}^{(k)}$ may be obtained by equating $\partial f(\underline{x})/\partial(\Delta \underline{x})$ to zero. Hence (Himmelblau, 1972),

$$\Delta \underline{x}^{(k)} = -\left[\nabla^2 f\left(\underline{x}^{(k)}\right)\right]^{-1} \nabla f\left(\underline{x}^{(k)}\right) \tag{54a}$$

or

$$\Delta \underline{x}^{(k)} = -H^{-1}\underline{g} \tag{54b}$$

where H^{-1} is the inverse of the Hessian matrix and $\underline{g}$ is the gradient vector, both evaluated at $\underline{x}^{(k)}$.

Rewriting Eqs. (53) and (54) yields

$$\underline{x}^{(k+1)} = \underline{x}^{(k)} - H^{-1}\underline{g} \tag{55}$$

Equation (55) is the essential basis of Newton's method, which requires that at each interation, $\underline{g}$ and H^{-1} should be evaluated at $\underline{x}^{(k)}$. If the objective function is quadratic, only one step is required to reach the extremum. However, for general nonlinear objective functions more than one step is required.

Improved numerical accuracy may be obtained if the independent variables are scaled so that the Hessian matrix is approximately equal to the unit matrix. The latter is true for all the methods based on calculating the Hessian matrix. Moreover, scaling of variables is always required when the variables are very different in their magnitudes. Furthermore, to ensure convergence of Newton's method, the initial guess should be in the neighborhood of the extremum. Also, H^{-1} should be positive definite and not approaching singularity.

To avoid some of the time-consuming calculations associated with Newton's method, H and H^{-1} should not be recalculated after each iteration, but instead only after m iterations. Although this procedure would necessitate more iterations, a significant reduction in total computation time might be obtained (Box, Davies, and Swann, 1969). Since matrix inversion is a time-consuming computational process, in practice the Hessian may be estimated by a system of linear equations instead of being inversed directly.

The iteration procedure given by Eq. (55) can be changed to yield a limited-step Newton formula:

$$\underline{x}^{(k+1)} = \underline{x}^{(k)} - \theta H^{-1}\underline{g} \tag{56}$$

where θ is chosen so that $f\left(\underline{x}^{(k+1)}\right) < f\left(\underline{x}^{(k)}\right)$. This last change is used to ensure a decrease in the function value at each iteration. The value of θ is determined by a linear search from $\underline{x}^{(k)}$ in the direction $-H^{-1}g$.

Finally, it should be noted that Newton's method and the steepest descent method are identical when H^{-1} is a unit matrix.

C. Quasi-Newton Methods

Generally, the quasi-Newton methods (known also as variable metric) have now become the standard method for finding an unconstrained minimum of a differentiable function. Quasi-Newton methods approximate and update the inverse of the Hessian matrix at each iteration without the need for matrix inversion. This avoids the time-consuming calculations and difficulties associated with the evaluation of the inverse of a matrix. In these methods the matrix H^{-1} is replaced by a positive definite symmetric matrix which is updated in each iteration.

The mathematical complexity of quasi-Newton methods requires a more elaborate discussion which may be found elsewhere (Avriel, 1976; Fletcher, 1981; Gill and Murray, 1972, Gill, Murray and Wright, 1981; Murray, 1976; Broyden, 1967).

One of the best known quasi-Newton methods is the Davidon-Fletcher-Powell (DFP) method. Originally devised by Davidon (1959) and further improved by Fletcher and Powell (1963). The DFP method, like the steepest descent method, provides a rapid reduction of the objective function, yet in the neighborhood of the extremum it benefits from Newton-method fast convergence. The DFP thus overcomes the main drawbacks of these other two methods: slow convergence near the optimum in the method of steepest descent, and failure to converge due to initial poor estimation of the optimal point in Newton's method.

The DFP method produces conjugate gradient directions by means of matrix formulas. The conjugate directions are generated by the minimization technique chosen to ensure that the optimum of the quadratic function is found in a finite number of iterations (Walsh, 1976). This is a very important property, since quadratic behavior is observed near the optimum of most functions (Beveridge and Schechter, 1970).

We mentioned that the DFP method produces conjugate directions by means of matrix formulas; this matrix is of the order n × n (n = number of variables). When the number of variables is quite high, the amount of computer storage required may be so large that problems concerning memory capacity may be anticipated. In such cases, the Fletcher-Reeves (1964) method may be more appropriate. The latter uses only a vector formula (of the order n); so there is a substantial reduction in storage requirements. The DFP method was further improved by Broyden (1967). The formula now commonly referred to as BFGS [Broyden (1970), Fletcher (1970), Goldfarb (1970), Shanno (1970)] provides one of the most effective, robust hard to beat method (Sargent, 1980).

V. CONSTRAINED MINIMIZATION

A more general optimization problem than previously considered is to find
the best conditions of an n-variable objective function:

$$J = f(x_1, x_2, \ldots, x_n) \tag{57}$$

subject to m inequality constraints which have the form

$$g_k(x_1, x_2, \ldots, x_n) \geq 0 \qquad k = 1, 2, \ldots, m \tag{58}$$

where m is no longer necessarily less than n.

Before reviewing the constrained optimization concept and solving
methods, we need to explain the terminology and special features of the
method.

In the nonlinear-constrained programming problems, the objective
function, or one or more of the constraints, assumes a nonlinear form.
Cases and problems defined by linear constraints and linear objective functions are discussed in a separate chapter (see Chap. 7). Inequality constraint described by Eq. (58) may be converted into equality constraints by
defining new variables, called "slack" or "surplus" variables. These may
be given as

$$g_i(x_1, x_2, \ldots, x_n) \leq b_i \qquad i = 1, \ldots, m_1 \tag{59}$$

$$g_i(x_1, x_2, \ldots, x_n) \geq b_i \qquad i = m_1 + 1, \ldots, m \tag{60}$$

where b_i is taken to be nonnegative.

The first m_1 inequalities [Eq. (59)] are rewritten as equalities:

$$e_i(x_1, x_2, \ldots, x_n, x_{s_i}) = g_i(x_1, \ldots, x_n) + x_{s_i} - b_i = 0 \tag{61}$$

for $i = 1, 2, \ldots, m_1$. The other $m_1 + 1$ to m constraints are transferred
to equalities:

$$e_i(x_1, x_2, \ldots, x_n, x_{s_i}) = g_i(\underline{x}) - x_{s_i} - b_i = 0 \tag{62}$$

for $i = m_1 + 1, \ldots, m$. It is convenient to make no distinction between the
real and the slack variables; thus,

$$x_{n+i} = x_{s_i} \qquad i = 1, 2, \ldots, m \tag{63}$$

A set of constraints, such as those imposed by the above conditions, defines a valid and admissible system region (still of n dimensions) within which all variables must lie. Obviously, the system region includes the boundaries. Furthermore, the optimum within the region can exist either at a stationary point or at a discontinuity in the function or its derivatives. A point $\underline{x}_p$ is said to lie on a constraint or to be "active" if $g(\underline{x}_p) = 0$.

Thus, a zero contour of the constraint function $g(x)$ is regarded as a barrier separating the positive value of $g(\underline{x})$ (i.e., feasible or permissible region) from the negative values (i.e., nonfeasible region). The barrier (zero contour of the constraint function) is denoted by shading to indicate the nonfeasible region (Box, Davies, and Swann, 1969).

Figure 12 shows a typical two-dimensional case, and illustrates that due to the constraint $g(\underline{x})$, point A (the global minimum) is not included in the feasible region. Thus, the constrained minimum occurs at point B. Furthermore, at point C the direction of the steepest descent is no longer directed toward the constrained minimum (point B). This latter deviation increases until the constraint starts to be active (point D). At this point, the progress toward the constrained minimum depends on the constraint as much as the nature of the objective function.

A. Unconstrained Approach

Several methods deal with constraints; however, the simplest is to treat the system as if it were unconstrained and to check the feasible solutions against the inequality constraints. If the derived values are not in the valid region, the desired optimum may be either on the boundary imposed by the conflicting restrictions (when the inequality is identically zero) or at one of the other local optimal locations within the bounded region (Beveridge and Schechter, 1970). Another useful way to handle constraints is, whenever possible, to use transformations that remove the constraints from the formulation of the problem. This scheme is possible for a constrained system expressed by the inequality constraints

$$g(\underline{x}) \geq \underline{b} \tag{64}$$

or simple gounds

$$\underline{\ell} < \underline{x} < \underline{h} \tag{65}$$

where $\underline{b}$, $\underline{h}$, and $\underline{\ell}$ are constant vectors.

Using several transformations (Box, 1966), the constraints may be removed, and an unconstrained optimization could be performed.

A different approach is to reduce the complexity of the problem by removing from the formulation any inequality constraints that are not effective at the optimum and/or to reduce the dimension of the problem by using the equality constraints to eliminate some of the variables.

B. Methods for Bounds-constrained Problems

The methods used for bounds-constrained cases are closely related to unconstrained minimization; the main difference is that only some of the variables
are free to change at each iteration. An active set strategy is used (NAG,
1977) where certain constraints hold as equalities (i.e., active). During the search, one of the free variables may encounter a bound, or
one of the variables currently on a bound may become a free variable.
The number of free variables changes until all free variables are identified. These methods reduce the dimension of the problem according
to the number of active-bound variables. A complete description of
methods for bounds problems is given in Gill and Murray's (1974)
book.

C. Hemstitching

Hemstitching was created by Roberts and Lyvers (1961). The procedure
may be applied to problems with any number of variables and constraints,
and is based on moving along the steepest descent direction toward the
minimum. If all constraints are satisfied, the procedure is repeated. If
however, a violation of one of the constraints occurred, a step orthogonal
to this constraint is taken toward the feasible region. The hemstitching
procedure for a two-dimensional problem is shown in Fig. 13. Steps in the
direction orthogonal to the constraint boundary (given by lines BC and DE)
are taken until a feasible point within the permissible region is reached
(points C and E). Since the steps toward the feasible region do not assure a reduction in the objective function value, convergence may be
limited. An improved method was suggested by Box, Davies, and Swann
(1969).

D. Penalty Function

A constrained problem may be transformed into an unconstrained one by
applying a penalty to the objective function at nonfeasible points.   A
typical inequality-constraint problem formulation would be for

$$g_i(\underline{x}) \geq 0 \qquad i = 1, 2, \ldots, m \tag{66}$$

a modified objective function F(x) is defined by

$$F(\underline{x}) = f(\underline{x}) + P(\underline{x}) \tag{67a}$$

where

$$P(\underline{x}) = \sum_{i=1}^{s} w_i [g_i(\underline{x})]^2 \tag{67b}$$

and s is the number of constraints violated at the current point, and w_i

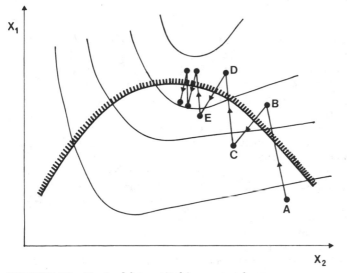

FIGURE 13   Typical homstitching procedure.

are positive weighting factors. As w_i increases, it can be shown (Fiacco
and McCormick, 1966) that the solution of the problem converges to the real
extremum. In practice, the choice of w_i is critical. The recommended
procedure is to solve the problem several times. In each iteration the
previous results are inserted back as the new starting point for the next
iteration, whereas the value of w_i is increased progressively. When
w_i becomes large enough, then optimality is achieved. Yet, practice showed
that as the constraint must be heavily weighted in relation to the objective
function, it yields an ill-conditioned unconstrained problem. An elaborate
scheme for calculating and updating the values of w_i is described by Leitmann
(1962). This procedure is not useful for on-line calculation.

E.   Sequential Unconstrained Minimization Technique (SUMT)

A highly developed form of the penalty functions method is called the sequential unconstrained minimization technique (SUMT). It is a different application of the penalty function concept which has proven to be more useful.
The main disadvantage of the penalty function is that it allows computation
of the function at nonfeasible points. This problem is avoided by restricting
the search to the feasible points only. Thus, all nonfeasible function evaluations are prevented. This method was proposed by Carroll (1961), who
called it the "created response surface technique," and further developed by
Fiacco and McCormick (1966). The method replaces the constrained problem with a sequence of unconstrained minimization problems by applying
different penalty functions to the given objective functions in such a way that
the optimal solutions of successive unconstrained problems approach the
optimal solution of the given constrained problem. For further details on
this method the reader should consult Carroll (1961), Fiacco and McCormick (1966), and Walsh (1976).

F. Complex Method

The techniques evolved from the simplex method for unconstrained minimization, (described in Sec. III.B.5) are capable of handling constraints by adjusting the nonfeasible vertex until it becomes feasible when a constraint is encountered. After many such adjustments, the flexible polyhedron used may collapse into n - 1 or fewer dimensions. This collapsed polyhedron may afterward not readily be expanded back into the original n + 1 dimensions. Box (1965) modified this method to handle constraints and termed his procedure the "complex method." This method is a sequential search technique which has been proven effective in solving problems with nonlinear objective functions, subject to nonlinear inequality constraints. The complex method provides one of the most promising methods of direct search for finite-space problems. The method is quite simple and offers easy handling of various inequality constraints (Umeda, Shinodo, and Ichikawa, 1972; Mishkin, Karel, and Saguy, 1982; Saguy, 1982). Furthermore, the procedure tends to find the global extremum since the initial set of points is randomly scattered throughout the feasible region. Because no derivative is required, the method is based on the evaluation and comparison of the objective function values at all vertices of a simplex and the replacement of the worst vertex by a new feasible point.

The complex method employs $k = n + p$ vertices, where $p = 1, \ldots, n$. Usually, $k = 2n$ is a good choice. This ensures that the polyhedron (which contains 2n vertices) will not collapse or flatten near the constraints. Box recommended from his empirical studies that the reflection factor α should be taken as 1.3 and $k = 2n$. This value of k may cause some difficulties when n is large (approaching a value of 10 or more).

This method is widely used for constrained optimizations, variational problems, and derivations of optimal temperature profiles (for further details, see the Appendix of Chap. 11, where a computer code and an example of its use are furnished). Although this method converges rather slowly, it may be applicable to many broad and different optimization problems.

Many other methods for constrained optimization are available. Among those frequently used, the following should be mentioned: Davidon-Fletcher-Powell (Fletcher and Powell, 1963), also called the "DFP" method; Rosenbrock (1960); Goldfarb and Lapidus (1968); and sequential augmented Lagrangian (Murray, 1976).

VI. DETERMINING A FEASIBLE
STARTING POINT

The iterative methods applied in solving a constrained problem require sequences of feasible points (i.e., within the permissible region). These feasible values could be found by using the following method (Walsh, 1976):

Starting at some nonfeasible point, write the m constraints in the form

$$g_i(\underline{x}) \geq 0 \qquad i = 1, \ldots, m_1 \tag{68}$$

$$e_j(\underline{x}) = 0 \qquad j = m_1 + 1, \ldots, m \tag{69}$$

and minimize the unconstrained function

$$Z = -\sum_{i=1}^{k} g_i(\underline{x}) + \sum_{j=k+1}^{m} [e_j(\underline{x})]^2 \tag{70}$$

where k is the number of inequality constraints that are violated at the current (nonfeasible) point ($k \leq m_1$). When the minimization of Z reaches zero, a feasible point has been located.

VII. AVAILABLE SOFTWARE

One of the most exciting opportunities presented by the growth of computers is the transfer of research results from one field to another via general-purpose software (Chan et al., 1978). Programs and subroutines are becoming the central channels between software, numerical analysis, and application. The principle criteria of choosing a subroutine is based generally on ease of use, machine efficiency, and availability. For more details on this subject, see Chap. 1.

Numerical analysts and computer software experts often remark (and argue) on the difficulty of comparing "codes," and only rarely provide potential users with clear-cut selection guidelines. We will focus, therefore, on providing information on the most sophisticated subroutines collected from commercial sources. Among the numerous sources, only a few are included: HAR (Hopper, 1981); IMSL (1977); NAG (1977); NPL (1978, 1980); ANL-U.S. Department of Energy (1982); and the National Bureau of Standards (1977) (for further detail, see Chap. 1.).

Also, most books written on optimization include several routines, e.g., Brent (1973), Forsythe, Malcom, and Moler (1977), and Kuester and Mize (1973). Current research and literature can be found in many journals, including ACM Transaction on Mathematical Software, Journal of Optimization Theory and Applications, and Computers and Chemical Engineering.

In our review of available software, no preference is shown for a specific library. The library chosen should be based on previous experience, special needs and requirements, availability, and implementation suitability. A technical report written by Gill et al. (1977) is an excellent source which details the design principles and structures of existing FOR-TRAN program libraries intended primarily to solve optimization problems. (For further information on FORTRAN libraries, see Chap. 1.) Remember, however, that no library or "code" can help in formulating your own problems.

A. Subroutines

The most comprehensive and sophisticated collection of optimization routines currently available was released by the National Physical Laboratory (NPL) adopted and recommended by the Stanford Center for Information Processing (Bolstad et al., 1978). Other users, such as The Weizmann Institute of Science, recommend the NAG library first, followed by IMSL, HAR and ANL (Jarosch and Druck, 1982).

Tables 5 to 8 summarize some of the available subroutines for non-constrained, simple bounds-constrained, linearly constrained and nonlinear general constrained minimization problems, respectively. In the above tables only the well-known and up-to-date subroutine versions are given. However, the potential user is warned that changes and improvements are frequently due. All routines are written in Standard FORTRAN (ANSI, 1966) and are available in single and double precision. Double precision is greatly recommended for most practical cases.

B. Subroutine Selection

Selecting the "right" algorithm is a complicated task and requires skill and experience. When this expertise is missing, the NAG and NPL libraries provide easy-to-use routines which include only those parameters absolutely essential to the definition of the problems, as opposed to the "normal" routines that have additional parameters. The latter allow the more experienced user to improve his or her program efficiency and to control some of the options offered, such as the maximum step size allowed, number of iterations, tolerances, and other decisions. These features are of little or no value to the inexperienced user; in some cases they may even cause failure or confusion. In the easy-to-use routines, these extra parameters are determined either by using fixed, efficient values or by providing auxiliary routines which generate a reasonable value automatically. Furthermore, the easy-to-use routines check numerically to see whether the user-supplied first or second derivatives are consistent with the objective function values. This option is strongly recommended when the "normal" subroutine versions are used. This is carried out by calling the "service" routines provided by these libraries.

The easy-to-use routines are, however, less flexible than their normal counterparts and generally less efficient. In Tables 5 to 8, the easy-to-use versions are given with their counterparts (separated by slash).

TABLE 5   Subroutines for the Optimization of Unconstrained Nonlinear
          Objective Function

Number	$\nabla f(\underline{x})$	$\nabla^2 f(\underline{x})$	Library	Subroutine code[a]	Algorithm method
One variable					
1	–	–	HAR	VD01A	Powell's method
2	–	–	IMSL	ZXGSP/ZXGSN	Golden section
3	–	–	NAG	E04ABF	Successive quadratic approx.
4	–	–	NPL	UNIFUN	
∇f-1	+	–	HAR	VD02A	Successive cubic approx.
∇f-2	+	–	NAG	E04BBF	Successive cubic approx.
∇f-3	+	–	NPL	UNIGRD	
N Variables					
1	–	–	HAR	VA03A	Powell's method
2	–	–	HAR	VA10A	Quasi-Newton
3	–	–	IMSL	ZXMIN	Quasi-Newton
4	–	–	NAG	E04CCF	Simplex method
5	–	–	NAG	E04CFF/E04CGF	Quasi-Newton
6	–	–	NPL	CNGRDF	Conjugate gradient
7	–	–	NPL	QNMDIF/UBNDQ1	Quasi-Newton
∇f-1	+	–	HAR	VA06A	Steepest descent and Newton's method
∇f-2	+	–	HAR	VA08A	Conjugate gradient
∇f-3	+	–	HAR	VA09A	Quasi-Newton
∇f-4	+	–	HAR	VA13A	BFGS Formula
∇f-5	+	–	HAR	VA14A	Conjugate gradient
∇f-6	+	–	NAG	E04DBF	Conjugate gradient
∇f-7	+	–	NAG	E04DDF/E04DEF	Quasi-Newton
∇f-8	+	–	NPL	CNGRDR	Conjugate gradient
∇f-9	+	–	NPL	QNMDER/UBFDQ2	Quasi-Newton
$\nabla^2 f$-1	+	+	HAR	VA11A	Newton's method
$\nabla^2 f$-2	+	+	NAG	E04EAF/E04EBF	Modified Newton
$\nabla^2 f$-3	+	+	NPL	MNA/UBSDN2	Modified Newton

[a] The second subroutine separated by a slash is the easy-t0-use counterpart
of the first.

TABLE 6 Subroutine for the Optimization of Simple Bounds Constraints Imposed on Nonlinear N-Variable Objective Function

Number	$\nabla f(\underline{x})$	$\nabla^2 f(\underline{x})$	Library	Subroutine code[a]	Algorithm method
1	-	-	NAG	E04JBF/E04JAF	Quasi-Newton
2	-	-	NPL	BCQNDF/BCNDQ1	Quasi-Newton
∇f-1	+	-	NAG	E04KBF/E04KAF	Quasi-Newton
∇f-1	+	-	NAG	E04KDF/E04KCF	Modified Newton
∇f-3	+	-	NPL	BCMNAF/BCFDN2	Modified Newton
$\nabla^2 f$-1	+	+	NAG	E04LBF/E04LAF	Modified Newton
$\nabla^2 f$-2	+	+	NPL	BCMNA/BCSDN2	Modified Newton

[a] The second subroutine separated by a slash is the easy-to-use counterpart of the first.

The choice of the appropriate routine is based on the following factors: the type of problem (number of variables, constrained or unconstrained), the level of derivative information available (using derivative values increases efficiency due to faster convergence), the experience of the user, storage demands, computational time, and availability. In most cases, availability seems to be the dominant reason for choosing an algorithm.

TABLE 7 Subroutines for the Optimization of Linear Constraint Imposed on Nonlinear N-Variable Objective Function

Number	$\nabla f(\underline{x})$	$\nabla^2 f(\underline{x})$	Library	Subroutine code[a]	Algorithm method
1	-	-	NPL	LCQNDF/LCNDQ1	Quasi-Newton
∇f-1	+	-	HAR	VE01A	Davidon's method
∇f-2	+	-	HAR	VE05A	Goldfarb's method
∇f-3	+	-	NPL	LCMNAF/LCFDN2	Modified Newton
∇f-4	+	-	NPL	LCQNDR/LCFDQ2	Quasi-Newton
∇f-5	+	-	ANL	KEELE	Variable metric
$\nabla^2 f$-1	+	+	NPL	LCMNA/LCSDN2	Modified Newton

[a] The second subroutine separated by a slash is the easy-to-use counterpart of the first.

TABLE 8 Subroutines for the Optimization of General Nonlinear
Constrained N-Variable Objective Function

Number	$\nabla f(\underline{x})$	$\nabla^2 f(\underline{x})$	Library	Subroutine code	Algorithm method
1	–	–	NAG	E04HAF	Sequential penalty function
2	–	–	NAG	E04UAF	Sequential augmented Lagrangian
3	–	–	NPL	SALQDF	Sequential augmented Lagrangian
∇f-1	+	–	HAR	VF01A	Penalty function
∇f-2	+	–	NAG	E04VAF	Sequential augmented Lagrangian
∇f-3	+	–	NPL	SALQDR, SALMNF	Sequential augmented Lagrangian
$\nabla^2 f$-1	+	+	NAG	E04WAF	Sequential augmented Lagrangian
$\nabla^2 f$-2	+	+	NPL	SALMNA	Sequential augmented Lagrangian

For unconstrained nonlinear minimization, the simplest routines are
VD01A (HAR), ZXGSN (IMSL), E04ABF (NAG), and UNIFUN (NPL). All
these routines require only function values. Bolstad et al. (1978) suggested
that when the derivative can be evaluated with less than four times the work
it takes to calculate the objective function, then a derivative routine such
as UNIGRD (NPL) should be used to find the minimum more efficiently.

For unconstrained nonlinear minimization of n-dimensional objective
functions, several routines are available (Table 5). In these cases the
first partial ∇f or the second partials $\nabla^2 f$ may be needed. Since no one
method for nonlinear optimization is universally applicable, we mention a
variety of routines with different characteristics so that the user may take
advantage of any special features and demands his or her problem may
have and of the software contained in his or her system.

Following the general recommendations given by Bolstad et al. (1978)
and Jarosch and Druck (1982), we may conclude with the following suggestions (when two subroutines are separated by a slash the second is the easy-
to-use version):

1. The modified-Newton algorithms such as E04EAF/E04EBF (NAG) or MNA/UBSDN2 (NPL) are rapidly convergent, but usually only if the initial guess is fairly close to the minimum. Furthermore, these require second derivatives either supplied in an explicit analytical form or evaluated numerically. In some cases this may be quite impractical.

2. If one has poor initial estimates or very difficult derivative evaluations, then a quasi-Newton routine may be used. The quasi-Newton method requires only first-derivative information; thus, an overall reduction in computational time may be obtained. Such routines are E04DDF/E04DEF (NAG) or QNMDER/UBFDQ2 (NPL).

3. If the calculation of the derivatives results in a time-consuming process, then another compromise may be taken. Routines E04CFF/E04CGF (NAG) or QNMDIF/UBNDQ1 (NPL) are a possible choice; these require function values only.

4. The use of explicit derivatives usually enhances both accuracy and convergence. However, such derivative calculations may be quite difficult to compute. Therefore, the derivative methods should be used with great caution.

Solving constrained nonlinear minimization problems can be complicated since constraints are often imposed on the solution, which increases the computational difficulty. For special requirements and features, see Table 6, which lists the routines available when the constraints are of simple bounds type. Linear constraints imposed on the system are solved with the aid of the subroutines summarized in Table 7; a general constrained nonlinear system may be approached with the routines given in Table 8.

All routines find local minima only, as we pointed out in the preceding section. We do not envision a "super"-routine in the near future that guarantees the location of the global minimum. Thus, attempts should be made to find a global minimum by using several different, widely separated starting points to see if the same solution is obtained and/or by excluding local minima, already determined by adding constraints to the problem.

Other problems concerned with convergence failure should be tested individually. For further information, see the NAG introduction (NAG, 1977), Chap. 7 of the book by Murray (1972) and Gill, Murray and Wright (1981).

C. Test Problems

A number of papers compare and evaluate which subroutine performs better in areas such as efficiency, reliability, and less CPU time. The ability to handle constraints and complex topography (i.e., curved valley, ridges, or saddle points) and minimize computation efforts (measured in terms of function evaluations, CPU, implementation time, etc.) are the topics of other papers [see, for example, Colville (1968); Dennis, Gay, and Welsh (1979); Himmelblau (1972); and Hillstrom (1977)]. Based on the results obtained from running these evaluations, numerical analysis experts can review and improve the available software.

After the problem has been formulated and subroutine chosen, the next step should be to study and check the algorithm. For this purpose, most software libraries include a typical example which users should study before running their own problems.

Another possibility is to use well-known test problems which have evolved from actual optimization situations and are reported in the litera- ture [see Gill and Murray (1976), Himmelblau (1972), and Lootsman (1972)]. Test-problem names come from their original source, for instance, Rosen- brock (1960):

$$f(\underline{x}) = 100\,(x_2 - x_2^{\frac{1}{2}}) + (1 - x_1)^2$$

Initial guess $\underline{x} = (-1.2, 1.0)$

Optimal results $\underline{x}^* = (1.0, 1.0)$ $f(\underline{x}^*) = 0$

You may find many other test problems in the literature.

VIII. CONCLUSION

For a food engineer or a food practitioner, the most crucial decisions are whether or not to use any optimization procedure, and how to formulate the objective function to represent the process, and what quantity is to be maxi- mized or minimized. Hence, in most cases we are engaged with the process itself rather than choosing the most efficient numerical method reported in the literature. Generally, the decision of which software to use has to be judged on reducing the time required for the implementation rather than which software is most sophisticated. Hence, it is our belief that whenever the decision is between an available and well-documented "slow" subroutine and a better code (which has to be purchased and implemented), the first choice should be taken.

Obviously, each case should be weighed separately according to the time, money, labor, and effort needed. The reader should consult Chap. 15 by Frank which further discusses this subject.

ACKNOWLEDGMENTS

We acknowledge, with appreciation, Professor Larry Evans for his remark- able and stimulating course (no. 10.38 entitled, "Analysis and Simulation of Chemical Processing Systems," given at the MIT Department of Chemical Engineering) which inspired this work.

    We are indebted to many people who have helped in the development
of this chapter. Special thanks are due to Marty Mishkin and Steve Haralampu, who carefully read this manuscript at various stages of its development and offered a number of very helpful suggestions and comments; and
to Mrs. Miri Rauchwerger for her graphic work. We would like to thank
Mrs. Kathy Blackfield for her dedication in editing this manuscript and Ms.
Ginger Burr for her typing.
    Finally, I wish to thank Professor Marcus Karel for his encourage-
ment, collaboration, and guidance.

REFERENCES

Aguilera, J. M. and Kosikowski, F. V. (1976). Soybean extruded product:
    a response surface analysis, J. Food Sci. 41:647-651.

ANSI (American National Standard Institute) (1966). American National
    Standard Fortran. Publ. X3.9. American National Standard Institute,
    New York.

Avriel, M. (1976). Nonlinear Programming Analysis and Methods. Pren-
    tice-Hall, Englewood Cliffs, N.J.

Bender, F. E., Kramer, A., and Kahan, G. (1976). Systems Analysis for
    the Food Industry. Avi Publishing Co., Westport, Conn., Chap. 5.

Beveridge, G. S. G. and Schechter, R. S. (1970). Optimization: Theory
    and Practice. McGraw-Hill, New York.

Bolstad, J. H., Chan, T. F., Coughran, W. M., Jr., Grosse, E. H.,
    Heath, M. T., Luk, F. T., Nash, S. G., and Trefethen, L. N. (1978).
    Numerical Analysis Program Library User's Guide NAPLUG, User
    Note 82. Computer Science Dept. and Stanford Accelerator Center,
    Stanford University, Menlo Park, Calif.

Box, G. E. P. (1957). Evolutionary operation: A method of increasing
    industrial productivity, Appl. Stat. 6:3-22.

Box, G. E. P. (1975). Evolutionary operation of a food plant, Food Manu-
    fact. 50(4):61.

Box, M. J. (1965). A new method of constrained optimization and a com-
    parison with other methods, Computer J. 8:42-52.

Box, M. J. (1966). A comparison of several current optimization methods,
    and the use of transformation in constrained problems, Computer J. 9:
    67.

Box, M. J., Davies, D., and Swann, W. H. (1969). Non-Linear Optimiza-
    tion Techniques. Imperial Chemical Industries Monograph No.5.
    Oliver and Boyd, Edinburgh.

Brent, R. P. (1973). Algorithms for Minimization without Derivatives.
    Prentice-Hall, Englewood Cliffs, N.J.

Broyden, C. G. (1967). Quasi-Newton methods and their application to
    function minimization, Math. Comp. 21:368.

Broyden, C. G. (1970). The convergence of a class of double-rank minimi-
    zation algorithms, J. Inst. Maths. Applic. 6:76-90.

Campey, I. G. and Nickols, D. G. (1961). Simplex Minimization. Imperial Chemical Industries Ltd., Central Instruction Laboratory Report (August) London.

Carroll, C. W. (1961). The created response surface technique for optimizing nonlinear restrained systems, Operation Research 9:169-184.

Chan, T. F., Coughran, W. M., Jr., Grosse, E. H., and Heath, M. T. (1978). A Numerical Library and Its Support. STAN-CS-78-673, Computer Science Dept., Stanford University, Menlo Park, Calif. November.

Coggins, G. F. (1964). Univariate Search Methods. Imperial Chemical Industries Ltd., Central Instruction Laboratory Research Note 64/11, London.

Colville, A. R. (1968). A Comparative Study of Nonlinear Programming Codes. IBM, New York Scientific Center Technical Report No. 320-2949, New York.

Cooper, L. and Steinberg, D. (1970). Introduction to Methods of Optimization. W. B. Saunders, Philadelphia, Pa.

Dantzig, G. B. (1963) Linear Programming and Extensions. Princeton University Press, Princeton, N.J.

Davidon, W. C. (1959). Variable Metric Methods for Minimization. AEC Research and Development Rep. ANL-5990 (December). Argonne Nation Laboratory, Argonne, Ill.

Dennis, J. E., Jr., Gay, D. M., and Welsch, R. E. (1979). An Adaptive Nonlinear Least-Squares Algorithm. Technical Report TR 77-321, Dept. of Computer Science, Cornell University, Ithaca, N.Y.

Edelbaum, T. N. (1962). Theory of maxima and minima in Optimization Techniques (G. Leitmann, ed.). Academic Press, New York, Chap. 1.

Fiacco, A. V. and McCormick, G. P. (1966). Extension of SUMT for nonlinear programming: equality constraints and extrapolation, Management Sci. 12:816-829.

Fletcher, R. and Powell, M. J. D. (1963). A rapidly convergent descent method for minimization, Computer J. 6:163-168.

Fletcher, R. and Reeves, C. M. (1964). Function minimization by conjugate gradients, Computer J. 7:149-154.

Fletcher, R. (1970). A new approach to variable metric algorithms, Comp. J. 13:317-322.

Fletcher, R. (1981). Practical Methods of Optimization. Vol. I and II, Wiley, New York.

Forsythe, G. E., Malcom, M. A., and Moler, C. B. (1977). Computer Methods for Mathematical Computations. Prentice-Hall, New York.

Gill, P. E. and Murray, W. (1972). Quasi-Newton methods for unconstrained optimization, J. Inst. Math. Appl. 9:91-108.

Gill, P. E. and Murray, W. (eds.) (1974). Numerical Methods for Constrained Optimization. Academic Press, New York.

Gill, P. E. and Murray, W. (1976). Nonlinear least squares and nonlinearity constrained optimization, in Lecture Notes in Mathematics No. 506 Numerical Analysis. Springer-Verlag, New York.

Gill, P. E., Murray, W., Picken, S. M., and Wright, M. H. (1977). The Design and Structure of a Fortran Library for Optimization. Technical Report SOL 77-7 (April), Systems Optimization Laboratory, Stanford University, Stanford, Calif.

Gill, P. E., Murray, W. and Wright, M, H. (1981). Practical Optimization, Academic Press, New York.

Goldfarb, D. (1970). A family of variable metric updates by variational means, Math Comput. 24:23-26.

Goldfarb, D. and Lapidus, L. (1968). Conjugate gradient method for nonlinear programming problems with linear constraints, Indust. Eng. Chem. Fund. 7:142-151.

Herika, E. G. (1972). Simple and effective system for use with response surface methodology, Cereal Sci. Today 17:309.

Hillstrom, K. E. (1977). A simulation test approach to the evalustion of nonlinear optimization algorithms, ACM Trans. Math. Software 3(4): 305-315.

Himmelblau, D. M. (1972). Applied Nonlinear Programming. McGraw-Hill, New York.

Hooke, R. and Jeeves, T. A. (1961). Direct search solution of numerical and statistical problems, J. Assoc. Compl Mach. 8:212-229.

Hopper, M. J. (1981). Harwell Subroutine Library: A Catalogue of Subroutines. Computer Science and Systems Division, Atomic Energy Research Establishment, Harwell, Oxfordshire, U.K.

IMSL (International Mathematical and Statistical Libraries) (1977). IMSL Edition 6 User's Manual, Vol. 2. Houston, Tex.

Jarosch, H. S. and Druck, S. J. (1982). Numerical Analysis User's Guide. The Weizmann Institute of Science, Rehovot, Israel. January.

Kiefer, J. (1953). Sequential minimax search for a maximum. Proc Amer. Math. Soc. 4:502-506.

Kissell, L. T. (1967). Optimization of white layer cake formulations by a multiple-factor experimental design, Cereal Chem. 44:253.

Korn, G. A. and Korn, T. M. (1968). Mathematical Handbook for Scientists and Engineers, 2d ed. McGraw-Hill, New York.

Kramer, A. and Twigg, B. A. (1966). Evolutionary operations—EVOP, in Quality Control for the Food Industry. Avi, Westport, Conn., Chap. 16.

Kuester, J. L. and Mize, J. H. (1973). Optimization Techniques with Fortran. McGraw-Hill, New York.

Kuhn, H. W. and Tucker, A. W. (1951). Nonlinear programming, in Proceedings of the 2d Berkeley Symposium on Mathematical and Statistics Problems (J. Neyman, ed.). University of California Press, Berkeley, Calif., pp. 481-492.

Leitmann, G. (ed.) (1962). Optimization Techniques. Academic, New York.

Liener, J. E. (1972). Nutritional value of food protein products, in Soybeans Chemistry and Technology, Vol. 1 (A. K. Smith and S. J. Circle, eds.). Avi, Westport, Conn., p. 203.

Lootsma, F. A. (ed.) (1972). Numerical Methods for Non-Linear Optimization. Academic, New York.

Maddox, I. S. and Richert, S. H. (1977). Use of response surface methodology for the rapid optimization of microbiological media, J. Appl. Bacteriol. 43:197-204.

Min, D. B. and Thomas, E. L. (1980). Application of response surface analysis in the formulation of whipped topping, J. Food Sci. 45:346-348.

Mishkin, M., Karel, M. and Saguy, I. (1982). Application of optimization in food dehydration, Food Technol 36(7):101-109.

Morgan, S. L. and Deming, S. N. (1974). Simplex optimization of analytical chemical methods, Anal. Chem. 46(9):1170-1181.

Moskowitz, H. R., Stanley, D. W., and Chandler, J. W. (1977). The eclipse method: Optimizing product formulation through a consumer generated ideal sensory profile, Can. Inst. Food Sci. Technol. J. 10(3):161-168.

Mullen, K. and Ennis, D. M. (1979). Rotatable designs in product development, Food Technol. 33(7):74-80.

Murray. W. (ed.) (1972). Numerical Methods for Unconstrained Optimization Academic, New York.

Murray, W. (1976). Methods for unconstrained optimization, in Optimization in Action (L. C. W. Dixon, ed.). Academic, New York, Chap. 12.

NAG (Numerical Algorithms Group) (1977). NAG Fortran Library Manual, Mark 6. Oxford, U.K.

National Bureau of Standards (1977). Computer Science and Technology: Guide to Computer Program Directories. U.S. Department of Commerce, Washington, D.C.

Nelder, J. A. and Mead, R. (1965). A simplex method for function minimization-errata, Computer J. 0.27.

NPL (National Physical Laboratory) (1978). Introduction to the NPL Numerical Optimization Software Library, Vols. 1-2. Division of Numerical Analysis and Computing, Teddington, U.K.

NPL (National Physical Laboratory) (1980). A Brief Guide to the NPL Numerical Optimization Software Library. Division of Numerical Analysis and Computing, Teddington, U.K.

Powell, M. J. D. (1964). An efficient method for finding the minimum of a function of several variables without calculating derivatives, Computer J. 7:155-162.

Ray, W. H. and Szekeley, J. (1973). Process Optimization. Wiley, New York, p. 34.

Roberts, S. M. and Lyvers, H. I. (1961). The gradient method in process control, Ind. Eng. Chem. 53:877-882.

Roozen, J. P. and Pilnik, W. (1976). Sensory evaluation of the protein enrichment of orange juice, Lebensm. Wiss. U. Technol. 9:329.

Rosenbrock, H. H. (1960). An automatic method for finding the greatest or least value of a function, Computer J. 3:175-184.

Routh, M. W., Swartz, P. A., and Denton, M. B. (1977). Performance of the super modified simplex, Anal. Chem. 49(9):1422-1428.

Saguy, I. (1982). Utilization of the "complex method" to optimize a fermentation process, Biotech. Bioeng. XXIV:1519-1525.

Saguy, I. and Karel, M. (cochairmen) (1982). Optimization theory, tech-
    niques and their implementation in the food industry, Food Technol
    36(7):87-113 (Symp. presented at IFT Meeting Atlanta, GA June 9,
    1981).
Sargent, R. W. H. (1980). A review of optimization methods for non-
    linear problems, in Computer Applications to Chemical Engineers,
    ACS Symp. Ser. No. 124, Amer. Chem. Soc. Washington D.C.
    pp. 37-52.
Shanno, D. F. (1970). Conditioning of quasi-Newton methods for function
    minimization, Math. Comput. 24:647-656.
Smith, H. and Ross, A. (1963). Subjective responses in process investiga-
    tion, Ind. Eng. Chem. 55:25.
Sood, V. K. and Kosikowski, F. V. (1979). Ripening changes and flavor
    development in microbial enzyme treated cheddar cheese slurries,
    J. Food Sci. 44:1690-1694.
Spendley, W., Hext, G. R., and Himsworth, F. R. (1962). Sequential
    applications of simplex designs in optimization and evolutionary
    operation. Technometrics 4:441-461.
Swann, W. H. (1964). Report on the Development of a New Direct Search
    Method of Optimization. I.C.I. Ltd. Central Instruction Research
    Laboratory Research Note 64/3, London.
Umeda, T., Shinodo, A., and Ichikawa, A. (1972). Complex method for
    solving variational with state-variable inequality constraints problems,
    Ind. Eng. Chem. Process. Des. Develop. 11(1):102.
U.S. Department of Energy (1982). Compilation of Program Abstract
    (ANL-7411), National Energy Software Center, Argonne National Lab-
    oratory, Argonne, Ill.
Walsh, G. R. (1976). Methods of Optimization. Wiley, New York.
Wilde, D. J. (1964). Optimum Seeking Methods. Prentice-Hall, Engle-
    wood Cliffs, N.J.
Wilde, D. J. and Beightler, C. S. (1967). Foundations in Optimization.
    Prentice-Hall, Englewood Cliffs, N.J.
Zahradnik, R. L. (1971). Theory and Techniques of Optimization for
    Practicing Engineers. Barnes and Noble, New York.

11

Optimization of Dynamic Systems Utilizing the Maximum Principle

ISRAEL SAGUY* The Volcani Center, Agricultural Research Organization, Bet Dagan, Israel

*Present affiliation:

Massachusetts Institute of Technology, Cambridge, Massachusetts

I. INTRODUCTION

The preceding chapter (Chap. 10) dealt with analytical and numerical methods frequently utilized in searching for minima (or maxima) of single or multivariable objective functions. In all the cases discussed thus far the objective function was given explicitly, and was a function of neither time nor distance. We now focus our attention on dynamic optimization problems, which deal with optimizing a continuous controllable process. A controllable process can be made to behave in different ways depending on the quantities controlling the system and often can be described by a system of first-order differential equations.

A dynamic optimization problem may be stated as follows: Given a process with all the performance equations (a set of differential equations describing the process) and the known initial and/or final values of some of the state variables, find a piecewise continuous decision control variable(s), subject to certain constraints in such a way that the objective function is minimized (or maximized).

These types of problems are called "trajectory optimization problems" since their solution requires that a complete decision trajectory along the path should be specified. In many engineering situations, trajectory optimization is often encountered. For instance, a typical problem of chemical engineering is the time-varying behavior of a batch or continuous reactor or the start-up of a process. A typical aerospace engineering problem is the search for the minimum amount of propellant consumption required to transfer a spaceship from one orbit to another. A typical food-oriented problem is to find the optimal retort temperature trajectory in a retort process required to maximize vitamin retention in canned foods.

There are several approaches to the solution of dynamic optimization problems. Among the most frequently applied we may include (1) calculus of variations; (2) Bellman's principle of optimality, dynamic programming; and (3) Pontryagin's maximum principle.

We shall restrict our treatment to Pontryagin's approach and a discretized decision approach based on the complex method. However, the reader should consult several textbooks on the other methods (Athans and Falb, 1966; Bellman, 1957; Bellman and Dreyfus, 1962; Denn, 1969).

II. THE MAXIMUM PRINCIPLE

The continuous maximum principle (Pontryagin et al., 1962) will be introduced by utilizing a study conducted by Saguy and Karel (1979). Its aim is to determine the optimum temperature profile of a retort process for a canned food in order to maximize its nutrient retention.

Consider a continuous process (represented schematically in Fig. 1) where $\underline{x}(t)$ is the s-dimensional vector of the state variables (x_1, x_2, ..., x_s), $\theta(t)$ is the r-dimensional vector of the control variables (θ_1, θ_2, ..., θ_r); known also as the "decision variable," representing the decision at

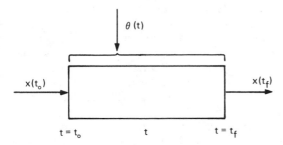

FIGURE 1 Schematic process diagram. (From Saguy and Karel, 1979. Reprinted from Journal of Food Science 44(5):1485-1490. ©1979 by Institute of Food Technologists.)

time t ($t_0 = t = t_F$, where t_0 and t_F indicate initial and final time, respectively).

The performance (rate) equations for the process have the form

$$\frac{dx_i}{dt} = f_i[x_1(t), x_2(t), \ldots, x_s(t); \theta_1(t), \theta_2(t), \ldots, \theta_r(t)] \qquad (1a)$$

$$(i = 1, 2, \ldots, s)$$

or in vector form,

$$\dot{\underline{x}} = \frac{dx}{dt} = f[\underline{x}(t), \theta(t)] \qquad (1b)$$

Thus, the system is described by s ordinary differential equations. The initial condition of the system is given by

$$\underline{x}(t_0) = \alpha \qquad (2)$$

where α is an s-dimensional vector $(\alpha_1, \alpha_2, \ldots, \alpha_s)$ representing the initial values of the state variables.

In the continuous maximum principle the objective function S (known also as a "functional") is defined as a linear combination of the final values of the state variables $\underline{x}(t_F)$. Thus,

$$S = \sum_{i=1}^{s} C_i x_i(t_F) = \underline{C} \, \underline{x}(t_F) \qquad (3)$$

where $\underline{C}$ denotes a vector of s constants.

To optimize the system, it is necessary to find a continuous control vector $\theta(t)$, possibly subjected to constraints which maximizes (or minimizes) S. The control vector function so chosen is called "optimal" and is denoted by $\theta^*(t)$.

We may add that if the objective function is an arbitrary function of the state variables at $t = t_F$. The object then is to select the $\theta^*(t)$ so as to generate the optimum value of S at t_F. In the general case the "best" value of t_F may be selected (e.g., minimum time problem), however, it is being assumed here that the value of t_F has been specified and is therefore constant and known.

The continuous maximum principle uses an s-dimensional adjoint vector $\underline{z}(t)$, $(z_1, z_2, \ldots, z_s)$, and a Hamiltonian function H, which satisfy the following relations:

$$H[\underline{x}(t), \ \underline{z}(t), \ \theta(t)] \ = \ \sum_{i=1}^{s} z_i(t) f_i[x(t), \ \theta(t)] \tag{4}$$

$$z_i(t) \ = \ - \frac{\partial H}{\partial x_i(t)} \ = \ - \sum_{j=1}^{s} z_j \frac{\partial f_j[\underline{x}(t) \ \theta(t)]}{\partial x_i} \qquad i = 1, 2, \ldots, s \tag{5}$$

The terminal values of the adjoint vector are given as

$$z_i(t_F) \ = \ C_i \qquad i = 1, 2, \ldots, s \tag{6}$$

It is worth noting that the initial conditions for the performance equations and the final conditions for the adjoint variables are given [Eqs. (2) and (6)]. This leads to a two-point boundary-value problem (TPBVP). To solve this type of problem, numerical methods are generally required.

When the time interval is fixed (i.e., t_F is specified) and the initial conditions (α) are given, there are two types of basic problems (Katz, 1960): those with free unassigned final values of the state variables and those with preassigned final values for some of the state variable. The first type is characteristic of processes such as batch fermentation (Constantinides, Spencer, and Gaden, 1970) and chemical reactors, where the second type has prerequisites on the final values of the state variables, as in distillation, drying, sterilization, and other processes.

The necessary conditions for the objective function S to be minimum (or maximum) with respect to $\theta(t)$ are

$$\frac{\partial H}{\partial \theta} \ = \ 0 \qquad \text{for unconstrained } \theta \tag{7}$$

$$H = \text{minimum (or maximum)} \qquad \text{for constrained } \theta \tag{8}$$

The optimal decision vector function $\theta^*(t)$ is determined by solving Eq. (7) for unconstrained $\theta(t)$ or by searching the boundary of the set to meet the requirement of Eq. (8).

The algorithm described above may be summarized as follows: In order for the scalar function S, given in Eq. (3), to be minimized (or maximized) for the process described by Eq. (1) with initial conditions given by Eq. (2), it is necessary that there exist a nonzero continuous vector function $\theta(t)$, such that H is a minimum (or maximum) for every t ($t_0 \leq t \leq t_F$). Furthermore, the minimum (or maximum) value of H, is constant for all t.

It is worth noting that in minimizing the functional S, the Hamiltonian H must be minimized as well.

To illustrate the maximum principle outlined above, we consider a simple example which has an analytical solution. This feature makes this example frequently used, e.g., Fan (1966) and Loncin and Merson (1979).

Example 1: Application of the maximum principle. Use the maximum principle to derive the optimal policy which minimizes the objective function S given as

$$S = \frac{1}{2} \int_{t_0=0}^{t_F} \left[x_1^2(t) + \theta^2(t) \right] dt \tag{9}$$

and the performance equation for this one-variable case (i.e., s = 1) is

$$\dot{x}_1(t) = -ax_1(t) + \theta(t) \tag{10}$$

The initial condition is given as

$$x_1(t_0 = 0) = \alpha_1 \tag{11}$$

To solve this problem we shall follow the maximum principle procedure. Hence, we introduce an additional state variable, $x_2(t)$:

$$x_2(t) = \frac{1}{2} \int_0^{t_F} \left[x_1^2(t) + \theta^2(t) \right] dt \tag{12}$$

and it follows that

$$\dot{x}_2(t) = \left(\frac{1}{2} \right) \left[x_1^2(t) + \theta^2(t_0) \right] \qquad x_2(0) = 0 \tag{13}$$

This transformation, obtained by introducing another state variable, allows us to express the problem with a new objective function S which is a linear combination of the state variables' final values [according to Eq. (3)]. Thus,

$$S = \underline{C} \ \underline{x}(t_F) = x_2(t_F) \tag{14a}$$

where

$$C_1 = 0$$
$$C_2 = 1 \tag{14b}$$

The Hamiltonian function is created [based on Eq. (4)]:

$$H\left[\underline{z}(t), \underline{x}(t), \theta(t)\right] = z_1(t)[-ax_1(t) + \theta(t)]$$
$$+ z_2(t)\left[\frac{1}{2} x_1^2(t) + \frac{1}{2} \theta^2(t)\right] \tag{15}$$

The adjoint vector is generated following the scheme of Eq. (5), and using the Hamiltonian function derived previously:

$$\dot{z}_1(t) = -\frac{\partial H}{\partial x_{1(t)}} = az_1(t) - z_2(t) x_1(t) \tag{16a}$$

$$\dot{z}_2(t) = -\frac{\partial H}{\partial x_{2(t)}} = 0 \tag{16b}$$

The final values of the adjoint variables are given [following Eq. (6)] and using Eq. (14b)] as

$$z_1(t_F) = C_1 = 0 \tag{17a}$$

$$z_2(t_F) = C_2 = 1 \tag{17b}$$

Solving the differential equation given by Eq. (16b) with the boundary condition expressed by Eq. (17b) yields

$$z_2(t) = 1 \qquad t_0 \leqslant t \leqslant t_F \tag{18}$$

Rearranging the Hamiltonian function and inserting $z_2(t) = 1$ into Eq. (15) yields

$$H = -az_1(t)x_1(t) + \frac{1}{2} x_1^2(t) + z_1(t)\theta(t) + \frac{1}{2} \theta^2(t) \tag{19}$$

According to the maximum principle, if S is to be minimized, then H must be at its minimum value. For an unconstrained problem it follows [Eq. (7)] that $\partial H / \partial \theta = 0$. Hence,

$$\frac{\partial H}{\partial \theta} = z_1(t) + \theta(t) = 0 \qquad \text{or} \qquad z_1(t) = -\theta(t) \tag{20}$$

Substituting Eqs. (20) and (18) into Eqs. (10) and (16a) yields

$$\dot{x}_1(t) = -ax_1(t) - z_1(t) \tag{21}$$

$$\dot{z}_1(t) = -x_1(t) + az_1(t) \tag{22}$$

This set of differential equations is solved simultaneously. First, $z_1(t)$ is expressed explicitly, using Eq. (21):

$$z_1(t) = -\dot{x}_1(t) - ax_1(t) \tag{23}$$

Differentiation of this last expression with respect to t yields

$$\dot{z}_1(t) = -\ddot{x}_1(t) - a\dot{x}_1(t) \tag{24}$$

By substituting Eqs. (23) and (24) into (22) and rearranging, we obtain

$$\ddot{x}_1(t) - (a^2 + 1)x_1(t) = 0 \tag{25}$$

The solution to this differential equation is

$$x_1(t) = A_1 \exp(bt) + A_2 \exp(-bt) \qquad b = \sqrt{a^2 + 1} \tag{26}$$

It follows from Eqs. (23) and (26) that

$$z_1(t) = -A_1(b + a) \exp(bt) + A_2(b - a) \exp(-bt) \tag{27}$$

The values of A_1 and A_2 are derived by applying the boundary conditions [Eqs. (11) and (17a)] to yield

$$A_1 = (b - a) \frac{\exp(-bt_F)}{D} \tag{28a}$$

$$A_2 = (b + a) \frac{\exp(bt_F)}{D} \tag{28b}$$

where

$$D\alpha_1 = (b + a) \exp(bt_F) + (b - a) \exp(-bt_F)$$

Finally, the optimal policy, $\theta*(t)$ is obtained by substitution of Eqs. (20) and (28) into Eq. (27); thus,

$$\theta*(t) = \left(\frac{1}{D}\right)\left\{\exp[-b(t_F - t)] - \exp[b(t_F - t)]\right\}$$

and the minimum objective function is

$$S = \left(\frac{\alpha_1}{2D}\right) [\exp(bt_F) - \exp(-bt_F)]$$

III. EXTENSION OF THE MAXIMUM PRINCIPLE

Before introducing numerical methods which are frequently required for deriving the optimal process policy when analytical solutions are impractical, we shall focus our attention on several cases which are encountered in real engineering situations.

A. Process with Constraints Imposed on the Final Values of the State Variables

In optimizing a dynamic process and especially food processes, some pre-assigned values imposed on the system are to be met. For instance, in the sterilization of a food product, the number of log cycle reductions of the initial spore population is defined by health requirements. Another example is the maximization of nutrient retention during a drying process. In this case, a desired degree of dryness in the product at the end of the process is required. Hence, some of the final state variables are preassigned, say $x_j(t_F)$ and $x_k(t_F)$. The objective function will read

$$g_i[\underline{x}(t_F)] \geqslant 0 \qquad i = j, k \tag{29}$$

$$S = \sum_{i=1}^{s} c_i x_i(t_F) \qquad i \neq j, k \tag{30}$$

In these cases the basic optimization formulation presented previously still holds, but with several changes:

$$z_i(t_F) = C_i \qquad i = 1, \ldots, s \qquad i \neq j, k \tag{31}$$

Equation (31) specifies that for $i = j$, k the adjoint variables' final values are no longer known. An iterative procedure is then to be applied to derive these required values.

Several methods for solving this constrained problem are known. We shall, however, describe only two of them, namely,

1. Lagrange multipliers (maximum principle).

2. The "complex method" (Box, 1965). This method helps solve the problem defined by the maximum principle; however, it is a completely different approach which utilizes a direct search method.

These methods are to be utilized with further discussion in Sec. IV.

B. Nonautonomous Systems

A system is defined to be "nonautonomous" if the performance equation depends explicitly on the time t as follows:

$$\dot{\underline{x}}(t) = f[\underline{x}(t), \theta(t), t] \tag{32}$$

Solving these types of problems requires an introduction of a new state variable $x_{s+1}(t)$ to satisfy

$$x_{s+1}(t) = t \tag{33}$$

and following the procedure for solving s + 1 state variables as outlined previously in Eqs. (1) to (8). For further details see Fan (1966) and Rozonoer (1959).

C. Process with Memory in Decisions

A process which depends not only on the decision vector $\theta(t)$ but also on the slope of decision vector $d\theta(t)/dt$ is termed (Fan, 1966) "process with memory in decision." The performance equation may be written in this case as

$$\dot{x}(t) = f[\underline{x}(t), \theta(t), \frac{d\theta(t)}{dt}] \tag{34}$$

Solving this type of problem is somewhat more difficult. The reader should consult Fan (1966) or other textbooks.

IV. NUMERICAL MINIMIZATION TECHNIQUES

As pointed out previously, when analytical solutions are impractical, numerical minimization techniques are necessary. Several different techniques have been developed for finding the minimum (or maximum) of the hamiltonian. General descriptions of a number of these minimization techniques have been compiled by many authors, e.g., Assad (1971) and Rudd (1966), and have been published in the literature. For special purposes and sophisticated methods the reader should consult several publications and reports by Miele and co-workers (1974, 1975, 1979), Gonzalez and Miele

(1979), and numerous others which may be found in the <u>Journal on Control and Optimization</u> (see, for instance, the November 1978 and January 1979 issues). Other sources are the <u>Journal of Optimization Theory and Applications</u> and the <u>IEEE Transactions on Automatic Control</u> published by IEEE Control Systems Society.

We shall illustrate three methods which have been utilized for several years in the chemical industry to optimize dynamic processes, whereas their use in the food industry has had only recent application (Mishkin, Karel, and Saguy, 1982; Saguy and Karel, 1979).

A. Optimal Temperature Profile in a Tubular Reactor

We use an example to illustrate the following three different numerical minimization techniques:

1. Guessing final values for the state vector

2. Gradient method in function space

3. Complex method

The first two techniques are based on the maximum principle, while the third is a discretized-decision direct search method.

Consider a plug-flow tubular reactor in which the chemical reactions are

$$A \xrightarrow{k_1} B \xrightarrow{k_2} C \tag{35}$$

Both reactions are first order, and the product B is to be maximized. This popular problem of deriving the optimal temperature profile which maximizes the production of B in a tubular reactor has been solved by numerous methods (Aris, 1961; Assad, 1973; Bilous and Amundson, 1956; Fan, 1966; Fine and Bankoff, 1967; Jaspan and Coull, 1971; Lee, 1964; Levine, 1966; Umeda, Shindo, and Ichikawa, 1972).

Before focusing our attention on the different numerical methods, we formulate the maximum principle for this problem. The performance equations describing the rates of change of components A and B (denoted 1 and 2, respectively) are

$$\dot{x}_1 = -k_1 x_1 \tag{36a}$$

$$\dot{x}_2 = k_1 x_1 - k_2 x_2 \tag{36b}$$

where k_1 and k_2 are the rate constants. Temperature dependence of the rate constants follows the Arrhenius equation:

$$k = k_0 \exp \frac{-Ea}{R\theta(t)} \tag{37}$$

where

k_0 = constant, min^{-1}

Ea = activation energy, cal/mol

R = ideal gas constant, 2 cal/mol - K

$\theta(t)$ = absolute temperature at any given time, K

According to the maximum principle, the objective function is

$$S - -x_2(t_F) \tag{38}$$

and

$$C_1 = 0 \qquad C_2 = -1 \tag{39}$$

The negative sign was introduced to convert a maximization problem into a minimization one. The Hamiltonian function is expressed

$$H[\underline{x}(t), \underline{z}(t), \theta(t)] = z_1 \dot{x}_1 + z_2 \dot{x}_2$$

$$- z_1(-k_1 x_1) + z_2(k_1 x_1 - k_2 x_2) \tag{40}$$

The adjoint variables are given by

$$\dot{z}_1 = \frac{-\partial H}{\partial x_1} = k_1 z_1 - k_1 z_2$$

$$\dot{z}_2 = \frac{-\partial H}{\partial x_2} = k_2 z_2 \tag{41}$$

The boundary conditions are given as follows: For the state variables,

$$x_1(0) = \alpha_1 \qquad x_2(0) = \alpha_2 \tag{42a}$$

and for the adjoint variable,

$$z_1(t_F) = C_1 = 0 \qquad z_2(t_F) = C_2 = -1 \tag{42b}$$

If we assume that no constraints are imposed on $\theta(t)$, the minimum of H occurs at the stationary point. Hence,

$$\frac{\partial H}{\partial \theta^*(t)} = 0$$

$$[z_2(t)x_1(t) - z_1(t)x_1(t)] k_{01} Ea_1 \exp \frac{-Ea_1}{R\,\theta^*(t)}$$

$$- z_2(t)x_2(t)k_{02}Ea_2 \exp \frac{-Ea_2}{R\theta^*(t)} \qquad (43)$$

and after rearrangement, $\theta^*(t)$ is derived:

$$\theta^*(t) = \frac{1}{R} (Ea_1 - Ea_2) \ln \frac{Mx_1(t)[z_2(t) - z_1(t)]}{x_2(t)z_2(t)} \qquad (44)$$

where

$$M = \frac{k_{01}Ea_1}{k_{02}Ea_2}$$

Equation (44) gives implicitly the optimal temperature at any given time, as a function of the values of the state and adjoint variables. Therefore, it follows that only when x and z are known, $\theta^*(t)$ may be derived. We would like to emphasize that the necessary condition for optimality, is generally a transcendental equation with respect to the control variable, and therefore cannot be solved explicitly. Nevertheless, one should first check this equation; otherwise, a most valuable piece of information may be wasted (see Algorithm 1, Sec. IV.B).

Generally, for this sort of problem, it is recommended that the time be transformed into a dimensionless number:

$$\bar{t} = \frac{t}{t_F} \qquad (45)$$

Also, it should be pointed out that in this example, for θ much greater than 360K, which corresponds to large rate constants, the numerical approach could become unstable. Thus, in any method which is to be applied, large perturbations of the optimal control are to be avoided. An alternative would be to solve a constrained problem [i.e., $290K \leqslant \theta(t) \leqslant 360K$], which could be handled easily by the "complex method."

Finally, before turning to the numerical algorithms, the values used in all the examples are (Lee, 1964)

$$k_{01} = 0.535\ 10^{11}\ min^{-1} \qquad\qquad k_{02} = 0.461\ 10^{18}\ min^{-1}$$

$$Ea_1 = 18,000\ cal/mol \qquad\qquad Ea_2 = 30,000\ cal/mol$$

$$x_1(0) = 0.53 \text{ mol}/1 \qquad x_2(0) = 0.43 \text{ mol}/1$$

$$t_F = 8 \text{ min}$$

B. Algorithm 1: Guessing Final Values for the State Vector

Generally this method is not recommended due to severe problems related to numerical instability projected during the integration. Nevertheless, when $\partial H/\partial \theta$ yields an implicit relation for $\theta^*(t)$, it is a reasonable approach. This procedure is also known as the "shooting-point method" and is defined by Roberts and Shipman (1972) as follows:

After guessing the missing conditions at one end of the problem, the equations are integrated straightforwardly, and the boundary conditions are checked at the other end of the integration interval. The magnitude of the error obtained is then used "in a good way" to correct the guessed boundary conditions.

The algorithm is as follows:

1. Guess the final values for $\underline{x}(t_F)$. Note that $\underline{x}(0) = \alpha$.

2. From the assumed and known values of $\underline{x}(t_F)$ and $\underline{z}(t_F)$, calculate $\theta^*(t_F)$ from Eq. (44).

3. A numerical integration scheme is carried out backward starting at $t_F(t = 1)$. For each new value of $\underline{x}(t)$ and $\underline{z}(t)$ a new value of $\theta^*(t)$ is derived until $t = 0$.

   The result of this integration is the optimal solution for those values of $\underline{x}(t_0)$ derived by the backward integration. Repeating this process many times with different $\underline{x}(t_F)$ yields a range of optimal solutions.

4. New guess for $\underline{x}(t_F)$ is made then the integration procedure is repeated until the initial conditions of $\underline{x}(t_0)$ are satisfied, e.g., Eq. (42a), or the results are within the desired accuracy. The latter is expressed by the tolerance. Various methods of improving the approximation of the initial guess of $\underline{x}(t_F)$ are available. The reader should consult Fan (1966), Roberts and Shipman (1972), and Wahrman and Dayan (1978) for some of these methods.

Obviously the initial guess influences the number of iterations necessary for convergence. When the initial guess is closer to the optimum, fewer iterations are required.

The algorithm described previously was programmed on an IBM 370/168 computer at the MIT Information Processing Center. Some of the results are given in Table 1 [each unit of dimensionless time corresponds to 0.8 min; the latter follows Eq. (45)].

TABLE 1 Typical Computational Results Obtained by Utilizing Algorithm 1 for the Problem A → B → C

No.	$\bar{t}$	$\bar{x}_1(t)$ (mol/liter)	$\bar{x}_2(t)$ (mol/liter)	$\bar{z}_1(t)$	$\bar{z}_2(t)$	$\theta*(t)$ (K)	H
ITERATION NO.:	1	INITIAL GUESS FOR X1 AND X2 :		0.1700	0.6637		
1	1.0000E 00	1.7000E-01	6.6370E-01	0.0000E-01	-1.0000E 00	3.3628E 02	-8.6656E-03
2	9.0000E-01	1.8840E-01	6.5578E-01	-9.6894E-02	-9.8240E-01	3.3649E 02	-8.6656E-03
3	8.0000E-01	2.0919E-01	6.4565E-01	-1.8425E-01	-9.6827E-01	3.3673E 02	-8.6654E-03
4	7.0000E-01	2.3277E-01	6.3289E-01	-2.6284E-01	-9.5201E-01	3.3703E 02	-8.6654E-03
5	6.0000E-01	2.5968E-01	6.1700E-01	-3.3338E-01	-9.3537E-01	3.3728E 02	-8.6652E-03
6	5.0000E-01	2.9061E-01	5.9729E-01	-3.9652E-01	-9.1826E-01	3.3776E 02	-8.6655E-03
7	4.0000E-01	3.2647E-01	5.7287E-01	-4.5284E-01	-9.0049E-01	3.3824E 02	-8.6654E-03
8	3.0000E-01	3.6854E-01	5.4247E-01	-5.0285E-01	-8.8186E-01	3.3880E 02	-8.6656E-03
9	2.0000E-01	4.1867E-01	5.0425E-01	-5.4703E-01	-8.6201E-01	3.3949E 02	-8.6655E-03
10	1.0000E-01	4.7981E-01	4.5533E-01	-5.8576E-01	-8.4037E-01	3.4048E 02	-8.6656E-03
11	-2.3842E-07	5.5712E-01	3.9058E-01	-6.1936E-01	-8.1583E-01	3.4190E 02	-8.6656E-03
ITERATION NO.:	5	INITIAL GUESS FOR X1 AND X2 :		0.1701	0.6800		
1	1.0000E 00	1.7013E-01	6.8004E-01	0.0000E-01	-1.0000E 00	3.3584E 02	-8.3710E-03
2	9.0000E-01	1.8787E-01	6.7242E-01	-9.3696E-02	-9.8515E-01	3.3600E 02	-8.3710E-03
3	8.0000E-01	2.0780E-01	6.6276E-01	-1.7848E-01	-9.7011E-01	3.3628E 02	-8.3711E-03
4	7.0000E-01	2.3030E-01	6.5070E-01	-2.5505E-01	-9.5482E-01	3.3649E 02	-8.3710E-03
5	6.0000E-01	2.5582E-01	6.3578E-01	-3.2407E-01	-9.3920E-01	3.3685E 02	-8.3709E-03
6	5.0000E-01	2.8495E-01	6.1744E-01	-3.8613E-01	-9.2318F-01	3.3717E 02	-8.3709E-03
7	4.0000E-01	3.1846E-01	5.9491E-01	-4.4175E-01	-9.0662E-01	3.3761E 02	-8.3711E-03
8	3.0000E-01	3.5738E-01	5.6717E-01	-4.9142E-01	-8.8934E-01	3.3810E 02	-8.3711E-03
9	2.0000E-01	4.0319E-01	5.3278E-01	-5.3557E-01	-8.7109E-01	3.3872E 02	-8.3710E-03
10	1.0000E-01	4.5808E-01	4.8955E-01	-5.7458E-01	-8.5146E-01	3.3957E 02	-8.3710E-03
11	-2.3842E-07	5.2567E-01	4.3391E-01	-6.0876E-01	-8.2974E-01	3.4074E 02	-8.3709E-03

Results presented in Table 1 show that the approximation process was
repeated until the computed values of $x_1(0)$ and $x_2(0)$ were approximately
equal to α_1 and α_2, respectively. To meet this requirement a tolerance of
0.01 was applied for both $x_1(0)$ and $x_2(0)$.

The program utilized a Runge-Kutta integration method [subroutine
DVERK (IMSL, 1977) and the CPU time for these five iterations was 1.6
sec.

C. Algorithm 2: Gradient Method in Function Space

This algorithm was first described by Storey and Rosenbrock (1964), and it
is based on the steepest descent method. As we saw in the previous chapter
the steepest descent technique involves calculating the gradient direction
$\partial H/\partial\theta(t)$ in which the Hamiltonian decreases most rapidly with respect to
$\theta(t)$, then adjusting the decision variable by a successive improvement
scheme.

The algorithm is as follows:

1. Guess an initial trajectory of the decision variable $\theta(t)$. Let $\theta^{(n)}(t)$
denote the nth guess for the decision variable at any given time, $t(t_0 \leqslant t \leqslant
t_F$, or $t_0/t_F \leqslant t \leqslant 1)$. To guess this trajectory, some approximations are
required. Obviously, we cannot choose an infinite number of points along
the independent variable, and make the calculations at every point, as this
would be an enormous task. Therefore, the independent variable is discretized into m intervals. The values of $\theta(t)$ as a function of time are assumed to be composed of m straight-line segments connecting the points.
The number of intervals, m, that the independent variable is to be discretized has to be selected carefully. Too large a number would result in
a time-consuming calculation, whereas too small a number would cause
severe numerical instability and decrease the accuracy obtained. A very
rough estimation of m would be 10 to 20 segments.

2. Integrate numerically the state equations $\underline{x}^{(n)}(t)$ forward in time
from $t = t_0$ to $t = t_F$.

3. Using the values of $\underline{x}^{(n)}(t)$, integrate numerically backward in time
(from $t = t_F$ to $t = t_0$) the adjoint function $\underline{z}^{(n)}(t)$. This backward integration
was found to be necessary to avoid numerical instability encountered if the
integration started at $t = 0$. Also note that unless the final values of z are
utilized, integration from $t = t_0$ to $t = t_F$ results in a trial-and-error calculation until the derived values of $z(t_F)$ coincide with the given values.
Based on experience, this bidirectional integration method is very useful
and is recommended for improved stability.

It is worth pointing out, that for most cases, a fourth-order Runge-
Kutta integration method could be used, while for high accuracy, Krogh's
method (Krogh, 1973) seems a better choice. If, however, the system of
differential equations is stiff (i.e., a system with rapidly decaying transient solutions), special methods are required. To deal with stiff systems,

the Gear's method (Gear, 1971) may be utilized. These methods are normally provided as subroutines in most computer libraries.

    4. Calculate the Hamiltonian $H^{(n)}$.

    5. Compute the value of the analytical derivation of H with respect
to θ. If differentiation of the Hamiltonian with respect to the decision variable is not simple, one can apply numerical differentiation techniques,
noting that

$$\frac{\partial H}{\partial \theta} = \frac{\partial H}{\partial t} \frac{\partial t}{\partial \theta} \tag{46}$$

where $\partial H/\partial t$ and $\partial t/\partial \theta$ are evaluated numerically [for instance, subroutine
DTE 4, IMSL (1977)]. It should be emphasized however, that analytical dif-
ferentiation should be used where possible, to alleviate inaccuracies generated by the numerical methods.

    6. Determine the (n + 1)st guess for the decision variable trajectory
using the following improvement scheme:

$$\theta^{(n+1)}(t) = \theta^{(n)}(t) - e \frac{\partial H^{(n)}}{\partial \theta(t)} \tag{47}$$

where ϵ is a small positive number. The value of ϵ is to be determined for
each system. ϵ should be small enough so that no instability will result,
yet large enough so that convergence will not be too slow. Appropriate
values for ϵ were found as 0.01 to 0.2 for this system.

    This steepest descent method converges very slowly toward the opti-
mum, and the converged objective function value may differ from the "real
one" by a small amount. The speed of convergence may be increased by
adjusting ϵ during the iterative descent. Also, as we pointed out previously,
the choice of the initial trajectory influences the number of iterations, as
can be expected.

    Table 2 shows a typical output obtained from utilizing the algorithm
outlined above. In this particular case, the dimensionless time was discretized into 20 sections (i.e., each one corresponds to t = 0.05 or t = 0.4
min). The optimal temperature profile obtained agreed very well with the
results derived using algorithm 1.

    The method of evaluating the derivative of the Hamiltonian with numeri-
cal scheme [Eq. (46)] was compared to the analytical approach [Eq. (43)].
The values obtained by these two approaches were in good agreement
(Graves, 1979). This indicates that for complicated cases where the analytical derivative is complicated the numerical approach may be utilized.

D. Algorithm 3: The Complex Method

As we have pointed out, most food processes lead to constrained problems.
The constraints imposed on the system include (among others) health

TABLE 2 Typical Computational Results Obtained by Utilizing Algorithm for the Problem A $\longrightarrow$ B $\longrightarrow$ C

No.	$\bar{t}$	$\theta^*(\bar{t})$ (K)	$z_1(\bar{t})$	$z_2(\bar{t})$	$\bar{x}_1(\bar{t})$ (mol/liter)	$\bar{x}_2(\bar{t})$ (mol/liter)	H	$\partial H/\partial\theta$
1	0.00	340.423	6.0369E-01	8.3565E-01	5.3000E-01	4.3000E-01	9.5876E-03	1.1787E-04
2	0.05	339.460	5.8699E-01	8.4630E-01	4.9516E-01	4.5921E-01	9.4653E-03	1.3767E-04
3	0.10	339.314	5.6921E-01	8.5633E-01	4.6395E-01	4.8486E-01	9.4474E-03	1.0735E-04
4	0.15	338.403	5.5040E-01	8.6580E-01	4.3585E-01	5.0750E-01	9.3362E-03	1.3755E-04
5	0.20	338.296	5.3061E-01	8.7476E-01	4.1045E-01	5.2757E-01	9.3224E-03	1.1530E-04
6	0.25	338.195	5.0923E-01	8.8368E-01	3.8673E-01	5.4584E-01	9.3119E-03	9.5146E-05
7	0.30	338.099	4.8618E-01	8.9257E-01	3.6455E-01	5.6247E-01	9.3037E-03	7.6729E-05
8	0.35	338.010	4.6137E-01	9.0145E-01	3.4379E-01	5.7759E-01	9.2974E-03	5.9896E-05
9	0.40	337.184	4.3551E-01	9.0989E-01	3.2488E-01	5.9106E-01	9.2295E-03	1.0450E-04
10	0.45	337.116	4.0867E-01	9.1792E-01	3.0761E-01	6.0307E-01	9.2229E-03	9.1378E-05
11	0.50	337.052	3.8002E-01	9.2595E-01	2.9135E-01	6.1404E-01	9.2174E-03	7.9202E-05
12	0.55	336.991	3.4948E-01	9.3398E-01	2.7602E-01	6.2402E-01	9.2128E-03	6.7875E-05
13	0.60	336.185	3.1793E-01	9.4163E-01	2.6197E-01	6.3293E-01	9.1385E-03	1.1610E-04
14	0.65	336.139	2.8545E-01	9.4892E-01	2.4908E-01	6.4092E-01	9.1332E-03	1.0703E-04
15	0.70	336.094	2.5104E-01	9.5622E-01	2.3586E-01	6.4820E-01	9.1285E-03	9.8517E-05
16	0.75	336.052	2.1461E-01	9.6353E-01	2.2528E-01	6.5482E-01	9.1247E-03	9.0550E-05
17	0.80	336.012	1.7605E-01	9.7086E-01	2.1430E-01	6.6081E-01	9.1211E-03	8.3023E-05
18	0.85	335.973	1.3528E-01	9.7820E-01	2.0389E-01	6.6622E-01	9.1182E-03	7.5973E-05
19	0.90	335.937	9.2180E-02	9.8557E-01	2.9401E-01	6.7108E-01	9.1155E-03	6.9280E-05
20	0.95	335.902	4.6646E-02	9.9295E-01	1.8464E-01	6.7543E-01	9.1132E-03	6.3010E-05
21	1.00	335.125	0.0000E-01	1.0000E 00	1.7600E-01	6.7928E-01	9.0435E-03	1.1554E-04

requirements, biological tolerance (e.g., temperature, oxygen), and quality demands.

Once the system is constrained, gradient methods require the introduction of penalty functions to transform the constrained problem into an unconstrained one. To illustrate this approach let us consider maximizing nutrient retention during an air-drying process, and assuming that the only state variables are x_1, the nutrient retentions, and x_2, the moisture content. From quality requirements the final moisture content of the product should be below 5%. Hence the objective function may be formulated to transform this constrained problem into an unconstrained one by using the Lagrange multiplier method:

$$S = -x_1(t_F) + \lambda[x_2(t_F) - \beta]^2$$

where

λ = Lagrange multiplier

β = desired dryness (e.g., 5%)

One method recommended to solve this type of problem has been outlined in Chap. 10. Basically it requires solving the optimization problem several times with increasing values of λ, until the constraint is satisfied. This method, although simple, requires a successive guessing of λ which results in an inefficient algorithm.

With this type of problem, a much easier and straightforward approach is the utilization of the "complex method" of Box (1965). The method is quite simple and offers easy handling of various inequality constraints and a nonlinear objective function (more details are given in Chap. 10).

Utilizing the complex method for the optimization of a dynamic system requires that the decision variable be divided into m discrete intervals over which it is constant (or a linear interpolation may be applied). This discretized decision is varied via the complex method of minimization in order to minimize the objective function. A more complete and elaborate discussion of this method was outlined by Umeda, Shindo, and Ichikawa (1972) and Mishkin, Karel, and Saguy (1982).

All that is needed for utilizing this method for dynamic systems is the complex-method minimization routine, an integration routine, and a subroutine to evaluate the state equations. One must also specify any constraints on the state variables. In order to obtain the desired degree of dryness in our example of the drying process, one would specify the value required as an upper constraint which would, of course, result in an active constraint.

Figures 2 and 3 depict the optimal temperature profiles derived by the complex method ($290K \leqslant \theta(t) \leqslant 360K$) and the Pontryagin maximum principle (algorithm 1) for the tubular reactor problem. In these figures the normalized time, $\bar{t} = t/t_F$ was discretized into 5 and 10 sections, respectively.

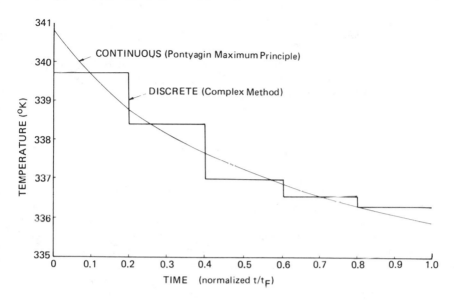

FIGURE 2   Optimal temperature profile obtained by the Pontryagin maximum principle (algorithm 1) and the complex method with five intervals. (Adopted from Mishkin, 1980.)

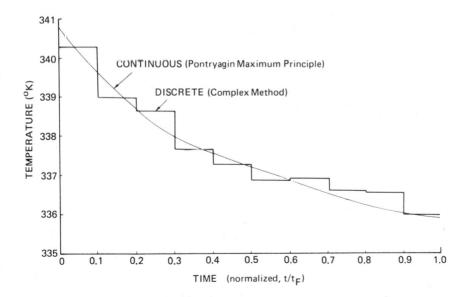

FIGURE 3   Optimal temperature profile obtained by the Pontryagin maximum principle (algorithm 1) and the complex method with 10 intervals. (Adopted from Mishkin, 1980.)

It is obvious that increasing the numbers of intervals improves the accuracy of the derived results.

The work by Mishkin, Karel, and Saguy (1982) indicates that in terms of practicality the complex method yields reliable results with considerably less mathematical complication than Pontryagin's maximum principle. Further discussion and comparison between these two methods is given in Sec. V.

V. OPTIMIZATION OF FOOD PROCESSES
UTILIZING THE MAXIMUM PRINCIPLE

The algorithms described previously were applied successfully in optimizing several food processes. These attempts include the optimization of thiamin retention during sterilization of a conduction-heating canned food (Saguy and Karel, 1979) using algorithm 2 (entitled "gradient method in function space"). Also, the latter algorithm and the complex method (algorithm 3) were applied to an air-drying process by Mishkin, Karel, and Saguy (1982). A similar approach was applied in the fermentation industry, for instance, Constantinides, Spencer, and Gaden (1970) and Rai and Constantinides (1973), used the maximum principle to determine the optimal temperature and pH profiles in batch fermenters. Also worth noting is Meo and Friedly's (1973) research, in which they applied variational analysis to derive a set of necessary conditions, yielding optimal control of a radiant-heat freeze-drying process, with the overall objective of minimizing the total drying time.

As full details on the application of the maximum principle to optimize the retort process have been described elsewhere (Saguy and Karel, 1979), we shall now focus our attention on some of the results derived from this research.

Figure 4 shows the optimum retort temperature $(\theta *)$ profile which maximizes thiamin retention in an A/2 can of pork puree. The resultant can-average (θ_m) and mid-center (θ_c) temperatures are also shown. Thiamin retention and microorganism destruction corresponding to the above values are given in Fig. 5.

The optimal retort temperature profile determined by this procedure improved thiamin retention, as compared with other methods, and showed that a single solution for the temperature profile exists (according to the maximum principle).

Since the research undertaken at the time was to investigate the application of modern optimization techniques, it was beyond the scope of the research to test the industrial applicability of the optimal temperature profile derived. However, the results show that retention of thiamin in the sterilization process used was limited by heat conduction factors and the dimensions of the cans used.

The gradient method in function space (algorithm 2), in which the control policy is corrected after each iteration using the maximum principle, is recommended. It is found to be fast and reliable when a good initial

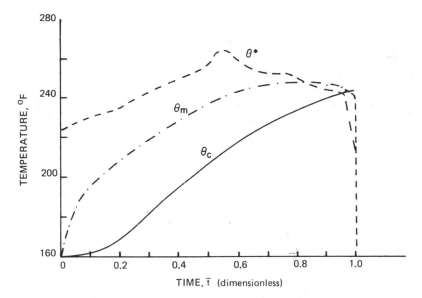

FIGURE 4 Optimal retort temperature (θ^*), mass average temperature (θ_m), and central point temperature (θ_c) during the sterilization process (A/2 can). (From Saguy and Karel, 1979. Reprinted from Journal of Food Science 44:1485-1490. © 1979 by Institute of Food Technologists.)

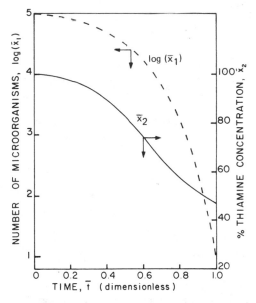

FIGURE 5 Optimal profiles of thiamin retention and microorganisms survival during the sterilization process (A/2 can). (From Saguy and Karel, 1979. Reprinted from Journal of Food Science 44(5):1485-1490. © 1979 by Institute of Food Technologists.)

guess for the profile has been made. However, when the initial estimate of the profile is far from the optimum, some difficulties are encountered, such as slow convergence and instability during the numerical integration. The instability is caused by the high sensitivity of the destruction rate (stiff equation) for spores at high temperatures and by high values of ϵ [see Eq. (47)]. The latter source of instability may be overcome either by decreasing the value chosen for ϵ or by making ϵ a function of time.

Less favorable results were obtained by applying algorithm 1, in which the initial values of the adjoint variables are guessed and the integration is performed only in a forward direction. This procedure resulted in severe instability, caused by the stiff nature of the system. Therefore, algorithm 1 was precluded and not utilized.

Research on the feasibility of applying the maximum principle algorithm 2) and the complex method to derive the optimal temperature profile during air drying was carried out by Mishkin, Karel, and Saguy (1982). Their main conclusions were as follows:

1. Both the complex method and the maximum principle could be applied to establish the optimal temperature profile.

2. These two methods differ significantly in ease of application. The complex method can be applied to any process which can be simulated, with only minor modification of the simulation routine. In contrast, implementation of the maximum principle often represents a formidable task, both in the mathematical analysis and numerical solution using the gradient method. This distinct contrast in tractability should greatly boost future use of the discretized-decision complex method in food processing analysis.

3. The resolution of the complex method may be increased by increasing the number of discrete decision points, at the expense of substantially increasing in computational time. This improved resolution may be unnecessary in most food engineering applications.

4. When no constraints are imposed on the final conditions, the system of equations resulting from applications of the maximum principle can be solved using the gradient method quite rapidly. But, for the fixed right-end problem, e.g., specifying the final values of the state variable (e.g., moisture content desired in the drying process), computational time is greatly expanded due to the iterative search for the Lagrange multiplier required to satisfy the constraint. Thus, the major advantage of applying the maximum principle, i.e., ability to apply rapid gradient methods to maximize the Hamiltonian (and therefore the objective function), is overshadowed by the difficulty in satisfying the constraint, resulting in increased computer run time.

Therefore, it is our recommendation that the complex method with discretized decision should prove to be an exyremely useful tool for improving food processes which can be approximated by mathematical modeling techniques. Also it may be applicable to many other engineering applications. Hence, we are including the computer program in the Appendix.

Optimization can be undertaken with respect to several objectives, for example, maximizing quality retention and/or minimizing energy expenditures ultimately yielding maximized profit. The latter is one of the most important reasons for justifying its use.

VI. APPENDIX: OPTIM

These notes describe a computer program OPTIM for solving general nonlinear optimization problems by use of the complex method.*

A. Formulation of the Problem

In order to use OPTIM, the user must first cast his or her problem into the following format:

Select a set of n <u>independent decision variables</u>

$$x = (x_1, \ldots, x_n) \tag{48}$$

Define an <u>objective function</u>

$$f = f(x_1, \ldots, x_n) \tag{49}$$

which is a function of the decision variables.

Define m additional <u>dependent variables</u>

$$y_1 = y_1(x_1, \ldots, x_n)$$
$$\ldots\ldots\ldots\ldots\ldots\ldots\ldots \tag{50}$$
$$y_m = y_m(x_1, \ldots, x_n)$$

which are also a function of the decision variables.

Establish <u>upper</u> and <u>lower bounds</u> on the independent and dependent variables of the form

$$(XL)_i \leq x_i \leq (XU)_i \qquad i = i, \ldots, n \tag{51}$$

$$(YL)_j \leq y_j \leq (YU)_j \qquad j = 1, \ldots, m \tag{52}$$

---

*Developed by Professor Lawrence B. Evans, Department of Chemical Engineering, Massachusetts Institute of Technology, Cambridge, Mass.

The optimization problem, then, is to find the values of the independent decision variables $(x_1, \ldots, x_n)$ which minimizes the value of the objective function [Eq. (49)] while satisfying all of the explicit constraints [Eq. (51)] and the implicit constraints [Eq. (52)].

B. Example Problem

To clarify some of the concepts, consider the following simple optimization problem: Minimize

$$f = (x_1 - a)^2 + (x_2 - b)^2 + 3 \tag{53}$$

with

$$0 \leqslant x_1 \leqslant 10 \tag{54}$$

$$0 \leqslant x_2 \leqslant 10 \tag{55}$$

$$0 \leqslant (x_1)^2 + (x_2)^2 \leqslant 16 \tag{56}$$

and

$$a = 5 \quad b = 5 \tag{57}$$

The constraint of Eq. (56) would be incorporated by defining a dependent variable

$$y_1 = x_1^2 + x_2^2$$

and requiring

$$0 \leqslant y_1 \leqslant 16$$

The objective function and constraints are shown in Fig. 6. In the absence of constraints, the minimum value of f would obviously occur at $x_1 = a$, $x_2 = b$. Because of the constraint, however, the minimum lies on the constraint at $x_1 = \sqrt{8}$, $x_2 = \sqrt{8}$.

C. Use of OPTIM

The user of OPTIM must perform two tasks:

1. Write a simple FORTRAN subroutine, called MODEL, to evaluate the objective function and dependent variables.

2. Prepare a set of data cards providing pertinent data for each optimization run.

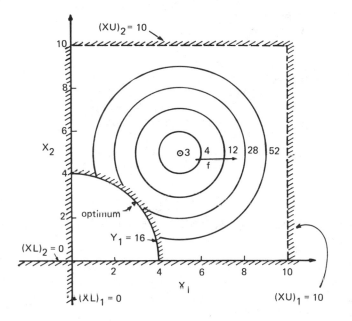

FIGURE 6   Diagram showing values of objective function and constraints for the example problem: $f = (x_1 - 5)^2 + (x_2 - 5)^2 + 3$; $y_1 = x_1^2 + x_2^2$.

1. Subroutine MODEL

The structure of the user-supplied subroutine MODEL is indicated in Fig. 7. The subroutine communicates with the OPTIM program through the variables in COMMON defined as follows:

$$F = \text{objective function}$$

$X(1), \ldots, X(NX)$ = an array containing values of the independent decision variables

$Y(1), \ldots, Y(NY)$ = an array containing values of the dependent variables

$P(1), \ldots, P(NP)$ = an array containing values of the parameters in the model which may be varied from one optimization run to the next

In the above definitions, NX denotes the number of independent decision variables, NY denotes the number of dependent variables, and NP denotes the number of parameters.

SUBROUTINE MODEL
COMMON F, X(5), Y(10), P(20)

F = . . .   Sequence of statements to evaluate
Y(1) = . . .   the objective function F and the
Y(2) = . . .   dependent variables Y(1). . .Y(NY)
                 as functions of current values of
                 the independent decision variables
etc.   X(1) . . . X(NX) and parameters
                 P(1) . . . P(NP)
RETURN
END

FIGURE 7   Structure of the user-supplied MODEL subroutine, exclusive
of system control cards.

The MODEL subroutine must simply evaluate the objective function and
dependent variables as functions of the current values of the independent decision variables and the problem parameters.

The COMMON statement in the subroutine must appear exactly as is
indicated. OPTIM is dimensioned to allow a maximum of five decision variables, 10 dependent variables, and 20 parameters. These limitations
coulb be removed readily by making minor changes in OPTIM.

2. Data Cards

In addition to providing the subroutine MODEL, the user must also supply
a set of data cards for each optimization run. The set of data cards is organized as follows (the FORTRAN formal specification for each type of data
card is indicated in parentheses):

1. A single title card containing the problem title expressed as 72
   alphanumeric characters (in the format 72 A1).

2. A single control card containing the following five integer parameters (in the format 5 I5):

   NX = number of independent decision variables

   NY = number of dependent variables

   NP = number of parameters

   MAXIT = maximum allowable number of iterations to be performed
           (a nominal value of 200 is typical)

NFREQ = iteration frequenty at which intermediate printing of the current simplex is to be performed to monitor progress toward solution (a nominal value of either 0 or 20 is typical)

3. A set of <u>independent variable cards</u> containing one data card for each independent decision variable. On each card, the following information will be provided (in the format 5 A1, 5 X, 3 F10.5):

NAMEX = name of variable, expressed as five alphanumeric characters

XL(I) = lower bound on variable (real)

XU(l) = upper bound on variable (real)

X(I) = initial value of the variable corresponding to a feasible point (real)

4. A set of <u>dependent-variable cards</u> containing one data card for each dependent variable (if any). On each card the following information is to be provided (in the format 5 A1, 5 X, 2 F10.5):

NAMEY = name of variable, expressed as five alphanumeric characters

YL(I) = lower bound on variable (real)

YU(I) = upper bound on variable (real)

5. A set of <u>parameter cards</u> containing one card for each parameter (if any). On each data card the following information is provided (in the format 5 A1, 5 X, F 10.5):

NAMEP = name of parameter, expressed as five alphanumeric characters

P(I) = value of parameter (real)

6. A single <u>objective-function card</u> containing the following information (in the format 5 A1, 5 X, 2 F10.5):

NAMEF = name of objective function, expressed as five alphanumeric characters

RDEV = allowable absolute deviation in objective function value to be used in convergence test (a nominal value of 0.001 is typical)

ADEV = allowable absolute deviation in objective function to be used
in convergence test (a nominal value of 0.01 is typical)

A coding form for case in preparing input data cards is shown in Fig. 8.
A series of optimization runs may be carried out by providing additional sets of data cards. Each new set must contain values for all of the data

FIGURE 8   Input-data coding form for OPTIM.

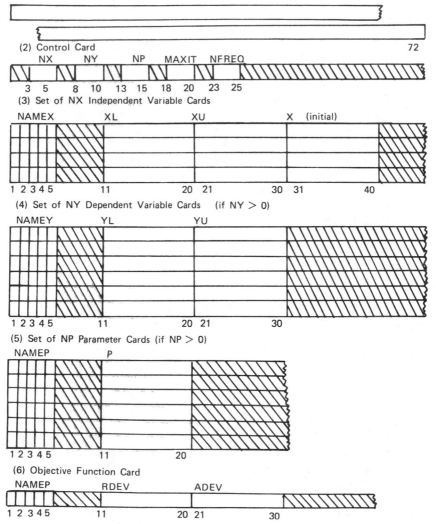

items, even if the values have not changed from the preceding run. The present version of OPTIM treats each set of data cards as defining a brand new problem.

D. Computer Solution of the Example Problem

The example problem described earlier (Sec. VI.B) was solved using OPTIM. The complete deck setup for submission to the IBM 1130 is shown in Fig. 9. It includes all of the necessary system control cards, the listing of the FORTRAN subroutine MODEL, and a set of data cards for two optimization runs. The results printed by OPTIM for the first run are shown in Fig. 10.

E. The OPTIM Program

The main program OPTIM is listed in Fig. 11. The main program calls the following subroutines:

OPTRD reads input data for OPTIM (Fig. 12).

OPTPR prints results for OPTIM (Fig. 13).

FIGURE 9   Sample deck setup for solving an example optimization problem using OPTIM on the IBM 1130.

```
// JOB  T                                             ⎫ system
// FOR                                                ⎬ control
*LIST SOURCE PROGRAM                                  ⎭ cards
*ONE WORD INTEGERS
      SUBROUTINE MODEL                                ⎫
      COMMON F,X(5),Y(10),P(20)                       │
      A = P(1)                                        │
      B = P(2)                                        ⎬ Fortran
      F = (X(1)-A)**2 + (X(2)-B)**2 + 3.0             │ source
      Y(1) = X(1)**2 + X(2)**2                        │ program
      RETURN                                          │
      END                                             ⎭
// DUP
*STORE      WS  UA   MODEL                            ⎫ system
// XEQ OPTIM                                          ⎬ control
FIRST OPTIMIZATION RUN--MINIMIZATION OF SIMPLE FUNCTION ⎭ cards
    2     1    2   200    10
   X1     0.0          10.0       2.0
   X2     0.0          10.0       2.0
   Y1     0.0          16.0
   A      5.0
   B      5.0                                         ⎫ OPTIM
   F      0.001         0.01                          ⎬ data
SECOND OPTIMIZATION RUN--MINIMIZATION OF SIMPLE FUNCTION ⎭ cards
    2     1    2   200     0
   X1     0.0          10.0       2.0
   X2     0.0          10.0       2.0
   Y1     0.0          16.0
   A      2.0
   B      1.0
   F      0.001         0.01
// *ENDJOB                              system control card
```

```
FIRST OPTIMIZATION RUN--MINIMIZATION OF A SIMPLE FUNCTION
INDEPENDENT VARIABLES

NAME      LOWER BOUND      UPPER BOUND          VALUE

  X1      0.00000E 00      0.10000E 02      0.20000E 01
  X2      0.00000E 00      0.10000E 02      0.20000E 01
DEPENDENT VARIABLES

NAME      LOWER BOUND      UPPER BOUND          VALUE

  Y1      0.00000E 00      0.16000E 02      0.80000E 01
PARAMETERS

NAME           VALUE

  A       0.50000E 01
  B       0.50000E 01
OBJECTIVE FUNCTION

NAME           VALUE          REL DEV          ABS DEV

  F       0.21000E 02      0.10000E-02      0.10000E-01
ITERATION   0

VARIABLES IN SIMPLEX (CENTROID IS VERTEX   5)

VERTEX              1                2                3                4                5
  X1      0.20000E 01      0.27439E 01      0.30424E 01      0.33694E 01      0.27889E 01
  X2      0.20000E 01      0.20321E 01      0.21332E 01      0.18576E 01      0.20057E 01
  Y1      0.80000E 01      0.11659E 02      0.13807E 02      0.14804E 02
  F       0.21000E 02      0.16898E 02      0.15051E 02      0.15534E 02
ITERATION   10  ENTERING VERTEX   2  FDEV=   0.3579E 00  FMIN=   0.1254E 02

VARIABLES IN SIMPLEX (CENTROID IS VERTEX   5)

VERTEX              1                2                3                4                5
  X1      0.29159E 01      0.28609E 01      0.29084E 01      0.30186E 01      0.29476E 01
  X2      0.27214E 01      0.27942E 01      0.26986E 01      0.26243E 01      0.26814E 01
  Y1      0.15908E 02      0.15993E 02      0.15741E 02      0.15998E 02
  F       0.12536E 02      0.12441E 02      0.12671E 02      0.12570E 02


          FIRST OPTIMIZATION RUN--MINIMIZATION OF A SIMPLE FUNCTION
          PROCEDURE HAS CONVERGED IN   15 ITERATIONS
          THE SOLUTION IS AS FOLLOWS
          INDEPENDENT VARIABLES

          NAME      LOWER BOUND      UPPER BOUND          VALUE

            X1      0.00000E 00      0.10000E 02      0.27998E 01
            X2      0.00000E 00      0.10000E 02      0.28562E 01
          DEPENDENT VARIABLES

          NAME      LOWER BOUND      UPPER BOUND          VALUE

            Y1      0.00000E 00      0.16000E 02      0.15997E 02
          PARAMETERS

          NAME           VALUE

            A       0.50000E 01
            B       0.50000E 01
          OBJECTIVE FUNCTION

          NAME           VALUE          REL DEV          ABS DEV

            F       0.12437E 02      0.10000E-02      0.10000E-01
```

FIGURE 10   Computer output from OPTIM generated in solving example problem with first set of data.

```
*NAME OPTIM
C      COPYRIGHT (C) 1975 BY LAWRENCE B. EVANS
       COMMON F,X(5),Y(10),P(20)
       COMMON NX,NY,NP,XL(5),XU(5),YL(10),YU(10)
       COMMON XC(5),XX(5,10),YY(10,10),FF(10),JG,NIT
       COMMON NTITL(72),NAMEF(5),NAMEX(5,5),NAMEY(10,5),NAMEP(20,5)
       COMMON ALPHA,BETA,KMAX,MAXIT,FR,FA,FDEV,FMIN,NFREQ
C      ...PRESET PARAMETERS OF OPTIM...
       ALPHA = 1.3
       BETA = 0.5
       GAMMA = 0.10
       NMAX1=50
       IX = 1357
       MAXIT = 100
C      ...READ BASIC DATA FOR OPTIMIZATION RUN...
99     CALL OPTRD
       CALL OPTPR(1)
       DO 100 I = 1,NX
100    XC(I) = X(I)
       CALL IMTST(N1,1,IFLAG)
       CALL OPTPR(2)
       IF (IFLAG) 500,501,500
501    CONTINUE
C      ...ESTABLISH INITIAL SIMPLEX...
       NIT = 0
       KMAX = 2*NX
       K = 1
104    FF(K) = F
       DO 102 I = 1,NX
102    XX(I,K) = X(I)
       IF (NY) 120,120,121
121    CONTINUE
       DO 105 I = 1,NY
105    YY(I,K) = Y(I)
120    CONTINUE
       DO 103 I = 1,NX
103    XC(I) = (XC(I)*(K-1)+X(I))/K
       IF (K-KMAX) 110,300,300
110    K = K+1
       DO 106 I = 1,NX
       CALL RANDU(IX,IY,YFL)
       IX = IY
106    X(I) = XL(I) + YFL*(XU(I)- XL(I))
       CALL IMTST(N1,NMAX1,IFLAG)
       IF (IFLAG) 502,503,502
503    CONTINUE
       GO TO 104
```

FIGURE 11 Listing of OPTIM (the main-line program).

IMTST tests for violation of implicit constraints in OPTIM (Fig. 14).

FAIL prints error messages for OPTIM (Fig. 15).

Another subroutine titled MODEL which is a user-supplied routine was described previously (Figs. 7 and 9).

```
C       ...BEGIN ITERATIVE SEARCH FOR OPTIMUM...
300     CONTINUE
C       ...ESTABLISH COUNTER FOR INTERMEDIATE PRINTING...
        IF (NFREQ) 520,520,508
520     IPRT = MAXIT + 1
        GO TO 509
508     IPRT = NFREQ
        WRITE (3,1003)
        CALL OPTPR(3)
509     CONTINUE
C       ...FIND POINTS OF SIMPLEX WITH HIGHEST AND LOWEST FUNCTION VALUES.
317     NIT = NIT + 1
        FMAX = -1.0E10
        FMIN = +1.0E10
        JG = 0
        JL = 0
        DO 323 J = 1,KMAX
        IF (FF(J)-FMAX) 301,301,303
303     JG = J
        FMAX = FF(J)
301     CONTINUE
        IF (FF(J) - FMIN) 322,323,323
322     FMIN = FF(J)
        JL = J
323     CONTINUE
C       ...TEST FOR CONVERGENCE...
        FDEV = FMAX-FMIN
        FTEST = FDEV - FR*ABS(FMIN) - FA
        IF (FTEST) 400,400,401
C       ...TEST SATISFIED, PROCEDURE HAS CONVERGED...
400     CALL OPTPR(1)
        WRITE (3,1000) NIT
        CALL OPTPR(2)
        GO TO 99
C       ...TEST NOT SATISFIED, PROCEED FOR ANOTHER ITERATION...
401     CONTINUE
        IF (NIT-MAXIT) 402,402,403
C       ...MAXIMUM ALLOWABLE NUMBER OF ITERATIONS HAS BEEN EXCEEDED...
403     CALL OPTPR(1)
        WRITE (3,1001) NIT
        CALL OPTPR(2)
        CALL OPTPR(3)
        CALL OPTPR(2)
        GO TO 99
402     CONTINUE
C       ...COMPUTE CENTROID OF POINTS IN SIMPLEX, EXCLUDING ONE WITH
C           HIGHEST FUNCTION VALUE...
        DO 304 I = 1,NX
        XC(I) = 0.0
        DO 305 J = 1,KMAX
305     XC(I) = XC(I) + XX(I,J)
304     XC(I) = (XC(I)-XX(I,JG))/(KMAX-1)
C       ...COMPUTE NEW TRIAL POINT BY REFLECTING POINT OF HIGHEST
C           FUNCTION VALUE THROUGH CENTROID OF REMAINING POINTS...
        DO 306 I = 1,NX
        X(I) = XC(I) -ALPHA*(XX(I,JG)-XC(I))
```

FIGURE 11   (continued)

```
C        ...TEST EACH EXPLICIT VARIABLE TO SEE IF IT VIOLATES BOUNDS.
C           IF SO, SET INSIDE BOUND BY A SMALL AMOUNT...
         IF (XU(I)-X(I)) 307,307,308
307      X(I) = XU(I) - GAMMA * (XU(I) - XC(I))
308      IF (X(I)-XL(I)) 309,309,306
309      X(I) = XL(I) + GAMMA * (XC(I) - XL(I))
306      CONTINUE
C        ...TEST TO SEE IF IMPLICIT VARIABLES VIOLATE BOUNDS...
         CALL IMTST(N1,NMAX1,IFLAG)
         IF (IFLAG) 504,505,504
505      CONTINUE
C        ...TEST TO SEE IF TRIAL POINT PRODUCES HIGHEST FUNCTION
C           VALUE IN NEW SIMPLEX...
         DO 312 J = 1,KMAX
         IF (J-JG) 316,312,316
316      IF (FF(J) -F) 312,312,313
312      CONTINUE
C        ...BECAUSE TRIAL POINT PRODUCES HIGHEST FUNCTION VALUE, MOVE TO
C           FRACTIONAL DISTANCE BETA FROM CENTROID OF OTHER POINTS...
         DO 314 I = 1,NX
314      X(I) = XC(I) + BETA*(X(I)-XC(I))
C        ...INSERT TRIAL POINT INTO NEW SIMPLEX...
313      CONTINUE
         CALL IMTST(N1,NMAX1,IFLAG)
         IF (IFLAG) 506,507,506
507      CONTINUE
         DO 315 I = 1,NX
315      XX(I,JG) = X(I)
         IF (NY) 320,320,321
321      CONTINUE
         DO 318 I = 1,NY
318      YY(I,JG) = Y(I)
320      CONTINUE
         FF(JG) = F
C        ...DO INTERMEDIATE PRINTING IF REQUIRED...
         IF (NIT-IPRT) 317,510,510
510      CALL OPTPR(4)
         CALL OPTPR(3)
         IPRT = IPRT + NFREQ
         GO TO 317
C        ...PRINT EROR MESSAGE AFTER CONSTRAINT VIOLATION IN IMTST...
500      WRITE (3,1002)
         GO TO 99
502      CALL FAIL(1)
         GO TO 99
504      CALL FAIL(2)
         GO TO 99
506      CALL FAIL(3)
         GO TO 99
C        ...FORMAT STATEMENTS...
1000     FORMAT ('0PROCEDURE HAS CONVERGED IN',I4,' ITERATIONS'/
     1   '0THE SOLUTION IS AS FOLLOWS')
1001     FORMAT ('0PROCEDURE HAS NOT CONVERGED IN',I4,' ITERATIONS'/
     1   '0THE CURRENT TRIAL SOLUTION AND SIMPLEX IS AS FOLLOWS')
1002     FORMAT ('0BASE SET OF VARIABLES VIOLATES SOME CONSTRAINT')
1003     FORMAT ('0ITERATION   0')
         END
```

FIGURE 11   (continued)

```
        SUBROUTINE OPTRD
C       COPYRIGHT (C) 1975 BY LAWRENCE B. EVANS
        COMMON F,X(5),Y(10),P(20)
        COMMON NX,NY,NP,XL(5),XU(5),YL(10),YU(10)
        COMMON XC(5),XX(5,10),YY(10,10),FF(10),JG,NIT
        COMMON NTITL(72),NAMEF(5),NAMEX(5,5),NAMEY(10,5),NAMEP(20,5)
        COMMON ALPHA,BETA,KMAX,MAXIT,FR,FA,FDEV,FMIN,NFREQ
C       ...READ BASIC DATA...
        READ (2,1000) (NTITL(I),I=1,72)
        READ (2,1001) NX,NY,NP,MAXIT,NFREQ
        DO 100 I = 1,NX
100     READ (2,1002) (NAMEX(I,J),J=1, 5),XL(I),XU(I),X(I)
        IF (NY) 112,112,113
113     DO 101 I = 1,NY
101     READ (2,1002) (NAMEY(I,J),J=1, 5),YL(I),YU(I)
112     CONTINUE
        IF (NP) 114,114,115
115     DO 102 I = 1,NP
102     READ (2,1002) (NAMEP(I,J),J=1,5),P(I)
114     CONTINUE
        READ (2,1002) (NAMEF(J),J=1, 5),FR,FA
        RETURN
C       ...FORMAT STATEMENTS...
1000    FORMAT (72A1)
1001    FORMAT (7I5)
1002    FORMAT (5A1,5X,3F10.5)
        END
```

FIGURE 12   Listing of OPTRD to read input data for OPTIM.

```
        SUBROUTINE OPTPR(IARG)
C       COPYRIGHT (C) 1975 BY LAWRENCE B. EVANS
        DIMENSION NINT(10)
        COMMON F,X(5),Y(10),P(20)
        COMMON NX,NY,NP,XL(5),XU(5),YL(10),YU(10)
        COMMON XC(5),XX(5,10),YY(10,10),FF(10),JG,NIT
        COMMON NTITL(72),NAMEF(5),NAMEX(5,5),NAMEY(10,5),NAMEP(20,5)
        COMMON ALPHA,BETA,KMAX,MAXIT,FR,FA,FDEV,FMIN,NFREQ
        DATA NINT/1,2,3,4,5,6,7,8,9,10/
        GO TO (1,2,3,4,5,6),IARG
1       CONTINUE
C       ...PRINT TITLE...
        WRITE (3,2000) (NTITL(I),I=1,72)
        RETURN
2       CONTINUE
C       ...PRINT TRIAL SOLUTION AND LIMITS...
        WRITE (3,2002)
        WRITE (3,2003)
        DO 200 I = 1,NX
200     WRITE (3,2004) (NAMEX(I,J),J=1, 5),XL(I),XU(I),X(I)
        IF (NY) 201,201,202
202     WRITE (3,2005)
        WRITE (3,2003)
        DO 203 I = 1,NY
203     WRITE (3,2004) (NAMEY(I,J),J=1, 5),YL(I),YU(I),Y(I)
201     CONTINUE
        IF (NP) 204,204,205
205     WRITE (3,2011)
        WRITE (3,2012)
        DO 206 I = 1,NP
206     WRITE (3,2004) (NAMEP(I,J),J=1,5),P(I)
204     CONTINUE
        WRITE (3,2010)
        WRITE (3,2006)
        WRITE (3,2004) (NAMEF(J),J=1, 5),F,FR,FA
        RETURN
3       CONTINUE
C       ...PRINT VALUES OF VARIABLES AT VERTICES OF CURRENT SIMPLEX...
        KMAX1 = KMAX + 1
        WRITE (3,2007) KMAX1
        WRITE (3,2008) (NINT(I),I=1,KMAX1)
        DO 301 I = 1,NX
        XX(I,KMAX1) = XC(I)
301     WRITE (3,2004) (NAMEX(I,J),J=1, 5),(XX(I,K),K=1,KMAX1)
        IF (NY) 303,303,304
304     DO 305 I = 1,NY
305     WRITE (3,2004)(NAMEY(I,J),J=1, 5),(YY(I,K),K=1,KMAX)
303     CONTINUE
        WRITE (3,2004) (NAMEF(J),J=1, 5),(FF(K),K=1,KMAX)
        RETURN
4       CONTINUE
C       ...PRINT RESULTS AT CURRENT ITERATION...
        WRITE (3,2009) NIT,JG,FDEV,FMIN
        RETURN
5       CONTINUE
        RETURN
6       CONTINUE
        RETURN
C       ...FORMAT STATEMENTS...
2000    FORMAT ('1',72A1)
2002    FORMAT ('0INDEPENDENT VARIABLES')
2003    FORMAT ('0',1X,'NAME',4X,'LOWER BOUND',4X,'UPPER BOUND',10X,
       1  'VALUE'/)
2004    FORMAT (' ', 5A1,6E15.5/( 6X,6E15.5))
2005    FORMAT ('0DEPENDENT VARIABLES')
2006    FORMAT ('0',1X,'NAME',10X,'VALUE',8X,'REL DEV',8X,'ABS DEV'/)
2007    FORMAT ('0VARIABLES IN SIMPLEX (CENTROID IS VERTEX',I3,')')
2008    FORMAT ('0VERTEX',I14,5I15/(7X,I14,5I15))
2009    FORMAT ('0ITERATION',I4,' ENTERING VERTEX',I3,' FDEV =',E12.4,
       1  ' FMIN =',E12.4)
2010    FORMAT ('0OBJECTIVE FUNCTION')
2011    FORMAT ('0PARAMETERS')
2012    FORMAT ('0',1X,'NAME',10X,'VALUE'/)
        END
```

FIGURE 13   Listing of OPTPR to print results for OPTIM.

```
      SUBROUTINE IMTST(N,NMAX,IFLAG)
C     COPYRIGHT (C) 1975 BY LAWRENCE B. EVANS
C
C     SUBROUTINE TESTS TO SEE IF IMPLICIT CONSTRAINTS HAVE
C     BEEN VIOLATED.  IF SO, EXPLICIT DECISION VARIABLES ARE MOVED
C     HALF WAY TOWARD CENTROID.
C     N = COUNT OF NUMBER OF TIMES TRIAL POINT IS MOVED TOWARD CENTROID
C     NMAX = MAXIMUM ALLOWABLE NUMBER OF MOVES
C     IFLAG = 0 IF CONSTRAINTS SATISFIED AND 1 IF VIOLATED
C
      COMMON F,X(5),Y(10),P(20)
      COMMON NX,NY,NP,XL(5),XU(5),YL(10),YU(10)
      COMMON XC(5),XX(5,10),YY(10,10),FF(10),JG,NIT
      COMMON NTITL(72),NAMEF(5),NAMEX(5,5),NAMEY(10,5),NAMEP(20,5)
      COMMON ALPHA,BETA,KMAX,MAXIT,FR,FA,FDEV,FMIN,NFREQ
      IFLAG = 0
      N = 1
C     ...EVALUATE OBJECTIVE FUNCTION AND DEPENDENT VARIABLES...
108   CALL MODEL
      IF (NY) 300,300,301
300   RETURN
301   CONTINUE
C     ...TEST TO SEE IF ANY IMPLICIT CONSTRAINT HAS BEEN VIOLATED...
      DO 103 I = 1,NY
      IF (Y(I)-YL(I)) 101,102,102
102   IF (YU(I)-Y(I)) 101,103,103
103   CONTINUE
      RETURN
C     ...BECAUSE TRIAL POINT VIOLATES IMPLICIT CONSTRAINTS, MOVE TO
C        FRACTIONAL DISTANCE BETA FROM CENTROID OF OTHER POINTS...
101   DO 104 I = 1,NX
104   X(I) = XC(I) + BETA*(X(I)-XC(I))
      IF (N-NMAX) 106,107,107
106   N = N+1
      GO TO 108
C     ...TRIAL POINT DID NOT SATISFY IMPLICIT CONSTRAINT AFTER NMAX
C        MOVES TOWARD CENTROID OF OTHER POINTS...
107   IFLAG = 1
      RETURN
      END
```

FIGURE 14   Listing of IMTST to test for violation of implicit constraints in OPTIM.

```
      SUBROUTINE FAIL(NARG)
C     COPYRIGHT (C) 1975 BY LAWRENCE B. EVANS
      WRITE (3,1000) NARG
      CALL OPTPR(2)
      CALL OPTPR(3)
      RETURN
1000  FORMAT ('0ERROR ENCOUNTERED IN OPTIM'/'0TYPE',I3,' CONSTRAINT VIOL
     1ATION')
      END
```

FIGURE 15   Listing of FAIL to print error message for OPTIM.

ACKNOWLEDGMENTS

We are indebted to many people who have helped in the development of this chapter. Special thanks are due to Marty Mishkin and Steve Haralampu, who carefully read this manuscript at various stages of its development and offered a number of helpful suggestions and comments, thus improving the quality of the material included.

We thank the Institute of Food Technologists for the permission to cite extensively from the paper of Saguy and Karel (1979) which appeared in the Journal of Food Science, Ms. Miri Rauchwerger for her graphic work, Ms. Judith Alkali for typing the first draft, and Ms. Janet Copley for her efforts and enthusiasm in editing and typing the manuscript.

Finally, we were very fortunate to have had the opportunity to collaborate in research on dynamic optimization carried out at MIT under Professor Marcus Karel and supported in part by the National Science Foundation Grant ENG-7824342 and CPE-8104582 from the Division of Engineering.

REFERENCES

Aris, R. (1961). The Optimal Design of Chemical Reactions. Academic, New York.

Assad, A. A. (1973). An evaluation of Computational Techniques for Solving Problems in Dynamic Optimization. M.S. Thesis, Department of Chemical Engineering, MIT, Cambridge, Mass.

Athans, M. and Falb, P. T. (1966). Optimal Control: An Introduction to the Theory and its Application. McGraw-Hill, New York.

Bellman, R. (1957). Dynamic Programming. Princeton University Press, Princeton, N.J.

Bellman and Dreyfus, S. E. (1962). Applied Dynamic Programming. Princeton University Press, Princeton, N.J.

Bilous, O. and Amundson, N. R. (1956). Optimum temperature gradients in tabular reactors, Chem. Eng. Sci. 5(I and II): 81, 115.

Box, M. J. (1965). A new method of constrained optimization and a comparison with other methods, Computer J. 8(1):42.

Constantinides, A., Spencer, J. L., and Gaden, E. L. (1970). Optimization of batch fermentation processes. II Optimum temperature profiles for batch penicillin fermentations, Biotechnol. Bioeng. 12:1081.

Denn, M. M. (1969). Optimization by Variational Methods. McGraw-Hill, New York.

Dingler, J. R. (1968). An Evaluation of Computational Techniques for the Optimization of Dynamic Systems. M. S. Thesis. Department of Chemical Engineering, MIT, Cambridge, Mass.

Fan, L. T. (1966). The Continuous Maximum Principle. Wiley, New York.

Fine, F. A. and Bankoff, S. C. (1967). Control vector iteration in chemical plant optimization, Irad. Eng. Chem. Fund. 6(2):288.

Gear, C. W. (1971). Numerical Initial Value Problem in Ordinary Differential Equations. Prentice-Hall, New York, Chap. 9.

Gonzalez, S. and Miele, A. (1979). Sequential gradient-restoration algorithm for optimal control problems with nondifferential constraints and general boundary conditions, Aero-Astronautics Rep. Nos. 142 and 143. Rice University, Houston, Texas.

Graves, R. E. (1979). Optimization of Ascorbic Acid Retention During Air
    Drying. B.S. Thesis, Department of Chemical Engineering, MIT,
    Cambridge, Mass.
IMSL (1977). International Mathematical Statistical Library Reference
    Manual. IMSL Lib. 0006. Houston, Texas.
Jaspan, R. K. and Coull, J. (1971). Trajectory optimization techniques
    in chemical engineering. I. Boundary condition iteration methods,
    A. I. Ch. E. J. 17(1):111.
Katz, S. (1960). Best temperature profiles in plug-flow reactors: methods
    of the calculus of variations, Ann. N.Y. Acad. Sci. 84(12):441.
Krogh, F. T. (1973). Algorithms for changing the step size used by a
    multistep method, SIAM, J. Num. Anal. 10:949-965.
Lee, E. S. (1964). Optimization by Pontryagin's maximum principle on
    the analog computer. A. I. Ch. E. J. 10(3):309.
Levine, M. D. (1966). Trajectory optimization using the Newton-Raphson
    method, Automatica 3:203.
Loncin, M. and Merson, R. L. (1979). Food Engineering Principles and
    Selected Applications. Academic, New York, Chap. 11.
Meo, D. and Friedly, J. C. (1973). Optimal control of a radiant heat
    freeze drying process. Paper presented at A. I. Ch. E. National
    Meeting, Detroit, Mich., June 3-6.
Miele, A.(1975). Recent advances n gradient algorithms for optimal con-
    trol problems, J. Optim. Theory Appl. 17:361.
Miele, A. and Cloutier, J. R. (1974). New transformation technique for
    optimal control problems with bounded state. Part I. Theory. Part
    II. Examples. Aero-Astronautics Rep. Nos. 122 and 123. Rice Uni-
    versity, Houston, Texas.
Miele, A., Wu, A. K., and Liu, C. T. (1979). A transformation technique
    for optimal control problems with partially linear state inequality con-
    straints, J. Optim. Theory Appl. 28(2):185.
Mishkin, M. (1980). Unpublished data.
Mishkin, M., Karel, M., and Saguy, I. (1982). Applications of optimiza-
    tion in food dehydration. Food Technol. (Presented at the IFT 41st
    Annual Meeting, Atlanta, Ga. June 1981.)
Pontryagin, L. S., Boltyanskii, V. G., Gamkrelidze, R. V., and Mish-
    chenko, E. F. (1962). The Mathematical Theory of Optimal Processes
    (English translation by K. N. Trirogoff). Wiley-Interscience, New
    York.
Rai, V. R. and Constantinides, A. (1973). Mathematical modeling and op-
    timization of the gluconic acid fermentation, Amer. Inst. Chem. Eng.
    Symp. Ser. 132(69):114.
Roberts, S. M. and Shipman, J. S. (1972). Two-Point-Boundary-Value
    Problems Shooting Methods. American Elsevier, New York
Rozonoer, L. I. (1959). L. S. Pontryagin's maximum principle in the
    theory of optimum system I and II, Automat. Remote Control 20:1288,
    1405.

Rudd, D. F. (1966). Mathematical optimization techniques, in <u>Computers in Engineering Design Education</u>. Vol. I (D. L. Katz, Ed.). University of Michigan, Ann Arbor, Mich.

Saguy, I. (1982). Utilization of the "complex method" to optimize a fermentation process, <u>Biotechnol. Bioeng.</u> 24:1519.

Saguy, I. and Karel, M. (1979). Optimal retort temperature profile in optimizing thiamin retention in conduction-type heating of canned foods. <u>J. Food Sci.</u> 44:1485.

Storey, C. and Rosenbrock, H. H. (1964). On the computation of the optimal temperature profile in a tabular reaction vessel, in <u>Computing Methods in Optimization Methods</u> (A. V. Balakrishnan and L. W. Neustat, eds.). Academic, New York, p. 23.

Umeda, T., Shindo, A., and Ichikawa, A. (1972). Complex method for solving variational problems with state-variable inequality constraints, <u>Ind. Eng. Chem. Process. Des. Dev.</u> 11(1):102.

Wahrman, S. and Dayan, J. (1978). Solving two point boundary value problems for the optimal control of simple batch reactor by several methods: A comparison study, <u>Proceedings of the Joint Automatic Control Conference</u> (JACC), Philadelphia, Pa., p. 179.

12

Process Control

MALCOLM C. BEAVERSTOCK   The Foxboro Company, Foxboro,
Massachusetts

I. INDUSTRY NEEDS

Advances in food technology has led to the study and consideration of many new and exotic processing areas such as single-cell protein production from petroleum and cellulose, synthetic fuels, synthesis of complex organic molecules, alcohol from black liquor, and new waste treatment processes. As a result, food technologists have had to incorporate new technologies including industrial process control.

Historically food processing has been one of the slower industries to adopt modern control instrumentation. However, several factors are at work to change that situation. Like other industries, the food industry is facing the problems of energy, quality, material costs, labor costs, and extensive record keeping. The need for increased production capacity in the face of higher operating costs has resulted in a productivity crisis. Energy costs and availability are major concerns in an industry that makes extensive use of cooking operations. Government regulations have re- quired a more extensive level of record keeping than is accomplished by operators writing entries on a clip board. These problems have resulted in new perceptions that include considering food processing more of a sci- ence than an "art" and accepting the idea that control methods may be of value.

Today, many large food processors realize that instrumentation and process control must be integrated into every process system design. One highly computerized food plant reported extremely high levels of operator- computer-process efficiency that assured consistently high quality product and energy savings that were expected to reach 30% compared to conven- tional facilities (Homan, Brennecke, and Forwalter, 1978). Even with such successes, acceptance of control and automation remains low (Beaverstock and Bernard, 1977). One important contributing factor is the increase in complexity that occurs when control systems have to take into account the characteristics of operating plants.

Operating procedures in a plant are all based on some aspect of the control system design. Every time the operator goes to the control panel to make a change he or she is following the directions established by the control system designer. Consequently, the operating procedures and even the economics of the plant are being significantly affected by the person who ends up specifying the control system. Coming to grips with this level of "designer power" requires a procedure that addresses both technical and operational issues in a consistent manner.

II. CONTROL ANALYSIS

Before the control problem can be analyzed, it has to be understood. Usually the problem isn't a control problem but a manufacturing one. A control problem involves the inability to maintain a particular temperature, pressure, level, flow, or other operating characteristic. The

TABLE 1 Control System Design
Considerations

Automation level

Plant operating procedures

Process management level

Personnel requirements

Reliability and security levels

Control performance criteria

Project scope and time frame

Operator job description

Accuracy requirements

System responsibility

Process dynamics and economics

---

manufacturing problem involves the inability to produce a product profitably.
The primary mission of any control system is to help the major manufacturing goal of producing acceptable, consistent-quality product at a profit,
If it doesn't do that, other goals such as maintaining safe operating conditions, complying with government regulations, and finally, reducing cost
will not be worth considering. To serve manufacturing, the control system
has to first work technically, and that means having a sound basic control
design.

That basic control design begins with recognizing all the factors that
tie control systems and plant operations together. A list of such factors is
shown in Table 1. Knowledge of the process and plant operating objectives
are key elements of the list along with the standard control system characteristics. Obtaining the information required by the factors is not an easy
task and usually requires more than what is normally acquired from pilot
or minimally controlled plant operations.

For example, scheduling, sequencing, and feedback control are areas
not normally considered in any detail at the experimental-process level.
Yet they predominate in a more automated environment such as the retort
control application shown in Fig. 1. Economic surveys show that the major
portion of the payback on automated systems in the food industry comes
from sequencing and scheduling. Such plant control systems in a brewery,
for example, can involve 500 or more contact inputs and outputs associated
with sequential control.

Simple feedback control, (e.g., single-variable flow, temperature,
pressure, and level etc.) which represents a small part of the experimental

FIGURE 1 Industrial retort processing area. (Courtesy of the Foxboro Co, Foxboro, Mass.)

effort, is a major control task in a plant. Two hundred or more feedback controllers and perhaps only five advanced control loops in a fermentation facility would not be unusual. In addition, any advanced control application such as optimization requires that the basic feedback loops be correctly operating.

These information requirements mean that any instrumentation and process control design must be integrated into every process system design. This integration in turn, requires a combination of talents (Stults, 1978). Food technologists have the responsibility to define process parameters necessary to produce specific product characteristics in each stage of the process. These include raw-materials requirements, formulation, blending procedures, and primary process variable specifications such as time, temperature, pressure, and process media. They also specify product control parameters such as moisture, texture, color, shape, bulk density, flavor, vitamin fortification, and additive ratios.

Process engineers are responsible for developing process system designs which involve selecting process equipment that fits the process theory. This includes specifying vessel characteristics such as agitation level, temperature limitations, and measurement/control accuracy requirements. Process engineers and plant production staff normally collaborate on such items as operational procedures and review process piping and equipment flow sheets. Major construction would also require central engineering personnel to purchase production equipment and controls, prepare installation drawings and specification, negotiate construction contracts, and supervise installation.

Engineers with control background take the responsibility for designing process control systems to control process variables; developing sensing devices and control instrumentation systems to measure and control product parameters; assembling and testing equipment and control system designs; and most importantly, working with all the other team members to make sure the control system faithfully represents the operating objectives of the plant. Because of its coordinative nature, one would expect that the function of control system design would be of prime importance in any food industry plant design—but it isn't. In almost every industry a cost-effective plant design must be frozen as much as possible once detailed engineering is ready for field construction. Unfortunately, control-system decisions are normally carried out at this time, and therefore, the impact of the decisions have little chance to feed back to the process design and plant operations issues (Beaverstock and Trearchis, 1978). All the average control engineer can do at this point is design conventional control schemes, usually based on the individual plant unit operations (i.e., filtration, drying, heat transfer). Such scheduling does not allow for any tailoring or innovation to better meet food industry control needs.

A recommended plant design format would allow for an integration of process, plant operation, and control goals early in the design stage when a free exchange of concepts among vendors and users can be accomplished (Beaverstock and Trearchis, 1978). Within this structure, questions concerning process requirements, plant operation, and process economic sensitivity as well as the technical areas of measurement, control, communications, and degree of control system distribution can be considered by the design team in an iterative fashion. However, carrying out a meaningful discussion about a complex subject still requires a methodology. Bernard and Howard (1968) suggest a method which involves a horizontal breakdown based on the concept of "operating units" instead of the standard unit operations. The result is a structure with many design units which interact only minimally. Carefully defined purposes, goals, and broad functional requirements are required. A key in the analysis is the willingness to write down purposes for different parts of the system and functions that seem obvious and simple. A good measure of the structure is how little the whole system must be changed to accommodate changes in any of its parts. Such segmentation results in smaller subsystems which are easier to analyze.

An example of how a brewery can be studied in this way is shown in Fig. 2. This figure gives a starting point for the plant study but doesn't begin to contain the detail that is required. The plant breakdown should conform to control-board layout and responsibilities of various organizations.

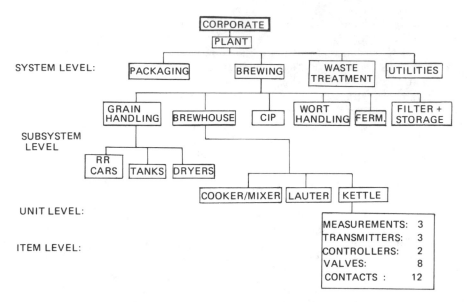

SYSTEM LEVEL:

SUBSYSTEM
LEVEL

UNIT LEVEL:

ITEM LEVEL:

FIGURE 2   Brewery structure by function.

In the example, the brewing operation is separated from the fermentation
and filtration sections by both control-board location and operator assignment.

Each plant area shown is broken down into its equipment, measurements, valves, and other operating parameters. The detail continues to the
point of individual input/output assignments for control devices. The resulting documentation of the plant becomes functional in nature. Consequently, information concerning a temperature controller on a dryer is found in
the grain-drying section of the plant rather than as part of a long list of controllers used throughout the plant. This sort of structure is also more compatible with newer, data base oriented system methods.

III.   CONTROL ALTERNATIVES

The methodology outlined in Sec. II provides the structure to study how control strategies interact with plant design and operation. Such a structure is
required to consider the control system options that are either presently
available or will be in the future. However, an understanding of the "control-
loop" concept is also required.

The control loop is the basic unit of any control system. It is made up
of four elements that may interact with the process in order to control it.
These elements include the measurement, display, controller, and actuator.

The display is used to monitor and interact with the operation of the loop and involves establishing a control target or "set point" as well as selecting automatic or manual modes of control. "Closed-loop" control implies that there is a continuous flow of information around the loop and through the process. These various elements may be physically separated or combined in various ways.

A single loop involves just one part of each function: one measurement, one control algorithm, one actuator, one process variable. Food-plant automation involves closed-loop control over many simple loops and some complex ones. A single loop performs a particular function such as maintaining a specified temperature, flow, pressure, or level. Figure 3 shows the single-loop concept as applied to flow control.

Some terms regarding these elements are worth mentioning. A closed loop implies that information from the process is used to manipulate an actuator which produces changes which are then used to further manipulate the actuator. This cycle is called "feedback" control and depends on an "error" or difference between the measurement and set-point values to operate. The changing of the set-point value is termed "supervisory" control and can be accomplished either manually by an operator or automatically by other control equipment or computers. Finally, "feedforward" control involves

FIGURE 3 Control loop concept.

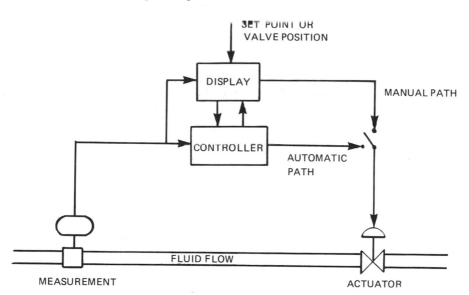

monitoring process values and then, on the basis of some model, anticipating process performance. If the predicted performance deviates from a desired level, then action can be taken either through a supervisory mode or changing an actuator directly. Feedforward control usually involves the use of a computer system to carry out the calculations.

A. Non-computer-based Systems

Discussing a "non-computer-based" system may seem out of place in a book dedicated to the use of computers. However, the vast majority of control systems in the food industry today do not have computers associated with them. Pneumatic instrumentation remains a major factor in plant control equipment. The two basic reasons are their high reliability factor and the lack of an immediate replacement for the pneumatic valve. Advances in pneumatics continue as evidenced by the recent development of an all-pneumatic, on-line process chromatograph (Waters, 1978).

Non-computer-based systems are considered to have a high reliability level. From an equipment point of view, a device failure in this configuration only affects the single loop. The single-loop concept also becomes part of the plant operating philosophy which is viewed as a combination of many independent control loops. Control equipment is either placed on the individual pieces of process equipment or collected on local control panels. Plant production personnel decides where equipment will be located. These decisions generally reflect staffing requirements.

Non-computer-based systems continue to form the basis for many advanced control structures in the food industry. Dryer control systems using an inferential moisture measurement technique and energy conservation systems are two examples (Shinskey and Myron, 1975).

Basically, non-computer-based systems represent a philosophy that is based on using single-loop control throughout a plant. Even if a microprocessor were used in place of pneumatics to produce a single-loop controller, the characteristics of the overall system would not change.

B. Computer-based Systems

This control system alternative makes use of more complex control loops and generally requires the use of a computer. A possible control structure is shown in Fig. 4. In this configuration the computer carries out control and supervisory calculations. Although a computer may be used to control one section of the plant, other sections may still utilize the non-computer-based alternative philosophy. Computer-based systems usually do require more design-level decisions; however, the additional work is justified by the degree of flexibility such a system provides.

Some of these design issues involve a choice between a supervisory control strategy and sending control signals directly to an actuator, how much back-up equipment is required in the event of a computer failure,

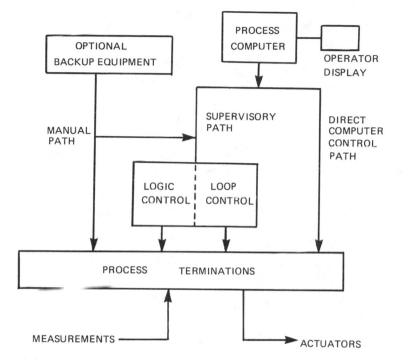

FIGURE 4 Centralized computer system.

and how long information will be stored in the computer before being printed as reports. The scope of the computer control is also important. How large a section of the plant is involved will greatly effect the complexity, size, and cost of the system.

Data becomes a key issue when computers are involved, and a consistent approach is critical. Security for the information in the system is also important, and access must be defined to prevent illegal or unauthorized use of the control system. Job descriptions may also be involved as an operator may not be required to make changes to the control parameters. If the computer services more than one area of the plant operation, then formats for data must be consistent so that the information is meaningful to everyone using the system. Obviously this means a higher level of documentation must be created and maintained.

Just as the previous section reflected a plant operational philosophy based on a single-loop concept, a computer-based system reflects an operating philosophy that involves centralized functions. In this respect the control system considers an operation such as cooking as an operating unit rather than a series of individual loops. Communication among such

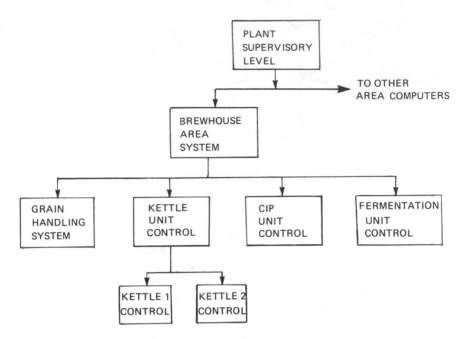

FIGURE 5 Computers for distributed/hierarchical control.

operating units, which is necessary to operate an entire plant, is normally carried out manually.

C. Multiple Computer-based Systems

Multiple computer-based systems are economically possible as a result of the rapidly growing microprocessor technology. This has resulted in new control structures that make use of distributed and hierarchical concepts. The technical reasons for such systems are reliability and flexibility which are the best features of the two alternative structures mentioned previously. A possible distributed structure for a brewery is shown in Fig. 5.

The general philosophy behind distributed systems is to consider the entire plant as a single unit. This requires the greatest interaction of process, production, and control personnel because of the necessity to connect all control and production elements together. An electronic data link establishes communication between sections of the plant. For example, in a fermentation plant, a communication link will connect computers in production control, fermentation, product recovery, and sterile operation. A problem arising in sterile operations can be quickly communicated to product recovery for data analysis and problem solving. If the problem is analyzed and found to originate in fermentation, proper action will be taken.

Once again data considerations are an important factor. For messages to be meaningful as they are passed from computer to computer, a plant-wide consistent format is a must. This is not easy to accomplish since variables describing an automatic potato peeler are different from those describing a controller, dryer, or feed storage tank. The location of data within the network, as well as when that data should be updated, must also be determined. Information concerning plant operations such as feed preparation, cooking, and sterilization procedures may be contained in various sections of the plant. Consequently, sufficient design effort is required to ensure that all records are updated correctly and error-recovery procedures are defined. The design objective is to have the resulting data and hardware distribution match the operational structure of the plant.

Distributed systems can be very reliable since the effects of failures are localized and easily corrected. A well-designed network or hierarchy will result in reducing system-wide data communication rates by localizing function and data requirements. Additionally, when the system matches the plant operation, problems can be recognized in terms of errors of function rather than abstract errors of computation. This reduces the number of confusing interactions among various parts of the control system. A functional structure also corresponds to traditional analog control structures and familiar human organization patterns. In summary, the distributed system puts hardware elements more thoroughly under the control of the operating people who have analogous roles in plant operations.

Distributed systems offer the greatest flexibility of any control alternative but also the greatest complexity in design. New concepts are being developed to cope with this problem. One approach is to design each element of the system to be "self-aware" so that it can function independently in a community of other control elements (Bristol, 1979). Each self-aware element will be autonomous and allow communications only in functional terms. A clean-in-place processor will not allow a change in its procedures or data base except by a message from a unit giving the correct format at the correct time.

While distributed systems represent a new control alternative for food plants, its philosophy contains many of the elements of the so-called non-computer-based approach and even depends on the correct application of single-loop control to successfully operate.

D. Benefit Potentials

The food processing industry, like other manufacturing operations, has to show a profit for the product they produce. Therefore any control alternative selected must have some benefit associated with the cost of implementation. The general rule is that any benefit figures must be developed with

the cooperation of the particular plant personnel in order to have proper
credability. Benefits are as varied as the number of plants in existence,
and only an individual study can accurately project figures. The following
discussion of benefits in a fermentation plant are typical of those found in
other industrial applications.

Sequential control typically provides a large portion of the benefits from
computer applications. The most common benefit comes from the reduced
time and increased accuracy involved with charging and emptying of tanks.
In sterilization procedures the computer ensures accurate heating cycles,
minimizes contamination, reduces the time between operations, and produces processing consistency where repeatable procedures are involved.
New control methods are permitting a reduction in energy costs, and the
flexibility of computer systems reduces the cost for future processing
changes.

Computer control of the fermentation operation itself provides easy
formulation of data logs and reports for management and FDA requirements.
Additional benefits result from maximizing yields through optimization routines, closer inventory control, reduced operator error, diagnostic analysis,
and energy reduction from close control of agitation and aeration. Engineers have also found that the data available from continuous monitoring of
fermentation operations have provided a better understanding of the metabolic process. Perhaps one of the greatest benefits comes from the greater
reproducibility of process conditions that is obtained from close control of
important parameters.

Formal procedures for determining the economic level of control for
plants have been developed and are available in the literature (Beaverstock,
Bernard, and Evans, 1979).

IV. USER-MACHINE INTERFACE

A. Background

The interface between the user and the processes that he or she controls
has undergone extensive changes over the past 20 to 30 years. The most
dramatic is the use of electronic display devices such as the cathode-ray
tube (CRT) (Fig. 6), which provides an infinitely variable medium for data
presentation and manipulation. Additionally, many of the human operator's
mental processes have been taken over or augmented by control systems
resulting in modifications of his or her tasks. This is especially significant to the food processing industry since its procedures have traditionally
depended on operators to use their sense of smell, sight, and touch to determine the quality characteristics of the product as it moves through the
plant. Automation has, therefore, required new thinking concerning the
operator's role.

Functionally, the process operator's job is to exchange information
with the process. The objective of the operator-process interface, then,

FIGURE 6 User-process interface. (Courtesy of the Foxboro Co., Fox-boro, Mass.)

is to implement this exchange as shown in Fig. 7. Any control system in-stallation in the food industry must consider and define the operator inter-face in terms of the following:

1. Content: What information is available and how much should be dis-played at any time?

2. Presentation: How should the information be displayed to the oper-ator (visually, audibly, etc.)?

3. Interaction: How should the operator transmit information and ac-tions to the process?

4. Configuration: How should the information be physically arranged (location, grouping of displays, size, etc.)?

5. Job responsibility: How is a good operator differentiated from a poor performer (criteria for advancement, salary action, or per-formance appraisal)?

A successful design in terms of these topics is especially difficult in the food industry since, historically, the operator's responsibilities have always required considerable judgment based on years of training. For

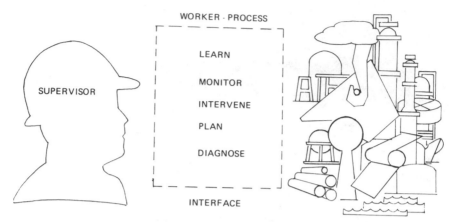

FIGURE 7 Process operator's tasks.

example, the brewmaster in a brewery is an operator, production chief, and
quality control representative all rolled into one. Additionally, product
characteristics based on texture, color, taste, or consistency cannot easily
be measured and therefore depend on human judgment for control. In spite
of these problems, newer production facilities are centralizing controls for
more efficient operation and therefore are having to cope with the operator-
process interface issues.

B. Methods

Just as preceding sections in this chapter have indicated the need for a combination of talents, designing an adequate operator-process interface also
requires a combination of expertise. The basic team still comprises the
food technologist, who supplies the processing requirements, and the engineer responsible for the control system design. However, individuals with
technical background in display technology, industrial design, human factors,
psychophysics, anthropometrics, and ergononmics are now also contributing
to the design effort.

   Ergonomics is the study of the way humans work and the assignment of
tasks based on an individual's background and training. Anthropometrics
involves the detailing of human body measurements such as height, reach,
angle of vision, etc. It provides information to industrial designers and
display technologists for use in correctly packaging display equipment to
meet the operator's personal and functional requirements. The study of
how an operator interprets information, even in the presence of stress,
fear, or other physicial or emotional conditions, involves the use of psychophysics.

Information from these technical areas is gathered together for the operator interface design effort by an individual with some background in human factors. This field of study developed from the aerospace industry and the design of aircraft instrument panels. The human-factor effort also contributes to the design work in areas such as human color perception, tactile feedback requirements, and blink rate or refresh times for CRT displays.

Recent studies of problems in nuclear plant control-room design have included theories on human error as well as typical problem areas (Sheridan, 1980). The results of such studies are equally applicable to the food industry. Examples of poor design include the use of large, mirror-image control boards which confuse operator work patterns. Accidental activation of switches can occur because of inadequate labels and inconsistent coding of equipment controls. Complex alarm annunciators resulted in confusing patterns when an emergency actually occurred and allowed for many "normal" alarms which were illuminated even though the plant was operating according to standard procedures. Some guidelines for avoiding many design errors are currently being developed by the International Purdue Workshop committee on user-process interface (Williams, 1978) and the Electric Power Research Institute (1979).

Some of the newer approaches to understanding how operators in process plants carry out their assignments involve the development of a model of the operator's actions (Beaverstock, Stassen, and Schneider, 1979). This technique assumes that the operator possesses knowledge of the dynamics of the process he or she is controlling, has some insight into the properties of the noise acting on the measurements he or she sees, and has knowledge to build an internal (mental) model of the process dynamics, noise statistics, and task requirements. The mental model is then used to control the process in some optimal fashion (in other words, the operator "does his or her best"). Current research is aimed at understanding the attributes of the operator's mental model.

It must be recognized that the operator's internal model of the process may not be perfect. Food processes are usually very complex and of high dimension. Consequently the operator's model may only include a simplified, lower dimensional (reduced order) representation of the actual process dynamics. The operator's model of the noise statistics may not be perfect either, and the operator will perform tasks according to his or her subjective interpretation of the task requirements.

However, the concept that the skilled operator possesses and applies an internal model seems reasonable. The fact that the operator's performance improves as he or she gains experience in controlling the plant indicates that control actions are based on knowledge about the process and his or her skill level.

In summary, work aimed at analyzing the human operator's ability to control a process plant by considering the basic structure of goals, assigned tasks, performance measure, knowledge of the process, degree of interaction with the control equipment, and the information displayed is just beginning to be included in control system design. The need for understanding the operator's role and behavior will become more important as it is recognized, considered, and then used effectively in the food industry during the next decade.

V. SENSORS

Advanced control systems in the food industry today are limited by the state of the art in sensors and on-line measurement. Generally, present practice involves simple feedback control using those few direct measurements which are currently available (Dobson and Trearchis, 1976). Unfortunately, the food industry often needs unique equipment unused by other industries. A canner wants to measure color of fresh tomatoes and finished sauce. A brewer needs to know how much protein has been removed from beer, or a soft drink manufacturer must know the carbon dioxide level (Fig. 8). Sweetener sensors are needed for syrup and carbohydrate measurements for cereal. Other instruments are required for acidity, alkalinity, the content of oils, amino acids, and microbiological activity. Not only is developing these specific measurements a problem, but each has to meet sanitary specifications as well.

A. Measurement Approaches

Measurement devices, like control systems, begin with a few very basic components and then get more complex and sophisticated. There are four basic measurements: flow, temperature, pressure, and level. These, when included in simple control loops, provide the control base needed for advanced systems. Often they alone can solve many processing problems. For example, one food processor desired a complex, on-line specific-gravity measurement on raw materials to accurately adjust cooking times. However an investigation determined that the consistency problems could be solved simply by using regular laboratory measurements combined with accurate and repeatable temperature and cook-time control on the cooker itself.

The most common measurement is flow. Magnetic flow meters, which produce a measurement signal proportional to volumetric flow rate, remain a popular device. The flow meter has no obstruction or moving parts and its accuracy is unaffected by changes in type of product or changes in fluid characteristics. A new flow meter gaining rapid acceptance is the vortex meter. This meter (Fig. 9) is highly accurate over wide flow ranges, has excellent reliability with no moving parts in the stream, and has a low installed cost. The vortex meter uses the principle of vortex shedding to transmit an output signal which is linearly proportional to the volumetric flow rate.

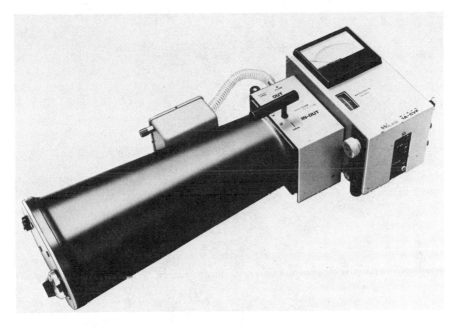

FIGURE 8 On-line infrared analyzer. (Courtesy of the Foxboro Co., Foxboro, Mass.)

The need for accurate temperature in heated vessels, heat exchangers,
fermentation tanks, and storage tanks is currently being met with a resistance temperature detector (RTD) as the primary temperature sensor. The
RTD provides a measurement with exceptional stability, accuracy, and resistance to long-term drift. The high-level output signals can be transmitted over ordinary, unshielded pairs of wires to a remote receiver. They
are most often used to control temperature in heated or cooled vessels or,
in conjunction with alarm relays, to sense input conditions as interlock
safeguards or to activate alarm annunciators.

Level transmitters are used where it is desirable to transmit inventory
information on tank contents to a remotely located receiver. They may also
be used to control level or work with alarm and interlock systems. The
diaphragm in the sanitary transmitter version is entirely crevice free.
This construction meets most international requirements for clean-in-place
installations and eliminates the "pockets" where accumulated materials can
remain to become contaminated with bacterial growth. The materials of
construction allow hot cleaning solutions to be used. Pressure sensors are
very similar to level devices and operate on the same principles.

Variables such as pH and conductivity are more difficult to handle than
the basic measurements described above. These are then followed in

FIGURE 9   Vortex flow meter.   (Courtesy of the Foxboro Co., Foxboro, Mass.)

complexity by composition and quality measurements.   Development of the electrodeless conductivity monitor has improved measurement of caustic concentration in clean-in-place (CIP) systems for the food industry.   This sensor element virtually eliminates the basic measurement difficulties from the buildup of materials on the sensor as well as electrode polarization experienced with direct-contact electrodes.   The accurate control of caustic solution strength is extremely important in order to minimize the use of makeup caustic and ensure safe CIP operations.

Composition and product-quality measurements are usually not available or are expensive to install and maintain.   Therefore a technique known as "inferential" measurement is often used.   This method uses measurable variables along with knowledge of the process to infer the desired quantity. Computers are making this method more viable as the calculational ability of the microprocessor increases and its price decreases.   The bottleneck to the use of inferential measurements is the amount of information known about the process.

An early example of inferential control in the food industry involved controlling product moisture in the drying of spent grain and resulted in savings on solids giveaway as well as energy input (Fadum and Shinskey, 1980). In place of an on-line moisture-measuring device, the dryer inlet and outlet air temperatures as well as the feed rate and ambient dew point were used to infer the conditions under which product moisture would remain constant. The output of the calculation was then used to regulate fuel flow to the dryer.

Computers in fermentation process control are being used in the inferential mode (Cooney, 1978). These computerized systems monitor physical and chemical properties and use them in conjunction with computer techniques to calculate the physiological properties of the system. Physiological conditions are usually inaccessible to direct measurement and yet are the key to fermentation productivity. By inputting the current physical and chemical properties of the system into a particular computer algorithm, physiological properties of the organism may be calculated. A decision with regard to maximizing production by using the trend of physiological variables and their relation to changes in the chemical and physical variables may then be accomplished by analog control with computer manipulation of set points. These advanced systems are just beginning since some biological parameters are still unmeasurable and there are major limitations in the theory of biological processes that prevent complete interpretation of the information. Still, the combination of computer calculational capabilities, process knowledge, and basic sensors will continue to be an accepted method of monitoring variables that are difficult to measure directly.

B. Accuracy and Repeatability

Three generic characteristics of measurement devices that must be recognized and understood to complete a successful control system implementation are accuracy, repeatability, and linearity. Too often these terms are misused and lead to the development of a substandard control system base. "Accuracy" is a term normally used by process engineers when specifying a measurement for use in a system, while "repeatability" and "linearity" are terms that have more meaning to control engineers.

A food technologist might specify that a process requires controlling a particular variable to 1% accuracy. This specification is not directly translatable to equipment, however. Measurement-device accuracy is normally specified as either percent of measurement or percent of instrument span, and these terms have quite different meanings. Consider a flow-meter device measuring flow from 0 to 100 gal/min with a measurement accuracy of 1%. This specification implies that at a reading of 90, the actual flow will be within 0.9 gal/min and at a reading of 5, the actual flow will be within

0. 05 gal/min. On the other hand, a meter measuring flow over the same range but with an accuracy of 1% of span will have different characteristics. In this case the actual flow will be within 0. 1 gal/min at a reading of 90, or 5 on the instrument scale. Therefore, when specifying accuracy or selecting a measurement device, the process engineer must always check the type of accuracy required.

While absolute accuracy is important, control loops are designed to control at a particular point on a scale. To do this function, the control system assumes that the measurement device continually measures the process variable in a consistent manner. This attribute of a measurement device is termed its "repeatability." For example, consider the same flow device as that used in the accuracy discussion. Repeated measurements might result in values of 11.8, 11.7, 11.8, and 11.9 gal/min with an actual flow rate of 12.5 gal/min. In this case the measurement is repeatable to within 1%, although the absolute accuracy is only 5.6%. Repeatability, especially long-term repeatability, is extremely important to good control. The relationship between repeatability and accuracy is determined through the instrument calibration.

Even after accuracy and repeatability specifications are understood, not taking into account linearity characteristics of a measurement device can cause control problems. A linear measurement produces an output signal that is directly proportional to the value of the variable. Some measurements, however, produce an output signal that is not linear with the measured value. The common orifice flow meter is an important example, as it represents the most popular flow measurement device presently in use. In this meter, the measurement output (pressure differential across the orifice) is proportional to the square root of the flow rate.

This nonlinear characteristic means that the pressure differential is less sensitive to flow at low flow rates and that this sensitivity (i.e., measurement gain) changes with flow level. Consequently, for best performance, orifice flow meters should be used in the more-linear region from about 40 to 90% of their total flow range. A tour through a food plant, however, will reveal most orifice flow meters indicating rate at 20% of scale or lower. Expecting consistent flow control under these conditions is almost unrealistic. This condition can only be rectified by recognizing process measurement requirements and matching them with instrument characteristics during the design phase of control projects.

VI. APPLICATIONS

Control applications in the food industry are carried out primarily on an ad hoc basis using the rationalization that each application is different. Yet if one looked closely at the many variations of applications, a number of common components can be recognized. In fact, present control systems are actually built up out of nonlinear subsystems, each like the major components of an automobile or television set, and each with well-defined functions (Bristol, 1980). This trend will be amplified even more as control components are able to be expressed as software application packages.

The most common control component is the basic controller. These components are routinely purchased either as separate hardware devices or as part of a digital control package. Other standard components would include cascade, ratio, blending, selector, or auctioneering functions. Future components will be more application orientated and contain control systems for such areas as fermenter control and energy management. The availability of such components for easy use by process designers will go a long way toward clarifying complex control system design. This section will examine some of the functions that are of particular importance to the food industry.

A. Blending

The blending function first came into prominence through its use in the petrochemical industry. Its primary purpose was to control accurately the blending of gasoline components as they were mixed on their way to storage tanks. Since that time, blending has been applied in many other industries and is presently experiencing rapid growth in the food industry where blending components is a common process function.

A blending system is basically an advanced ratio control system. Individual flows are controlled using conventional, single-loop flow control. The set points of the controllers are manipulated so that a particular ratio of the ingredients is maintained. Simple ratio control selects one flow to be the "master" and all others to be "slaves" with their set points becoming a multiple of the master's. Problems arise if one or more of the slaves deviates from its set point (possibly caused by pressure loss or a slowly plugging line or valve). In this case no feedback is provided to the master, and therefore, ratio control can be lost.

Modern blending systems are based on microprocessor technology and include operator displays for entering the desired ratio of ingredients (Fig. 10). In the case of the blender, only the total flow and component ratios need be entered as the system feeds back information concerning the flow status of each component and adjusts the others accordingly to maintain the correct ratios. The effect is to make each individual flow the "master" flow.

Blending has been successfully applied to many diverse areas in the food industry. One such use involves the blending of high-gravity beer where streams are mixed with the beer flow for dilution, additives, and injection of carbon dioxide. This application provided the high accuracy and flexibility needed to meet special brand requirements. In the production of edible oil a blender provides a complete, compact control center for the instrumentation of the entire operation. It minimizes neutral oil losses and caustic usage, and significantly improves the refinery operating performance.

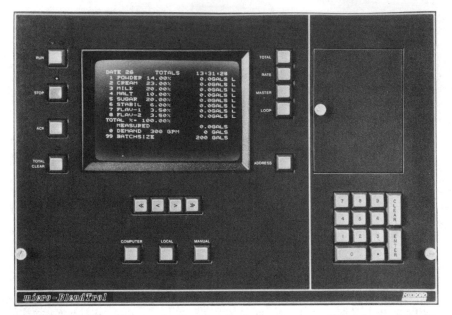

FIGURE 10 Blending controller for food processing. (Courtesy of the Foxboro Co., Foxboro, Mass.)

The dairy industry utilizes blending for standardization control and the coordination of the pasteurization process. In this case it can increase plant
production and lower operating costs by permitting faster and more efficient
product changes. Other applications include the mixing of ingredients for
bread manufacturing and for control of nutrient and substrate feeds to fermentation units.

B. Sequential Logic

Almost all food industry processing involves some aspect of sequential control as operating procedures call for operations such as fill, add ingredient,
cook for a time period, dump, and clean. Automation is particularly well
suited to such repetitive, well-defined procedures, and consequently sequential control is normally the first application area to be automated after simple loop control.

Sequential control need not require the use of computer systems. The
Foxboro Company manufactures solid-state, plug-in components for prepro-
grammed heating-cycle control in the brewing industry as part of its SPEC
200 analog equipment. The logic elements control rates of heating or cooling, holding temperature periods, and sequencing of heating and cooling
phases in the mash and cereal cookers. The resulting temperature control
is reproducible to ±0.5°C.

More complex applications which include time- or event-orientated stepping logic, interlock logic, control capability, as well as data storage for procedures and recipes require additional equipment—usually involving a computer. In some cases these components start with the ability to combine sequential logic with the blending component previously described (Fuller and Cook, 1979) to control one portion of a batch operation. More extensive automation applications in the pharmaceutical industry involve on-line computers to control processes, fermentation operations, lab instruments, and physiological testing. Special-purpose languages which allow engineers to easily design operating procedures have been developed by individual companies for their own use (Thome, Cline, and Grillo, 1979). The present trend, however, is for companies to utilize the packaged, batch control facilities provided by vendors rather than begin the expensive effort to develop and maintain batch software. Examples of plants who have successfully made use of such components are available in the literature (Vickers, 1975).

G. Energy Management

The food industry, with its emphasis on cooking, baking, sterilization, and drying, is a heavy user of energy. Consequently components emphasizing efficient use of energy are being included in automation plans. The key element to this control component is the conservation of available work (developed from a thermodynamic analysis of the process) from energy as it flows through the plant (Beaverstock and Cole, 1980).

New control techniques such as floating pressure control (Fauth and Shinskey, 1975) are being used to increase energy efficiency. Utility distribution optimization in steam systems can be achieved by floating the header pressures to meet the needs of the most demanding user. Regulation of steam-header pressures in cogeneration facilities can likewise be achieved through controlling turbine admission and extraction operating conditions. Components for further controlling the distribution of energy in food plants are also available. Tie-line control achieves a balance between plant-generated electricity and that purchased from an outside utility by maintaining electrical purchases at or slightly below a demand limit to avoid demand charges. Dispatching control looks at energy conversion costs and availability of equipment to maximize turbine shaft work and minimize inefficient pressure-relief-valve flows while satisfying process demands.

In the process areas, grain-drying control can yield savings of 15% in energy, increase potential dryer capacity, and reduce the amount of grain being given away in place of water. Balancing heat loads on cookers or evaporators is another aspect of energy management systems as multiple operations require phasing to minimize wasted energy.

The energy management component of an automation system becomes complex as its scope increases. However a common methodology is available (Shinskey, 1978) to design an energy control system that can effectively be integrated with traditional controls designed to achieve desired levels of quality, safety, and plant throughput.

VII. CONCLUSIONS

Computers for process control in the food industry will enable significant advances to be accomplished in the level of automation in many plants. However, a process plant is a complex entity with entirely different operating characteristics than a laboratory or pilot-plant environment. Because of these circumstances, a successful design of the control strategy goes beyond the selection of hardware depending on knowledge of the process and an understanding of the capabilities of advanced control tools. This can be accomplished by food industry personnel, based on applying available control components even without specialized control expertise. The following points, however, are crucial to any automation program:

1. A firm base of single-loop control must be established to assure consistent process operation before higher level automation is considered.

2. While the computer can be used to collect process data, knowledge of the process is required before such data can be used for control.

3. A top-down control strategy is required that defines overall short- and long-term objectives and implications of the control system.

REFERENCES

Beaverstock, M. C. and Bernard, J. (1977). Advanced control: Ready, able, accepted?, Fifth IFAC/IFIP International Conference on Digital Applications to Process Control. The Hague, The Netherlands, June 14–17.

Beaverstock, M. C., Bernard, J. W., and Evans, L. B. (1979). Economic modeling of process management and control systems, ISA Trans. 18:2.

Beaverstock, M. C. and Cole, W. (1980). Energy conservation control systems: A systematic and thermodynamic approach, Joint Automatic Control Conference, San Francisco, Calif., August.

Beaverstock, M. C., Stassen, H. G., and Schneider, H. W. (1979). Modeling the operator: A needed approach to interface design, InTech 26:4, April.

Beaverstock, M. C. and Trearchis, G. P. (1978). Fermentation control systems: A time for change, Second International Conference on Computer Applications in Fermentation Technology, University of Pennsylvania, Pa., August.

Bernard, J. W. and Howard, G. M. (1968). Organizing multi-level process control systems, IEEE System Science and Cybernetics Conference, San Francisco, Calif., October 16.

Bristol, E. H. (1980). After DDC: Idiomatic (structured) control, 88th National Meeting of A.I.Ch.E., Philadelphia, Pa., June.

Bristol, E. H. (1979). Organization and discipline for distributed process control, Instrum. Technol. January.

Cooney, C. L. (1978). Computer application in fermentation technology, Second International Conference on Computer Applications to Fermentation, University of Pennsylvania, Philadelphia, Pa., August.

Dobson, J. C. and Trearchis, G. P. (1976). The computer as a tool for fermentation control, American Chemical Society meeting, San Francisco, Calif., August.

Electric Power Research Institute (1979). Human factors methods for nuclear control room design, report NP-1118SY, June.

Fadum, O. and Shinskey, F. G. (1980). Saving energy through better control of continuous and batch dryers, Control Engineering, March.

Fauth, C. A. and Shinskey, F. G. (1975). Advanced control of distillation columns, Chem. Eng. Progr. 71:6, June.

Fuller, R. T. and Cook, R. E. (1979). Integrated process control systems, ISA Symposium, Wilmington Section, Wilmington, Del., May.

Homan, J., Brennecke, O., and Forwalter, J. (1978). Computer control, Food Processing June.

Sheridan, T. B. (1980). Human error in nuclear power plants, MIT Technol. Rev. February.

Shinskey, F. G. (1978). Energy Conservation Through Control. Academic, New York.

Shinskey, F. G. and Myron, T. J. (1975). Product moisture control for steam: Tube and direct fired dryers, MBAA Tech. Quart. 12:4.

Stults, B. (1978). Food process instrumentation and control, Food Technol. March.

Thome, R. J., Cline, M. W., and Grillo, J. A. (1979). Batch process automation, Chem. Eng. Progr. May.

Vickers, F. P. (1975). Application of a process control computer to a 26-reactor batch plant, Chemische Rundschau 28:48.

Waters, A. (1978). A pneumatic composition transmitter and its applications, Chem. Eng. Progr. June.

Williams, T. (ed.). Guidelines for the design of man/machine interfaces for process control, TC-6 committee report of the International Purdue Workshop on Industrial Computer Systems, Purdue University, West Lafayette, Ind. (1978).

13
Computer Analysis of Food Chromatographic and Electrophoretic Data of Protein

NICHOLAS CATSIMPOOLAS Boston University School of Medicine,
Boston, Massachusetts

I. INTRODUCTION

Food proteins play an important role in the formulation of new foods with desirable functional and nutritional characteristics. Research and analytical problems involving proteins require the use of modern separation and characterization techniques, the most important ones being electrophoresis and chromatography in their multiple variations. The increasing sophistication in the application of these methods necessitates the use of computers to perform the various statistical computations.

Computers are utilized to carry out three main functions: (1) acquisition and smoothing of data; (2) chromatographic or electrophoretic peak analysis, i.e., statistical characterization of the protein distribution in regard to position, width, and shape; and (3) incorporation of the peak analysis data into programs that provide physical characteristics of the proteins, e.g., molecular weight, diffusion coefficients, and test of identity between two molecular species. The peak analysis data can also be utilized in comparing the resolution provided by various separation methods and, therefore, choosing the best on a rational basis.

The chromatographic signal to be analyzed is usually provided by ultraviolet absorbance, fluorescence emission, and refractometric instruments which continuously monitor the effluent of the separation column. The separation of a mixture of proteins can be based on ion exchange, molecular sieving, hydrophobic interactions, partition, and biospecific affinity principles. Electrophoretic separation can be similarly monitored with the additional use of scanning densitometers which detect various stains (dyes) bound to the proteins after their separation.

The involvement of computers in all of the above steps for the final characterization of proteins will be outlined below, especially as used in the author's laboratory.

II. DATA ACQUISITION AND SMOOTHING

A. Data Acquisition

The analog signal from the photomultiplier during the monitoring of the chromatographic or electrophoretic column is a continuous current (or voltage) whose amplitude is proportional to the absorbance or fluorescence of the protein. In order to be processed by a computer, the analog voltage must be digitized to produce a word whose number of bits after conversion is equivalent ot the input voltage. To avoid large errors, the maximum input voltage has to be amplified to the maximum voltage that the analog-to-digital converter can accommodate, e.g., 10 V. A digital voltmeter can measure voltages as low as a few microvolts with as little as 0.01% error, but analog-to-digital conversion requires at least 10 msec for completion. Such slow converters act as a low-pass filter and thereby smooth the data to a considerable extent. However, fast analog-to-digital converters can process and 8-bit word in less than 10 μsec, which is indeed much faster than is normally required. In this case, amplification of the input signal is usually necessary.

The analog signal produced by the photometer is digitized by an analog-to-digital converter at preselected time intervals. Since the mechanical scanning speed v of the densitometer or the flow rate through the monitor is constant, the digitized values represent the photometer signal amplitude at equidistant points. The number of data points j obtained per peak depends on the velocity v (cm/sec) of the scanner or flow rate (ml/min), the digitizing rate g (data points per second), and the total width w_t (centimeters or ml) of the peak profile at the baseline, so that

$$j = \frac{gw_t}{v} \qquad \qquad (1)$$

To a first approximation, the narrower the slit width(s) of the scanner, or flow monitor, the more accurately the peak profile will be recorded. However, the narrower the slit width, the larger the random noise in the signal. A large slit width smooths the data by averaging the light absorbance over a larger window. In practice, the slit width should be approximately equal to the ratio w_t/j. Depending on baseline noise and peak shape (Catsimpoolas and Griffith, 1973) the desired value of j can be set between 10 and 100. If the peaks are of gaussian shape, 10 data points may be sufficient to obtain their statistical moments. The more asymmetric the peak, the more data points are needed for its characterization. The total number of points k to be acquired per scan depends also on the peak capacity n which is defined as the maximum number of peaks resolvable in the column, so that

$$k = jn \qquad \qquad (2)$$

The peak capacity depends primarily on the average standard deviation $\bar{\sigma}$ (centimeters or milliliters) of typical peaks. Since adjacent peaks can be considered completely resolved if the peak heights are at a distance of $4\bar{\sigma}$ from each other, the value of n can be approximated by $L/4\bar{\sigma}$, where L (centimeters or ml) is the length of the column, or total volume of usable effluent.

After acquisition, the digitized data can be stored on punched paper tape, magnetic tape, disks, or other data-storage devices. The speed at which these devices store data must be compatible with the data-acquisition rate.

B. Data Smoothing

The primary information from the data-acquisition step should be in the form of absorbance or percent transmittance of the compounds of interest versus distance of separation or volume. However, random errors which are characteristically described as noise are often superimposed on this information in an indistinguishable manner. Removal of the noise without degrading the underlying information can be performed by convoluted least-squares differentiation procedures (Grushka, 1975). Two important restrictions have to be taken into consideration in this type of analysis. First, the digitized data points must be at a fixed, uniform interval in the chosen abscissa (i.e., volume, distance, or time). In other words, each data point must be obtained at the same time interval from each preceding point. Second, the curves formed by graphing the points must be continuous and more or less smooth. All of the errors are assumed to be in the ordinate (absorbance, percent transmittance) and none in the abscissa.

TABLE 1 Smoothing (Zero–Derivative) Convolution Integers and Normalizing Factors (Quadratic or Cubic)

Data point	Fixed number of points					
	15	13	11	9	7	5
-07	-78					
-06	-13	-11				
-05	42	0	-36			
-04	87	9	9	-21		
-03	122	16	44	14	-2	
-02	147	21	69	39	3	-3
-01	162	24	84	54	6	12
00	167	25	89	59	7	17
01	162	24	84	54	6	12
02	147	21	69	39	3	-3
03	122	16	44	14	-2	
04	87	9	-9	-21		
05	42	0	-36			
06	-13	-11				
07	-78					
Normalizing factor	1105	143	429	231	21	35

Source: Catsimpoolas, 1975.

The least–squares calculations are carried out by convolution of the data points with properly chosen sets of integers. In this procedure, we take a fixed number of points and evaluate the central point by a convoluting function (i.e., quadratic) the next point at the other end is added, and the process is repeated. Table 1 contains the convolution integers and their normalizing factors for smoothing (zeroth derivative) polynomials of degrees 2 and 3 from 5 to 15 fixed numbers of points. As an example, the equation for smoothing a five-point quadratic or cubic convolute is

$$y = \frac{-3y_{-02} + 12y_{-01} + 17y_{00} + 12y_{01} - 3y_{02}}{35} \qquad (3)$$

The number of fixed points that may be used for the smoothing depends on how accurately the polynominal describes the experimental curve under examination. The number of points should be chosed so that no more than one inflection in the observed data is included in any convolution interval. Data digitized at high intensities (i.e., taken very closely together) offer more flexibility in the choice of the number of points to be used. Generally, the more data points included in the convoluting function, the less is the noise, since this is reduced approximately as the square root of the number points involved. Thus, a nine-point smoothing produces approximately a threefold improvement in the signal-to-noise ratio. Although there is no way to assess the degree of distortion of recorded data introduced by the smoothing procedure, experience gained with separation of known samples may suggest compromise conditions between rate of digitizing and number of data points involved in the convolution.

III. PEAK ANALYSIS

A. Slope Analysis

Since the distribution of a physically homogeneous population of proteins can be approximated by a gaussian curve, slope analysis can be very useful in detecting the baseline and peak shoulders and in estimating the resolution of strongly overlapping gaussian peaks. Slope analysis can be performed by convoluted least-squares differentitation procedures (Savitzky and Golay, 1964) similar to those used for smoothing. For most work it is adequate to estimate the first- and second-derivative convolutes. Tables 2 and 3 list the convolution integers and their normalizing factors for first- and second-derivative calculations for polynomials of degrees 2 and 3 from 5 to 15 fixed numbers of points. As an example, the equation for deriving the first derivative by a five-point quadratic convolute is

$$y' = \frac{-2y_{-02} - y_{-01} + 0y_{00} + y_{01} + 2y_{02}}{10} \tag{4}$$

and for deriving the second derivative by a seven-point convolute is

$$y'' = \frac{5y_{-03} + 0y_{-02} - 3y_{-01} - 4y_{00} - 3y_{01} + 0y_{02} + 5y_{03}}{42} \tag{5}$$

The first derivative of a gaussian curve has a maximum (+) and a minimum (−) at the inflection points of the curve and passes through zero (0) at peak maximum. Even in the presence of extreme baseline noise, the inflection points can be located. The distance between the maximum or minimum and the zero position corresponds to the standard deviation (σ_1, σ_2) on the front and back side of the curve, respectively. For a gaussian distribution, $\sigma_1 = \sigma_2$. Thus, the first derivative can be utilized to detect the presence of

TABLE 2 First-Derivative Convolution Integers and Normalizing Factors (Quadratic)

Data point	Fixed number of points					
	15	13	11	9	7	5
−07	−7					
−06	−6	−6				
−05	−5	−5	−5			
−04	−4	−4	−4	−4		
−03	−3	−3	−3	−3	−3	
−02	−2	−2	−2	−2	−2	−2
−01	−1	−1	−1	−1	−1	−1
00	0	0	0	0	0	0
01	1	1	1	1	1	1
02	2	2	2	2	2	2
03	3	3	3	3	3	
04	4	4	4	4		
05	5	5	5			
06	6	6				
07	7					
Normalizing factor	280	182	110	60	28	10

Source: Catsimpoolas, 1975.

an asymmetric peak ($\sigma_1 \neq \sigma_2$), where H is the peak height and X_m is the abscissa position of peak maximum from an arbitrary origin. In the case of a gaussian curve, the σ estimated by slope analysis is useful in obtaining the area A of the peak, where H is the peak height.

$$A = H\sigma(2\pi)^{1/2} \tag{6}$$

The second-derivative curve of a gaussian distribution produces a minimum (−) at the peak maximum of the original curve and two maxima (+). The magnitudes (ordinary values) of these extrema depend on the height of

TABLE 3 Second-Derivative Convolution Integers and Normalizing Factors (Quadratic or Cubic)

Data point	Fixed number of points					
	15	13	11	9	7	5
-07	91					
-06	52	22				
-05	19	11	15			
-04	-8	2	6	28		
-03	-29	-5	-1	7	5	
-02	-44	-10	-6	-8	0	2
-01	-53	-13	-9	-17	-3	-1
00	-56	-14	-10	-20	-4	-2
01	-53	-13	-9	-17	-3	-1
02	-44	-10	-6	-8	0	2
03	-29	-5	-1	7	5	
04	-8	2	6	28		
05	19	11	15			
06	52	22				
07	91					
Normalizing factor	6188	1001	429	462	42	7

Source: Catsimpoolas, 1975.

the peak and its variance (σ^2). However, the ratio of their maximum to the minimum (ordinates) is independent of the height and variance of the peak, and in fact, its value is $-2 \exp(-3/2) = -0.446$. The ratio between the two maxima is unity. Defining R_1 as the ratio of the maximum on the front side of the curve to the minimum, R_2 as the ratio of the maximum on the back side of the peak to the minimum, and R_3 as the ratio of the two maxima on the front and back side, Grushka (1975) has suggested the following applications. In a gaussian system, any deviation from the extrema ratios of $-2 \exp(3/2)$ and 1.0 immediately indicates the existence of double peaks. Second-derivative analysis is preferred to the first derivative because it is sensitive to slope changes when one of the two peaks in the composite is one-tenth or less than the height of the second. In the case of two overlapping, equal peaks, the first-derivative ratio (maximum to minimum) is always unity. While the ratio of two maxima (R_3) of the second derivative is unity when the overlapping peaks are identical, the two ratios R_1 and R_2 increase with the resolution between the two peaks. By employing calibration curves (Grushka, 1975) with suitable standards, the R_1, R_2, and R_3 ratios can indicate not only the resolution between strongly overlapping peaks in the composite but also the ratio of their heights. With single peaks, the standard deviation (σ) can be estimated from

$$\pm\sigma = \frac{x - x_0}{\sqrt{3}} \tag{7}$$

where the x's are the abscissa positions of the extreme and x_0 is the coordinate of the gaussian's maximum (center of gravity).

B. Peak Detection and Baseline Correction

First- and second-derivative slope analysis can be used to distinguish a peak maximum ($y' = 0$, $y'' < 0$) from a trough ($y' = 0$, $y'' > 0$) [baseline conditions ($y' = 0$, $y'' = 0$)], or a peak ascent ($y' > 0$, $y'' > 0$, then $y'' < 0$) from a peak descent ($y' < 0$, $y'' < 0$, then $y'' > 0$). In practice, "zero" for y' is represented by the range of two threshold values C_1 and C_2 and that for the y'' by two additional threshold parameters C_3 and C_4. Baseline conditions are met with $C_1 > y' > C_2$ and $C_3 > y'' > C_4$. The selection of the parameters C_1, C_2, C_3, and C_4 depends on the magnitude of the baseline noise in relation to the peak height. At first, a desired limit of integration (l) is selected such that it represents any percentage (e.g., 0.01, 1.0%) of the peak height H. This is called an H_l increment. C_1 and C_2 have the same value but exhibit positive and negative signs, respectively. The values of C_1 and C_2 for a defined H_l are obtained from the first derivative y' of consecutive data points 5 to 15 (see Table 2), increasing (C_1) or decreasing (C_2) by the H_l increment, the higher are the values of C_1 and C_2. For example, if H_l is set at 0.1% of the peak height, baseline conditions are met within the threshold window C_1 to $-C_2$ (computed for H_l of 0.1 H), even if the baseline ascends or decends by one H_l increment. Thus, the H_l values determine the sensitivity of detection of slope changes, which signify the start and end of peaks. It should be noted that detection of "peak start" by the slope increase method should be corrected by storing the data preceding the point of peak detection to set back the baseline. For a gaussian curve, the baseline should be set back approximately 0.3σ. Similar corrections apply to the back side of the peak where additional points are added on after the point of baseline detection. As an example, the equations for deriving the C_1 and C_2 parameters for a seven point moving average are

$$C_{1,2} = \frac{\pm[-3(H_1) - 2(2H_1) - 3(H_1) + 0(4H_1) + (5H_1) + 2(6H_1) + 3(7H_1)]}{28} \tag{8}$$

It can be seen that if the limit of integration is set at 0.1% of H, the parameters C_1 and C_2 have values of $\pm 1.0 \times 10^{-2}$. The second-derivative threshold parameters C_3 and C_4 are set empirically at such levels as to avoid false triggering of peak start and end by baseline noise. Some guidelines can be obtained by entering alternate or fluctuating values of H_1, $-H_1$ into the second-derivative (y'') equations of variable consecutive data points (see Table 3).

Once the baseline on both sides of the peak has been detected, the curve has to be corrected for an ascending or descending baseline by linear interpolation. The method is based on the assumption that the peak is superimposed on a baseline exhibiting a linear sloping continuum. Correction of the distribution is carried out according to the equation

$$y_{corr} = y_{obs} - (y_e - y_s)/(x_e - x_s) \, x_{obs} - \frac{y_s x_e - y_e x_s}{x_e - x_s} \tag{9}$$

where y_{corr} is the corrected ordinate, y_{obs} is the observed ordinate, x_{obs} is the corresponding coordinate, y_s and y_e are the ordinates at peak start and end, and x_s and x_e are the corresponding coordinates.

C. Moment Analysis

After baseline correction, the nth statistical moment (m'_n) of the concentration distribution $C(x)$ of a peak is estimated by

$$m'_n = \frac{\int x^n C(x) \, dx}{\int C(x) \, dx} \tag{10}$$

Subsequently, the second, third, and fourth central moments, measured relative to m'_1, are obtained by

$$m_n = \frac{\int (x - m'_1)^n C(x) \, dx}{\int C(x) \, dx} \tag{11}$$

where x is the distance and C is the concentration.

The first moment m'_1 is the center of gravity of the concentration profile. It coincides with the peak maximum only if the peak is symmetrical (Fig. 1). The second central moment m_2 is the peak variance σ^2. The square root of m_2 corresponds to the standard deviation σ of the concentration distribution which provides a measure of peak width. The third central

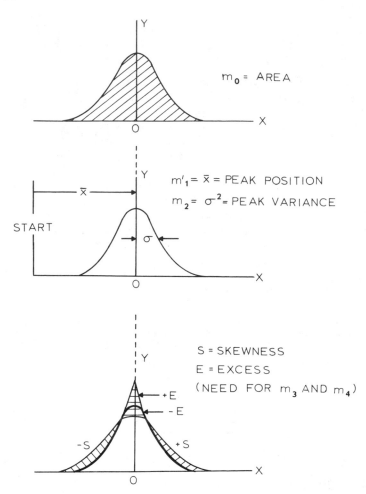

FIGURE 1 Statistical moments and their practical significance.

moment m_3 indicates the direction and magnitude of peak asymmetry, whereas the fourth central moment m_4 is a measure of peak flatness as compared to a gaussian shape. Negative values of m_3 indicate a fronting shape, and positive values indicate tailing peaks. The zeroth moment is the normalized area of the peak.

In addition, the second, third, and fourth central moments can be used to estimate the coefficients of skewness S and excess E by

$$S = \frac{m_3}{m_2^{3/2}} \tag{12}$$

$$E = \frac{m_4}{m_2^2} - 3 \tag{13}$$

The skew measures the asymmetry of the peak, while the excess indicates deviation from a gaussian shape in regard to flatness. The coefficients S and E are dimensionless quantities and therefore are independent of the size of the peak (Grushka, 1975). In practice, the moments are estimated by digitizing the absorbance A_1 at the fixed time intervals which correspond to a fixed distance Δx or volume in the column. Starting from an arbitrary origin, the coordinate at the ith interval is x_i. The moments are estimated by

$$m_0 = \sum_i A_i \, \Delta x \tag{14}$$

$$m_1' = \frac{\sum A_i x_i}{\sum A_i} \tag{15}$$

$$m_2 = \frac{\sum A_i (x_i - m_1')^2}{\sum A_i} \tag{16}$$

$$m_3 = \frac{\sum A_i (x_i - m_1')^3}{\sum A_i} \tag{17}$$

$$m_4 = \frac{\sum A_i (x_i - m_1')^4}{\sum A_i} \tag{18}$$

It is apparent that the exact description of the peak area, position, width, and shape by moment analysis has fundamental applications in protein analysis. The connection between the two is made by adopting mathematical models describing the separation process in regard to the protein transport as a function of time and also as affected by other physical factors such as molecular sieving.

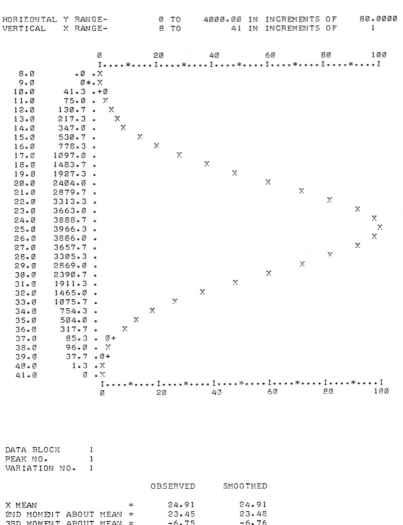

```
HORIZONTAL Y RANGE-          0 TO    4000.00 IN INCREMENTS OF    80.0000
VERTICAL   X RANGE-          8 TO      41 IN INCREMENTS OF       1


               0        20        40        60        80        100
               I....*....I....*....I....*....I....*....I....*....I
    8.0      .0  .X
    9.0       0*.X
   10.0      41.3 .+0
   11.0      75.0 . X
   12.0     130.7 .  X
   13.0     217.3 .   X
   14.0     347.0 .    X
   15.0     530.7 .      X
   16.0     778.3 .        X
   17.0    1097.0 .          X
   18.0    1483.7 .             X
   19.0    1927.3 .               X
   20.0    2404.0 .                  X
   21.0    2879.7 .                     X
   22.0    3313.3 .                        X
   23.0    3663.0 .                          X
   24.0    3888.7 .                            X
   25.0    3966.3 .                             X
   26.0    3886.0 .                            X
   27.0    3657.7 .                          X
   28.0    3305.3 .                       Y
   29.0    2869.0 .                    X
   30.0    2390.7 .                 X
   31.0    1911.3 .              X
   32.0    1465.0 .          X
   33.0    1075.7 .       Y
   34.0     754.3 .     X
   35.0     504.0 .   X
   36.0     317.7 .  X
   37.0      85.3 . 0+
   38.0      96.0 . X
   39.0      37.7 .0+
   40.0       1.3 .X
   41.0        0  .X
               I....*....I....*....I....*....I....*....I....*....I
               0        20        40        60        80        100


DATA BLOCK        1
PEAK NO.          1
VARIATION NO.     1

                            OBSERVED       SMOOTHED

X MEAN                =        24.91          24.91
2ND MOMENT ABOUT MEAN =        23.45          23.48
3RD MOMENT ABOUT MEAN =        -6.75          -6.76
4TH MOMENT ABOUT MEAN =      1533.09        1539.50
BETA 1                =         -.06           -.06
BETA 2                =         2.79           2.79
STANDARD DEVIATION    =         4.84           4.85
AREA                  =     49100.33       49109.16
```

FIGURE 2 Sample computer output of a hypothetical distribution and a printed report of the estimated moments.

In this laboratory, a computer program available in both FORTRAN and BASIC languages is routinely used for processing the data (Fig. 2). In the case of overlapping distributions, deconvolution techniques are applied (Cole, Shapiro, and Rodbard, 1978).

IV. PROTEIN CHARACTERIZATION

A. Gel Chromatography: Molecular-Weight Distribution Analysis

Methods for the determination of molecular-weight averages have been discussed in general references (Tanford, 1961; Morawetz, 1965) which should be consulted for detailed theoretical background information. The number-average molecular weight (M_n) can be determined by osmotic pressure measurements, the weight-average molecular weight (M_w) by light-scattering experiments, and the z-average molecular weight (M_z) from equilibrium sedimentation data. When colligative properties are utilized, each molecule, large or small, makes the same contribution to the observed averaging process. If the total weight were shared equally among the molecules in the mixture, the observed contribution will remain unchanged, thus leading to the determination of M_n. However, the large species are weighted more heavily in the weight average than in the number average, so that for a polydisperse sample $M_w > M_n$. The fraction M_w/M_n is used to characterize the "polydispersity" of the sample. Very high molecular weight species influence the n average more than the weight average. Gel-filtration chromatography can be used to determine simultaneously M_n, M_w, and M_z, which is a very attractive feature of the method considering the simplicity and rapidity of the experimental procedure (Catsimpoolas, 1975). However, GFC being a secondary method requires calibration against a primary method (e.g., osmometry, light scattering, sedimentation) of molecular-weight determination, and therefore, the use of molecular-weight markers.

The molecular-weight distribution is characterized by its principal statistical moments and is given by

$$\overline{M}_n = \frac{\sum\limits_{i=1}^{\infty} n_i M_i}{\sum\limits_{i=1}^{\infty} n_i} \tag{19}$$

$$\overline{M}_w = \frac{\sum\limits_{i=1}^{\infty} n_i M_i^2}{\sum\limits_{i=1}^{\infty} n_i M_i} \tag{20}$$

$$\overline{M}_z = \frac{\displaystyle\sum_{i=1}^{\infty} n_i M_i^{3}}{\displaystyle\sum_{i=1}^{\infty} n_i M_i^{2}} \tag{21}$$

where n_i is the concentration number of molecules of the ith kind per unit volume and M_i is their molecular weight. In addition, the partition coefficient (K_{av}) in rapid-gel-filtratrion experiments is given by

$$K_{av} = \frac{t_e - t_0}{t_n - t_0} \tag{22}$$

where t_0 is the minimum retention time for the column determined with a totally excluded compound (e.g., thyroglobulin), t_n is the maximum retention time measured with a totally included compound (e.g., dinitrophenyl alanine), and t_e is the retention time of any substance which satisfies the condition $t_0 < t_e < t_n$. Marker proteins and peptides obey the empirical relation

$$-\log K_{av} = RM^{2/3} + S \tag{23}$$

where M is the molecular weight of the marker, and R and S are constants obtained by linear regression analysis. Combination of Eqs. (22) and (23) produces.

$$M = \left(\frac{-\log [(t_e - t_0/t_n - t_0)] - S}{R} \right)^{3/2} \tag{24}$$

Considering t_i to be the ith time interval (of the molecular-weight distribution) at which M_i (the molecular weight at the ith interval) is measured, then Eq. (24) becomes

$$M_i = \left[\frac{(-\log [(t_i - t_0/t_n - t_0)]) - S}{R} \right]^{3/2} \tag{25}$$

Assuming the standard deviation (σ) of the concentration distribution of all species to be identical, the concentration of protein or peptide molecules in the ith interval is proportional to the absorbance (A_i) at 220 nm. Therefore, the proportion of the total weight present in the ith fraction is

$$n_i = \frac{A_i \, \Delta t}{\Sigma A_i \, \Delta t} \tag{26}$$

where Δt is the constant interval at which A_i measurements were taken. So that Eqs. (19) to (21) become

$$\overline{M}_n = \frac{\Sigma A_i}{\Sigma A_i / M_i} \tag{27}$$

$$\overline{M}_w = \frac{\Sigma A_i M_i}{\Sigma A_i} \tag{28}$$

$$\overline{M}_z = \frac{\Sigma A_i M_i^{\,2}}{\Sigma A_i M_i} \tag{29}$$

Combination of Eqs. (25) and (27) to (29) produces the final equations for calculation of M_n, M_w, and M_z. In practice, A_i at t_i can be obtained automatically by digitizing the absorbance at a constant time interval Δt. S and R are predetermined calibration constants obtained by linear regression from Eq. (22). The values of t_0 and t_n are obtained from the first statistical moment of the unretained and completely retained peaks, respectively, so that

$$t_0 = \left(\frac{\Sigma A_i t_i}{\Sigma A_i} \right)_{\text{peak 0}} \tag{30}$$

$$t_n = \left(\frac{\Sigma A_i t_i}{\Sigma A_i} \right)_{\text{peak n}} \tag{31}$$

This process assumes that the two marker peaks (distributions) are monodispersed (i.e., $M_w / M_n = 1$).

Equation (23) can assume other forms depending on the relationship between the partition coefficient and molecular weight of marker compounds, and whether this is assumed to be linear or polynomial. Third-degree-polynomial fitting techniques have been shown to offer a much greater degree of precision (i.e., as low as 0.1%) than linear equations in the analysis of retention data of globular proteins subjected to gel filtration.

B. TRANS Electrophoresis: Estimation of Molecular Weights and Diffusion Coefficients

Transient-state (TRANS) electrophoresis is a method which utilizes repetitive electrooptical scanning of the separation path (in the presence of the electric field) for the kinetic monitoring and computation of the distribution of charged species subjected to various forms of electrophoresis (Catsimpoolas, 1975). The electrophoresis support medium can be either a density gradient (e.g., of sucrose) or a polyacrylamide gel which exhibits molecular sieving properties. The experiments are performed in vertically oriented quartz cylindrical tubes with the electrical field applied at the top and bottom of the tubes. Thus, electrophoretic migration of the electrically charged proteins occurs downward.

The electrophoretic peak velocity can be measured from the slope of the plot of the mean peak position m'_1 versus time (t) by linear regression:

$$\bar{v} = \frac{d(\bar{m}'_1)}{dt} \tag{32}$$

If the velocity of the front $(\bar{v}')$ is measured (e.g., by using a tracking dye) simultaneously with the proteins of interest, the relative mobility value R_f of the latter is obtained from the ratio of the velocities:

$$R_f = \frac{\bar{v}}{\bar{v}'} \tag{33}$$

This measurement allows the determination of the retardation coefficient K_R of the protein from the slope of the plot of log R_f versus %T by linear regression analysis, where T denotes percent acrylamide concentration. In practice, each tube contains gels of different acrylamide concentrations and, therefore, have a different sieving properties. The molecular radius $(\bar{R})$ of the globular proteins is estimated from the K_R of "standard" proteins by linear regression between $(K_R)^{1/2}$ and $\bar{R}$. Subsequently, $\bar{R}$ is translated into molecular weight (MW) with its 95% confidence limits. For random-coil proteins, a linear relationship between MW and K_R is used (Chrambach et al., 1976).

The statistical analysis of the second central moment of a peak distribution provides a measure of its variance (σ^2). Changes in peak variance as a function of time can be utilized in the presence of the electric field $(E > 0)$ to determine peak spreading (Chen, Chrambach, and Rodbard, 1979) and in the absence of the electric field $(E = 0)$ to determine the apparent diffusion coefficient D. D can be estimated from the slope of the plot of σ^2 versus t using

$$D = \frac{1}{2} \frac{d(\sigma^2)}{dt} \tag{34}$$

Measurement of σ^2 (>0) in analytical polyacrylamide gel electrophoresis (PAGE) can provide useful information for the optimization of resolution in preparative PAGE (Rodbard et al., 1974). The apparent diffusion coefficients have to be corrected for current density, protein load, and ionic strength (Chen, Chrambach, and Rodbard, 1979).

Computer programs are available for performing the above statistical computations (Chrambach et al., 1976), as well as for estimating other parameters, i.e., electrophoretic mobility and valence. Recently, computer programs have been developed (Lutin, Kyle, and Freeman, 1978; Taylor et al., 1968; Alexander et al., 1980) for the analysis of two-dimensional maps of proteins in gels, which give an estimate of the isoelectric point and molecular weight of each protein species.

It can be safely predicted that food protein analysis in the not-so-distant future will be performed entirely by computer techniques.

REFERENCES

Alexander, A., Cullen, B., Enrighlolz, K., Norgard, M. V. and Monahan, J. J. (1980). A computer program for displaying two-dimensional gel electrophoresis data. Anal. Biochem. 103:176.
Catsimpoolas, N. (ed.) (1975). TRANS electrophoresis, in methods of Protein Separation, Vol. 1. Plenum Press, New York, p. 1.
Catsimpoolas, N. and Griffith, A. L. (1973). Transient state iselectric focusing: computational procedures. Anal. Biochem. 56:100.
Chen, B., Chrambach, A., and Rodbard, D. (1979). Continuous optical scanning in polyacrylamide gel electrophoresis: estimation of the apparent diffusion coefficient of beta lactoglobulin B. Anal. Biochem. 96:120.
Chrambach, A., Jovin, T. M., Svendsen, P. J., and Rodbard, D. (1976). Analytical and preparative polyacrylamide gel electrophoresis: an objectively defined fractionation route, apparatus, and procedures, in Methods of Protein Separation, Vol. 2 (N. Catsimpoolas ed.). Plenum, New York, p. 27.
Cole, B. R., Shapiro, M. B., and Rodbard, D. (1978). Chromatography and electrophoresis analysis system, in Electrophoresis '78 (N. Catsimpoolas, ed.). Elsevier North-Holland, New York, p. 79.
Gurshka, E. (1975). Chromatographic peak shape analysis, in Methods of Protein Separation, Vol. 1 (N. Catsimpoolas ed.). Plenum, New York, p. 161.
Lutin, W. A., Kyle, C. F., and Freeman, J. A. (1978). Quantitation of brain proteins by computer-analyzed two dimensional electrophoresis,

in <u>Electrophoresis '78</u> (N. Catsimpoolas, ed.). Elsevier North-Holland, New York, p. 93.

Morawetz, H. (1965). <u>Macromolecules in Solution</u>, Interscience, New York,

Savitzky, A. and Golay, M. J. E. (1964). Smoothing and differentiation of data by simplified least square procedures. <u>Anal. Chem. 36</u>:1627.

Tanford, C. (1961). <u>Physical Chemistry of Macromolecules</u>. Wiley-Interscience, New York.

Taylor, J., Anderson, N. L., Coulter, B. P. Scandora, A. E., and Anderson, N. G. (1968). Estimation of two-dimensional electrophoretic spot intensities and positions by modelling, in <u>Electrophoresis '79</u> (B. J. Radola, ed.), W. deGruyter, Berlin, p. 329.

14

Computer System Development: A New Approach

EDWARD W. BURNSIDE AND STEPHEN C. NICHOLSON   Reckitt and
Colman, Hull, England

I. INTRODUCTION

The microchip will change your lives! How constantly is this repeated in
today's scientific, commercial, and popular press? How many articles,
media slots, and study groups are devoted to exploring the implications of
this revolution which is just around the corner?

Well, is it?

Similarly strident claims were made in the early 1960s when large
computers first became available to the business community. These were
to inaugurate an age of scientific decision making. Stock control, material
requirements planning, production scheduling, vehicle fleet maintenance,
sales forecasting, and a host of other activities would be performed better
than ever before with the aid of these powerful machines. Everything would
be parameterized, prioritized, and then optimized!

It didn't happen—well certainly not on the scale envisaged in those early
days. The reasons for this failure need investigating because we may be on
the point of extracting equally disappointing performances from the new technology. The authors believe that the systems design methodology which was
necessarily adopted at that time, and which has unnecessarily been continued
to the present, is the fundamental cause of failure. With the help of a case
study* we present an alternative strategy for creating computerized decision
aids. As well as allowing us to illustrate the points we wish to make, the
case study, that of the blending of honey, should be of intrinsic interest to
food technologists.

The opinions stated below are derived from the authors' experience and
observations in a diverse range of industries, and from conversations, sem-
inars, and conferences where the subject of computer systems has been discussed. The views should not be taken as critical of one section of industry
or of one profession, but as a general exhortation to exploit the new tools
and technology which are becoming available in order to produce more dur-
able computerized decision aids.

We review the traditional approach to computer systems design.

A. Conventional System Development

A typical development would proceed something along the following lines:

1. A manager calls in the management services department to help
   solve a particular problem or is persuaded to seek their assistance.

2. A systems analyst then discusses the problem with the manager,
   and together they arrive at the relevant computer solution.

3. The systems analyst then goes away and translates these discussions into a proposal for a computer system, occasionally seeking
   further information from the manager. This proposal will be a detailed list of the human resources needed to develop and run the
   system, the demands that will be made of the computer, a cost-
   benefit analysis, where the data is to come from, what reports are
   to be produced, etc.

---

*This case study was first presented at IP Sharp's APL User Conference
Toronto 1980.

4. This proposal is then published, and the manager is invited to sanction the expenditure so that actual development may begin.

5. At this stage other departments start taking an interest in the new system. Perhaps having read the system proposal, they realize that with some extensions the system could provide them with useful information. It is therefore agreed that the system will be enlarged to cover the requirements of these other departments and that the costs of development will be shared.

6. Having consulted the interested parties, the systems analyst will design the system. This design will be embodied in a system specification which will describe technically what computer programs are needed and how they will work, the number and types of computer files, the exact layout of the reports, screen displays, etc.

7. This specification is then passed to computer programmers who are specialists at producing the logical step-by-step instructions required by the machine. They will write these instructions in one of the many commercially available languages such as COBOL, FORTRAN, PL/1, or BASIC. These programs will be individually tested on small amounts of data supplied by the systems analyst and then fitted together to form the full system. Each program will be documented for future reference by a written commentary explaining exactly how it has been written.

8. The whole system will then be tested using realistic amounts of data and producing a full set of reports and screen displays.

9. When the development team is satisfied that the system behaves as specified, they will invite the sponsoring managers to assist them in extended trials designed to test the system in its working environment. Instruction manuals explaining how to use the system will be produced for this purpose.

10. When discovered errors have been corrected and the users have been fully trained, the development project is considered finished. The management services department will make a commitment to maintain the system (that is, to correct any further errors which are uncovered), however, extensions or required changes will be treated as separate developments for which the above 10 stages will be repeated.

It must be admitted that a good development team will not slavishly follow the above prescription. For instance, an experienced systems analyst will naturally include other potentially interested managers in the initial discussions, gaining a fuller perspective before writing the system proposal. Similarly, it is not necessary for the system specification to be completed

before programming can begin. Thus it is usually possible to telescope the
above stages to some extent, and different companies will have slightly different procedures; however, when one uses the conventional approach, these
10 steps will always be present. Such a methodology is necessary in a
"batch" environment, where computer tasks are submitted on, say, punched
cards and 24 hr may elapse before the reports are printed, as was the case
with the early technology. Maximizing machine efficiency was of paramount
importance. Having looked at the typical development cycle of a computerized management aid, let us look at the typical outcome.

B. Typical Results

The first point to note is that the above procedure is very time consuming,
often taking 12 months to 2 years before anything useful emerges. Second,
the sponsoring managers rarely find that the system solves their problems.
Somehow the concepts have become diluted and compromises have been
made in order to accommodate their differing requirements and keep within
the cost and time constraints. Nobody is really happy.

Another source of frustration is that in order to get the system changed
a further development cycle has to be embarked upon and expenditure sanc-
tioned. Even then changes will not be made immediately as the members of
the development team will have been switched to other data processing assignments. Thus, managers have to persevere with an inflexible system,
which is already perhaps 2 years out of date, in a dynamic commercial
environment which demands flexibility. The depressingly frequent outcome
is that they stop using the system and revert to the manual methods which
they have been forced to devise whilst waiting for the computer solution!

We stated earlier our belief that this development procedure is the
main reason why computers have not made the impact on management decision making that was initially predicted. Why should it be, however, that
such a logical and methodical approach, applied by logical and methodical
people, fails on so many occasions? We attempt to answer this question.

C. Critique of Traditional Methods

Tasks involving repetitive low-grade clerical effort were the first to which
computers were applied. There was a simple tradeoff between an expensive,
but extremely flexible, manual system and a cheap, inflexible computer system. To run large, repetitive systems such as payroll or invoicing cheaply
demanded a "batch" environment and effective machine utilization. However
the early computers demanded great skill and knowledge to run them effectively and economically. Thus the computer professional was a person who
spent many years gaining such knowledge and who saw his or her main task
as maximizing the operating efficiency of the machine. In particular, training would not have equipped him or her to understand the problems of executive managers. Now, although the technology has changed over the years,

the training and outlook of the computer specialist has usually not done so. Bearing this in mind, let us reexamine the development cycle described above.

When the sponsoring manager first meets the systems analyst, he or she is invited to state the problem. This will usually be done by describing a collection of symptoms, e.g., carrying an excessive amount of stock and yet customers are still complaining of the service level, the sales forecasts are terrible, and they cannot revise them quickly enough to allow the factory time to change its production schedule. The manager will then explain to the analyst what he requires in terms of information to enable him to overcome these problems. He condenses this down to a series of forms, charts, and flow diagrams which he shows the manager for approval before writing the system proposal. The manager agrees. Already the seeds of failure have been sown.

What has happened is that the manager has been asked to describe the computer solution in his own language which the systems analyst then interprets into technical terms. However the manager is just not equipped to do this effectively, having no real knowledge of what is possible with the machine. Similarly, the analyst does not have the business experience to critically assess the commercial validity of what he is being asked to create. Consequently, inherent misunderstandings are built into the system design, which neither party is aware of and which will only come to light when the system is being tried and actual reports are being produced.

This in itself would not be too bad if only a short time passed between the initial discussions and the first output from the system and if the system was easy to change. Usually however a considerable amount of time passes before anything emerges from the computer. During this interval, the manager loses almost total contact with the construction of the system, being kept informed by progress reports detailing what programs have been written, which computer files have been created, what data have been accumulated, etc. Also, the programs will be written in one of the standard computer languages which are designed to make effective use of the machine resource, languages that require specialists to write and which need careful documenting so that similar specialists may later understand the intricate coding involved. Similarly, the system will have been designed with a great deal of thought about computer storage space, the number of records to be processed, lines of output required, etc., and the final product will form an integrated whole of which the development team will be justly proud. They will naturally feel strongly about adulterating this design with "dirty" changes subsequently requested by the sponsors. As well as incurring this emotional resistance, amending the system will be a tedious and time-consuming business, requiring that programmers be retained on the project when they are eager to start something new, and requiring all the documentation, training manuals, etc., to be rewritten to some extent.

Being aware of the difficulty of changing a supposedly finished system, the analyst will be keen to get the design "right" before any programming is attempted, thus his willingness to accommodate the requirements of any other departments which show interest. Better to do it at the start than be forced to alter the design later on. Consequently, the system is ultimately much bigger than the original manager expected, and he has to wait that much longer for his "bit" to be completed. He is usually pacified, however, with the thought that he is getting it cheaper than he would have done otherwise.

Admitting other sponsors into the project makes the proposed system much more complex, and in order to produce a neat and efficient design, the systems analyst recommends compromises, to which (not fully recognizing the implications) they agree. Also, implicit compromises will be made of which not even the development team may be aware. Consequently the original requirements become diluted and large areas may be ommitted in order that the time and cost quotes are acceptable. It is regrettable that the detail that gets pruned back in this way is the same detail which makes the original problem so difficult to solve for the individual manager. Without it he or she would easily devise a set of operating rules which subordinates could apply without any need for a computer. Consequently the system, if it does not handle this detail, does not solve the managers problems!

Even if this watering down does not take place, it is unlikely that the finished system will be acceptable. Over the length of time which we have stated as typical for development, things will change. The commercial environment in which most managers exist is extremely dynamic. In fact, it might be said that the prime function of an executive is to manage this continual change. Consequently a solution which takes 12 months to 2 years to develop will inevitably be out of date when it emerges and, unless given constant attention by the programming staff, will eventually become obsolete. Likewise, the personnel involved will change. It is not unknown for the manager who initiates a project to have left the company before the system is finished! The successor is unlikely to have exactly the same requirements.

Let us assume an ideal situation in which the systems analyst has correctly interpreted what the manager originally told him, i.e., no compromises have been made and no important changes have taken place in the environment or the people involved. Surely then the system will be acceptable to the sponsor? Alas, not always so.

In the initial exchanges between the analyst and manager, the latter describes the problem as he or she sees it, usually as a collection of symptoms. The manager in doing so states a solution, explicitly or otherwise, forming a starting point for the analyst's work. However it is almost always the case that when the manager sees the results issuing from the computer, his or her perception of the problem, and therefore of the required solution, changes. The act of analysis quantifies and clarifies some of the assumptions made and tends to modify ideas on what are the critical aspects. This, of course, only occurs when the first output from the system is available and should not be considered an oversight on the part of the manager, or of the analyst, but as a natural consequence of the way the human brain

conceptualizes problems. Failure to understand this process leads to the
remark, too frequently heard within computer departments, that "the sponsor never knows what he wants!"

D. A User-Written Alternative

An alternative approach is for the manager to dispense with the computer
department and write the system himself. Until recently this was an im-
possibility. Modern developments in computer languages, however, now
enable an intelligent person to pick up enough know-how to design and write
a modest computer system that works in a matter of weeks! Is this then the
new approach to computer system design we are advocating? No it is not,
for the following reasons.

Most managers will not have the luxury of a few spare weeks in which
to develop a computer system for themselves, and certainly that is not what
their company pays them to do. Thus a subordinate member of the department would be given the job of acquiring sufficient fluency in the chosen language to be able to create the system. Here we have an example of an application being developed by an experienced business person whith a smattering of computer knowledge, rather than the usual experienced computer
person with a smattering of business knowledge. In the authors' view a definite step in the right direction! However problems still remain.

Firstly, the programs which will be written will be highly individualis-
tic, full of "fixes," and not adequately documented. Consequently, if the
original programmer leaves the department, understanding how the system
works in order to change it will be very difficult. It is not unknown for a
person to be unable to grasp exactly how a program he or she wrote a few
months earlier actually works! In such situations the temptation is to throw
the whole thing away and start afresh. Modern computer languages in no
way discourage you from doing this.

Secondly, it is very easy to make programming errors which are not
readily detectable. Thus a user-developed system may well produce garbage under the guise of respectable statistics. This is an inherently dangerous possibility which is less likely if the system has been developed and
tested by an experienced team.

Finally, there is the question of data. While the manager is collecting
and inputting his or her own data, no real problems arise. However it is
more than likely that the manager will eventually wish to use data which
resides on some other computer file. As soon as this happens a professional systems analyst will have to be consulted for advice on how to get the external data into the system. The manager is then practically back where
he or she started—only worse. The last thing any self-respecting analyst
wants is to be associated with a user-written system! The minimum
amount of cooperation can be expected. Also there will be a real fear in
the computer department that the manager is going to "mess up" their data
files with highly suspect programs. The manager may find that the only

way to get at the required data is to have the system taken over by the computer department.

It is for these reasons that we do not believe that a user-written system
is currently a viable alternative to the traditional systems development
strategy. What we are about to propose is a methodology which combines
the strengths of both. This methodology, made possible by recent developments in computer technology, is based on the concept of a "prototype."

II. A NEW APPROACH

A. The Prototype Method

The case study which forms the core of this chapter describes an application of computer technology to aid decision making in an international company. The application, built around a linear programming model, was developed using the prototype approach. There is nothing radical about the
tactics employed; in fact, many managers may find them commonplace.
However, the prototype philosophy goes against the whole training and outlook of the traditional computer professional, who will regard it as experimental and dangerous. We give a brief outline of what we understand as the
prototype approach.

The traditional systems analyst tries very hard to get the design "right"
before any programming commences, knowing how difficult it is to amend a
completed program written in any of the standard languages such as COBOL,
FORTRAN, or PL/1. The above discussion has attempted to show that this
aim is impossible to achieve, not just practically but theoretically, due to
the change in perception that the construction of the system brings about.*
The prototype philosophy accepts this inability to get the initial design correct as its fundamental precept. Therefore it attempts to produce something realistic out of the computer as quickly as possible, no matter how
rough and ready, in order to focus the thinking of the sponsor and development team. This first attempt is then rapidly refined by suggestion and experimentation until a point is reached at which the manager feels he has a
usable tool. It will still be extremely primitive in its facilities, no attempt
having been made to supply professional "frills"; however, it will produce
information that the sponsor can act upon. The user of the prototype is then
allowed to operate the system in earnest for a reasonable interval during
which time no reprogramming takes place (unless serious program errors
are uncovered which would invalidate the results). At the end of this planned
interval, the prototype is reappraised, and the user is actively encouraged

---

*We are talking here of management information systems and decision aids
and not of standard applications such as payroll, invoicing, and order
processing.

to make suggestions for improving it. Any recommendations are incorporated into a second-generation prototype, and there follows a further trial
and appraisal period. This iterative cycle is repeated until the manager is
satisfied that the version of the system currently operating solves his or
her problems and does not wish to spend any more time or money extending
it. At this point the system is totally rewritten to professional standards,
with full documentation, error-handling and recovery procedures, operating
manuals, etc., and the finished system, now no longer a prototype, is implemented in the sponsors department.

The above is a very brief description of the prototype methodology
which is more fully illustrated in the case study. Some points are worth
noting however.

The first, and by far the most important, is that the sponsoring manager never loses control over the development of the solution. The user and
analyst are necessarily in close contact throughout the project, to their mu-
tual benefit, for any misunderstandings or potential opportunities will be
readily apparent to one or the other at an early stage.

Second, until the original manager is satisfied that the prototype solves
the problem, the requirements of other departments are totally ignored.
Thus we have the concept of a "problem owner," who is a single individual
or at most a single department for whom the prototype is being developed.
Only when the manager has a workable system, will the development team
shift its attention to a new problem owner and consider what extensions are
needed to accommodate him. At no time will the original sponsor be required to compromise his or her working prototype.

Finally, the user will identify closely with the finished product. He will
consider it "his" system and will have a detailed, although nontechnical,
knowledge of how it has been put together and will therefore be readily able
to guide amendments when the commercial environment forces change.

B. APL*

A prerequisite of the prototype approach is the availability of a modern interactive computer language. The standard languages, some of which have
already been mentioned such as COBOL or PL/1, are just not appropriate.
The accent is on speed of development which means that the language must
satisfy certain criteria.

Firstly, such a language should require no knowledge of how a computer
actually works. Secondly, the language should be interpretive. This means
that the programmer can sit at a teletype or visual display screen, type in

---

*See, for example, APL: An Interactive Approach by L. Gilman and A. J.
Rose, published by John Wiley & Sons, Inc., New York.

a statement equivalent to, say, "3 + 3 = ?" and immediately get back the answer 6. In other words, as the person types in each line of a program, the computer instantaneously checks it for simple errors such as grammatical inconsistencies. Thirdly, the language should be <u>interactive</u>. This means that the programmer can finish writing a program, run it immediately, and the computer will then stop at any programming error which has not already been discovered. The programmer will then be able to interrogate the machine about the nature of the error, quickly diagnose the trouble, rewrite the offending line, and restart the calculation from the point it stopped. The computer will then continue to the next error and so on. The twin features of being <u>interpretive</u> and <u>interactive</u> makes for speedy program writing and ease of debugging. Thus a system can be "thrown together" very quickly and be producing useful information within a matter of weeks.

The honey-planning system was developed using APL (this simply denotes A Programming Language) which is a typical modern interactive language, although by no means the only one. APL was originally invented as a convenient shorthand for writing down scientific calculations. It was intended to replace the standard mathematical notation which can be cumbersome and inconsistent. It was <u>not</u> initially developed as a computer programming language, although its potential in this field was quickly realized. Its main power lies in its conciseness and, therefore, its speed of writing, and its ability to handle large tables of data as if they were single numbers. Thus it is ideal for putting together decision aids, such as the linear program of the case study. It is also very easy to learn and can be learned in stages; a programmer only needs to know as much of the language as he or she currently requires, increasing this knowledge when it becomes necessary for the application in mind. In this respect, because of its interactive nature, the language forms its own teaching aid, allowing statements of the program to be tried out individually, and intermediate results displayed. Expertise in APL does not require the traditional computer professional's technical training and background. These are the typical features of a modern language.

The aspects of APL which make it so appropriate for developing prototypes also have their inherent dangers. The speed with which programs can be written and changed militates strongly against them ever receiving adequate documentation. If parts of a system cannot be subsequently understood the temptation is to discard them and rewrite them from first principles. This eventually leads to a totally unmaintainable set of programs. Secondly, because the language takes care of all the really technical aspects of computer use automatically, it is extremely prodigal of the machine resource. Consequently undisciplined use of APL can quickly overload the capacity of the computer. Thirdly APL, because of its philosophical differences, does not link well with traditional computer languages. Thus there are difficulties in passing data from a system developed in a standard language to one written in APL.

It is in order to overcome such potential dangers that the phase of the prototype methodology devoted to professionally rewriting the system is planned in from the start. It is also for the above reasons that there is a strong emotional resistance to using interactive languages amongst the professional computer fraternity. They see the rewriting as having to "rescue" a badly written system.

However it is not necessary for the final system to be written in the same language as the prototype. A machine efficient language can be chosen.

Having developed our arguments for a new approach to constructing computerized management aids, we present a case study to illustrate the points we have made.

III. AN INTERACTIVE AID TO PLANNING IN THE
      FOOD INDUSTRY

A. Introduction

This case history describes the development of an interactive computer planning model to aid in the development of a strategy and tactics for the purchasing, production, and blending of honey. The approach adopted by the project team was to use APL to quickly write prototype systems and to use these systems to help both the user and the project team develop an understanding of the user's problems and the interactions caused by these problems in other departments. During the development stages of the project, emphasis was placed on the following:

Providing a computational aid to create thinking time for the user

Not taking the decision making away from the user, but providing more information to enable better decisions to be made

Providing a means of communication between user departments where all assumptions have to be explicitly stated

Providing accurate and up to date information on stock levels, expected deliveries, etc., to all user departments

This study records how the original brief of solving a short-term blending problem was extended by phases into an integrated system to aid in the development of a long-term purchasing strategy. The system which finally evolved is given and details how the user, at run time, can dynamically redefine the criteria used to evaluate differing strategies and tactics. A brief description of how the system operates is also given.

B. Background

Honey is a product produced by bees from the pollen of flowers. Some honey is sold in the cones in which the bee produces it, but, more usually it is sold in liquid form in one of two states: clear honey, which as its name implies is transparent, and set honey, which is opaque. In this study demand for honey greatly outweighs the supply, thus large quantities need to be imported. These honeys are from many countries including Australia, China, Rumania, and Mexico. To ensure consistency of quality and taste these "raw" or source honeys need to be blended to give the final clear and set honey which are consumed at breakfast tables all over the country.

C. The Problem

The production of honey involves, in the main, five departments: production, purchasing, formulation, marketing, and finance.

The marketing department provides the sales forecast from which the production schedule is calculated within the inventroy level laid down by the financial department. The purchasing department then draws up a purchasing plan within the financial constraints imposed by the raw materials inventory levels, and with regard to the possible blends as supplied by the formulation department (Fig. 1.). Traditionally the blends used to make the final clear and set honeys followed a well-established recipe, which in times of difficulty in obtaining supplies of the required quality was replaced by a

FIGURE 1 Department involved in the blending process.

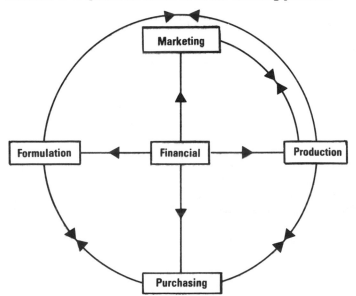

second-choice or third-choice recipe. This approach assumes that the quality of honey from a given location does not vary over time, and that supplies from traditional sources can always be obtained.

During 1978, a failure of the Australian honey crop (Australia is one of the world's largest producers of honey), coupled with an increase in world demand, led to severe difficulties in obtaining raw honeys of the required quality. The recipe approach to blending had to be abandoned as the traditional raw honeys could not be obtained. Honeys of unknown quality had to be purchased from hitherto unused sources, with the quality parameters only being established after the shipment had arrived. Frequent reblending therefore had to take place, as continuity of supplies and quality could not be guaranteed.

The coordination of honey production was carried out by a honey-working party which had representatives from all the five departments involved. This working party, consisting of about 15 people, met frequently to consider the changing situation and to revise the production blend to be used as information on the quality of the most recent shipment of raw honey became available. The computational complexity of calculating a new blend was very time consuming, and hence the first blend which satisfactorily met the quality requirements was accepted.

External consultants where called in by the research and development director (to whom the formulation department reports) to provide a computational aid to the blending of honey.

D. Development of the System

Phase 1. Acceptance of a Computerized Approach

From initial discussion with the formulation department it quickly became apparent that a computerized approach was viewed with something less than overwhelming enthusiasm. In fact, computer systems were considered to be big, very complex, and difficult to use; could not solve their problem; took a long time to develop; threatened their jobs; and would generally create a mess which they had to clear up. The first task of the consultant was to gain the confidence of the formulation department. This was achieved by quickly writing a simple program in APL to calculate the effect on the stock levels of any production schedule using a given blend for the clear and set honeys. This was an exercise which the formulation department performed every time there was a change in production schedule or blend, which during 1978 was very frequent. The stock-level program was designed to reproduce the manual calculation. The output exactly duplicated the then current stock-level documentation.

The program was written, tested, and being tried by the formulation department, within a week of discussing the problem by which time they had, fortunately, the need to produce a new stock level plan. The speed of the computer system response and the ability to produce exactly what they currently did manually greatly impressed them. Indeed within a few days they were requesting extensions to the system. At all times it was stressed that the system written was only a prototype system and would be modified to meet the blenders requirements.

Because the formulation department saw that modifications were quickly and easily carried out, they had no inhibitions in requesting them. The output from the stock-level program quickly replaced the original stock-level report.

Phase II. Development of a Blending Model

Phase I was very successful in gaining the formulation department's confidence in the ability of computer systems to help them in their work. They now became positively enthusiastic to try and extend the approach to solve the original problem, i.e., to provide a computational aid to honey blending. Together the consultant and a person from the formulation department formed a small project team which defined the working constraints when preparing an acceptable honey blend. These constraints were as follows:

1. Mass balance constraints: The production of clear and set honey should not exceed the opening stocks and the expected deliveries of raw honeys.

2. Legal constraints: Legislation specifies the minimum level of certain quality parameters which must be met. These are the soluble solid content, the reduced sugars, the sucrose level, and the hydrox F and diastase number.

3. Shelf-life requirements: To ensure the shelf life of the set and clear honeys are acceptable, the L/D ratio must be above minimum level. This level differs for clear and for set honey.

4. Consumer acceptability requirements: Two constraints are placed by the consumer acceptability requirements, the taste and the color. Discussion established that the taste of the raw honeys could be graded on a 1-5 scale. Furthermore the percentage of certain raw honeys and blended honeys from specified origins could not be exceeded in the final blend.

The primary objective is to meet all the above constraints using only raw honeys, any other additions being illegal. Because of the computational difficulties, the second stage of choosing a single blend from the many possible satisfactory blends was never encountered in the past, the first satisfactory blend being accepted. However, having seen the early results, the blenders expressed the desire to be able to test the effect of using differing criteria in choosing the "best" satisfactory blend, e.g., maximum color, minimum solids.

It was established that all the above constraints were linear; so a linear programming approach was used. In order to give the blenders the facility to test the effect of changing the criteria of selection of the best blend, it was decided to write the LP model so that it could be used interactively, the main advantage being that many blends could be reformulated within a very short time.

The computer program giving the blenders all the above facilities was written in APL using an IP Sharps* linear programming package. Development of this system was completed in 2 to 3 weeks.

Phase III. Testing the System

At all stages in the contact between the formalation department and the project team, it was stressed that the systems being written were prototype systems written by nonprofessional programmer and that these systems would be professionally rewritten when the user was more confident on the specification of the final system. User testing of the prototype systems was essential, and criticisms and suggested modifications positively welcomed.

The formulation department, therefore, was requested to test the systems over a 2-month period, for all possible combinations of events. Initial training on the use of the system was given, as well as background support if required.

During this 2-month period, the system was given a complete testing in a live environment, sometimes to destruction. The only modification made to the programs were those that were essential to the running of the system. At the end of the test period, a long list of modifications were suggested, concerned mainly with the ease of running the system rather than with the logic of the design. The overall conclusion of the formulation department was that the computerized aid was a major advance in reducing their time involvement but could be made easier to use.

All the modifications suggested by the formulation department, particularly regarding the machine-user interface, were written into the prototype system.

Phase IV. Extension to Other Departments

Section III.C describes how the production of honey, in the main, involves five departments: production, purchasing, formulation department, marketing, and finance. The project team so far had worked only with the formulation department. With their approval it was decided to extend the project brief to involve all the departments using the system and to provide an aid in the solution of longer term planning problems as well as the shorter term

---

*IP SHARPS ASSOCIATES, a world-wide computer bureau, based in Toronto, Canada, who specialize in the interactive language APL.

blending problem. This was necessary to include the constraints imposed on the formulation department by these other departments. To this end, a presentation was made to the honey-working party, where suggestions were made on how the system could be extended to answer the question of what honeys to purchase in addition to the question of what to do with that honey when it has been purchased (i.e., how to blend it).

The proposal to extend the system was accepted and a member of the purchasing department co-opted onto the project team. To answer his purchasing questions regarding

Quality of raw honey to purchase

Price to quote for raw honey

Contingency plans taking into account uncertainties on availability, quality, delivery, reliability, price, etc.

the model was extended to include the price of the blend as a constraint and also as a parameter in the selection of an acceptable blend.

Phase V. Testing the Modified System

The purchasing department was requested to test the modified system for a period of 3 months. Once again, it was emphasized that the system was still in prototype form and requests for modifications would be welcomed.

Phase VI. Rewriting the System

With both testing phases successfully completed, the programming department was called in to rewrite the system, giving all the facilities specified by the users but to professional standards, paying particular attention to the user-system interface, efficiency of coding, ease of maintenance, and the company's documentation standards. The users still had the prototype system to operate during this rewriting phase. When the rewriting was complete (2 months) full training of the user followed. The programming department retains the responsibility for maintaining the system.*

E. System Description

As stated above the total system consisted of two sections:

1. Stock-level calculation

2. Blending/planning model

---

*In this case the final system was rewritten in the same interactive language, APL, used for the prototype in order to retain certain specialized facilities.

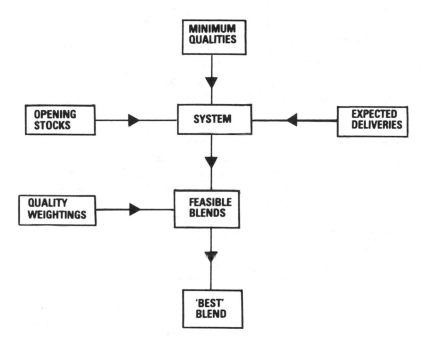

FIGURE 2 Flow sheet for blending model.

The two sections were designed to be integral parts of a total system, both
referencing the same data file and with the chosen blend being able to be
passed from the blending program into the stock-level program. To facilitate the understanding of the system, the blending/planning model will be
described first (even though it was written second), since the output from
this module can be fed into the stock-level program to check that adequate
stock cover is available throughout the whole of the planning period.

1. Blending/Planning Model

Schematically, the system can be reviewed in Fig. 2. The program plans
for a single period which can be as long or as short as the user wishes (it
is usually set at 1 year).

The system requires the following information which can be either via
the keyboard or taken from a data file.

1. Opening stocks of raw honeys

2. Expected deliveries of raw honeys

3. Quality parameters of these honeys

4. Minimum permissible closing stock level for each raw honey

5. The maximum percentage of each raw honey in the final clear and set formulations

6. The production required for clear and set honey

7. The weighting to be given to each clear and set quality parameter (including cost) when selecting the best blend from the set of feasible blends

Because the system is written in APL, it could become very expensive to run, the expense being geometric in proportion to the size of the problem. To reduce the size, and hence the cost to a minimum, the facility was built into the system to enable the user to specify the quality constraints which must be met. Thus, if the user believes that, say, the reduced sugar content will always be above the required quality, that quality constraint can be left out of the formulation, thus reducing the size of the problem. The value of the quality parameter will still be calculated and can be used in the selection of the best blend.

The first calculation the system does is to check that there is enough honey available (stocks plus expected deliveries less minimum required closing stock) to meet the production requirements. If production requirements are greater than expected availability, the user can choose between four options:

1. Make the requirement equal to the availability.

2. Make the availability equal to the requirement.

3. Allow system to suggest blends and deliveries.

4. Cancel the run.

The system will calculate the best blend displaying

1. Closing stocks

2. Quality parameters not met (if any)

3. Blend for set and clear

4. Quality achieved

5. Suggested deliveries, if requested

The user then has the option of changing any of the current setup values and rerunning. Output can also be stored on a file.

2. Stock-Level Calculation

Schematically, the system can be reviewed in Fig. 3. Given the specified blending program the system calculates the monthly movement in stock.

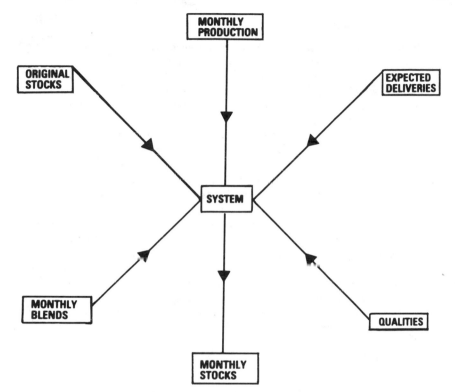

FIGURE 3   Flow sheet for stock movement model.

The blending program can be either that calculated by the blending/planning
model or any other specified at run time. This report is required because
it checks to see that the yearly schedule, as calculated by the blending/plan-
ning model, is feasible on a monthly basis. The same method of operation
applies to both sections of the system insofar as the user can change any of
the data and rerun at any time.

F. Summary

The case history given in this paper records the development of a blending/
planning model using a prototype system approach. This summary, therefore, will be divided into two sections: the first will describe the benefits
obtained in the construction and use of the planning model, and the second
section will list some of the perceived advantages of using a prototype system approach.

1. Benefits of the Planning Model

The ultimate criteria used to judge any system must be the acceptability of
the system to the user and the use he or she makes of it.

The planning system described in this study is used on a regular and
frequent basis by the formulation department to calculate blends and by the
purchasing department to evaluate planning strategies. The output from
both the planning/blending module and the stock-level calculation module
have now replaced the manually prepared reports. The honey working party
now meets very infrequently, all departments being prepared to accept the
calculations made by the system. In financial terms stock levels of raw
honeys have been reduced, because by better blending more of the source
honeys can be used. The system also leads to a more effective use of the
new honeys.

One of the major benefits of the system, however, is unquantifiable,
that is the understanding gained of the total blending problem and the interactions between the different departments involved. The user now has to
specify the criteria for the selection of the best blend and therefore has to
define and specify "best." This aids in the communication between departments as all assumptions have now to be stated explicitly.

Last, but not least, the computer is seen as being apolitical, and therefore its calculations are more readily accepted by all the user departments.

2. Benefit of the Prototype System Approach

The main benefit of a prototype approach is that the user develops and mod-
ifies the system and therefore identifies closely with it. His or her understanding of the problems will evolve during development, and the system
can quickly and easily be modified to accommodate these changes in understanding. At the end of the development phase, the user is left with a system with which he or she identifies, which more closely meets his or her
needs, and which he or she can more readily adapt to meet changing circumstances. A second major benefit is that throughout the development phase,
the designer of the system and the user are in close contact, the development phase usually being of short duration. The problem owner, therefore,
never loses touch with the development of the solution and, hence, will not
solve the problem in another way. If the problem is immediate and urgent,
the user cannot wait up to a year, which can be the usual time scale for the
development of a computer system by conventional methods. The final system may be the same, but throughout the "professional" writing phase, the
user will have a prototype system to use.

IV. CONCLUSIONS

It is firmly believed that in the development of this planning aid, the prototype system approach offered major advantages over conventional computer

development methods. The approach aided design, user acceptance of this system in particular and computer systems in general, communications between departments, and in the final analysis, the attainment of the financial benefits of stock reduction that the system could generate. The use of APL was fundamental in this approach, because it was only by the use of an interactive interpretive language that the flexibility of resizing the problem and the redefinition of the objective function at run time were possible; APL also aided the speed of development and modification.

V. FINAL REMARKS

The prototype system philosophy aims at exploiting the recent advances in programming languages to quickly alleviate the operational difficulties of a sponsoring manager, without leaving any long-term problems of system continuity and maintainability. The tactic is to use a speedily written program as a communication vehicle between user and analyst to clarify and sharpen their thinking. This is then rapidly refined by a series of suggestions and amendments, under the eyes and hands of the sponsor, into a usable prototype which will eventually be rewritten to professional standards. Now, however, the manager can afford to wait for the robust and streamlined system, having the prototype to plug the gap. Also the "professionals" have a working example of what they are attempting to produce, which greatly reduces specification and programming errors.

As we have stated, the prototype approach strikes at the fundamental philosophy of the traditional computer system specialist. He or she sees in it a vast duplication of effort and the need to set aside resources for reprogramming at some unspecified future time, which destroys his or her ability to plan effectively. We contend that with the conventional approach this duplication and reprogramming inevitably occurs, only in an uncontrolled fashion, with a consequent loss of confidence on the part of the sponsor and of morale on the part of the development team. We therefore exhort the specialist to embrace the prototype methodology as a means of forming close partnerships with executive managers, of enhancing their own commercial experience, and of producing management information systems and decision aids which will make a positive and lasting impact on the business. It has been shown to work, with benefits, in many areas—not only for interactive decision aids but also for conventional information producing tasks.

Similarly, we warn the potential recipient of a management system not to lose control over its development. If you do, and you allow the computer professionals to take your problem away to work on, the chances are you will not be satisfied with the results. You are paying for it, make sure you get what you want, even if at the outset you are not certain what it is!

15

The Changing Interface to Computation, Super-Routines, and Professional Responsibility

RONALD I. FRANK Cambridge Engineering/Scientific Support Center, Cambridge, Massachusetts

I. INTRODUCTION

This chapter is intended to orient the practicing professional or advanced
student of food science and engineering to the modes of using computers
and to encourage the learning and use of computation. This orientation
should enable the reader to better understand the changing modes of using
computers. Food science and engineering (FS&E) is not very different from
other technical disciplines as far as computation is concerned. The orientation to be derived from this chapter is valid for other fields. It seems
that food science and engineering professionals, as in other fields, need
encouragement in the application of computation to problems in their discipline. This is true more for older members of the field than for younger
members who are better trained in computation.

The reader's background is assumed to include some brush with computation, but not a long-standing in-depth experience. A historical survey
of the modes of using computation is presented to develop a common perspective from which one can view the present state of computation in FS&E.
This survey is in narrative form with some burden on the reader's recall
of computer terms. That means that the complete newcomer to computation, if such a person still exists, may find some terms undefined. Any
introductory text in data processing, such as Edwards and Broadwell
(1979), would be a suitable but not necessary background, in that the terms
used here are more completely defined there, but the reader does not need
the contents of that book to follow the contents here.

The specialized uses of computation called "sensor based" and "real
time" are not covered here. "Sensor based" means that data come into the
machine directly from remote sensors built for that purpose. "Real time"
means that the computational process must respond to external service requests in time to affect the process which generates the requests. Sensor-
based and real-time computation often appear in the same application such
as in on-line process control. Sensor-based computation occurs alone in
passive monitoring and logging of data from instrumentation. An example
where real-time computation appears alone is in airline reservation systems. In any case, the underlying tools and techniques of using computers
in these modes are sufficiently different and specialized to be beyond the
scope of this chapter. They are a small but not insignificant extra amount
to learn beyond those super-routine tools and techniques which this chapter
describes.

The main theme of the orientation provided by this chapter is that
powerful and effective use of computers is now within the grasp of any intelligent person without the necessity of taking a complete data processing
degree. However a time/effort investment decision is necessary. This
orientation provides the beginnings of an information base upon which that
decision can be founded. The orientation introduces "super-routines" (to
be defined), which provide a cost-effective and efficient mode of use within
the "interactive" environment (where the user employs a typewriter or

video terminal to interact with the computer). The orientation proposes that it is worth the investment of time and effort to learn the fundamental services offered by any modern interactive system available to the reader and to learn to use the super-routines available on that system. A harmony for this main theme is the call to FS&E practitioners to collaborate in the creation of super-routines (SRs).

Super-routines are computer programs organized as a complete set of coherently communicating routines that solve a small but important class of specific problems in a given discipline, such as the nutrition branch of food science and engineering. They usually employ a coordinated set of user aids, such as an easily learned command-and-control language (possibly mediated by a menu or dialogue handler), a system text editor and composer, a data manager for data files, a discipline-dependent application "kernel" of executable program code which "solves" the problem, and a report writer (possibly with graphic output capability). SRs carry the user completely through the solution of a problem (in the class of problems for which the super-routine is intended), from data entry to file maintenance, computational process, and output-report formatting and generation. They are characterized by being application specific. They often use the jargon of the discipline, as opposed to computer jargon, in order to present a friendly interface to the user. Good ones avoid unnecessary computer jargon.

In the future, one hopes that in a given field or in many fields, there will be only one set of SR components (except, of course, for the specific problem computational kernels). This will enable the learning of only one set of tools for most application development by professionals in a field such as FS&E. Also, users will have to learn only one set of commands for control of most of their discipline-dependent computing.

Creating an SR is the process of imbedding the discipline-dependent computational kernels into the user aids, which are often supplied by the machine vendor. The kernels are developed by professionals who know their computer system and a higher-level language. They also know how to solve a meaningful class of problems in their discipline. Their two prime contributions come from (1) stating the solution process precisely in an algorithmic way, so that a machine can execute the process correctly, and (2) organizing the entire solution process into a structure that colleagues can easily learn to use effectively.

What is "interfacing"? What is "computation"? Many words have both a vague general use and a technical use. We will need to go further into the technical meaning of these terms. Let us begin with the definition of these terms and follow our topic's history to the present so that we may see how to better make that time/effort investment decision for personal education in computational practice and to better apply that computational education by the use and creation of super-routines within the interactive environment.

II. DEFINITIONS

"Computation," in its basic meaning, is the application of the resources of a computer. These include the instruction execution cycles of the machine, the main storage of the machine, the printer/plotter of the machine, the auxiliary tape or disk storage of the machine, the operating system and languages supported for the machine, and whatever other output displays may be attached for communication to the user or operator. In its extended meaning, "Computation" means all the services provided by the original vendor of the machine, the operating department which runs the machine, and libraries of code available to the user including code distributed by societies of like-minded users.

A "user" is a person who uses a system. The term is most often employed to mean a programmer on a computer or a person using an application program written by someone else. It is a relative term. The relation is between a person and a system, which can be a machine, an operating system, or an application program. An "interface" is the boundary between two systems. "Interfacing" is the act and process of communication across an interface. "Interfacing to computation" is, therefore, the act and process of communication across the human-machine boundary by the user and the machine. It may involve communication with the components of the extended definition, i.e., the vendor, the operating department, or other users who have created code.

The changes in this communication and the changes in the way the user and the machine act toward each other are prime subjects of this chapter. The main point is that the act and process of communication to get a given result for the user are becoming more efficient and, hopefully, more human. On the other hand, since much more can be done overall and much more can be done per unit effort, the choice of what to do and how to do it becomes a significant investment decision for the individual. Therefore it pays to obtain an orientation within the act and process of communicating with a computer to better assess the important factors in the choices.

III. EARLY HISTORY (LATE 1940s TO LATE 1950s)

When computation in its modern programmable form first became generally available, it appeared to the user most often as a stand-alone machine that the user operated and programmed. The technically oriented user needed to invest considerable time and effort in learning to program the machine and to run it, via its console. In larger organizations the user might have had an operations staff available which would set up and run individual jobs one at a time, often under the direct supervision of the programmer/user. Production runs could be left to the operations staff to do by following a "run book," or set of printed instructions.

From our point of view, the important feature to remember of this era is the large personal investment of time and effort the individual user had to

make in order to get even the most trivial problems solved. In fact, the
effort was so large that many people actually spent so much time that they
switched out of their original line of work and became the first generation
of computer scientists and specialists. User libraries and societies to help
alleviate redundant effort were begun to share programs among users.

Machines were rare, access was limited and expensive, and funding
was difficult to get. Few people were properly trained, so that help was often a matter of going directly to the computer vendor for lengthy consultation.
The most a user could expect to do was some short series of independent
tasks.

An example, typical of scientific use in this era, starts with the sorting
of a set of cards onto which data had been manually punched, at large expense and with considerable error rate, and then resorting the cards either
on the computer or most likely on an off-line sorting machine (one not connected to the computer). The next step would be to feed the cards into the
computer which had been previously loaded with a statistical analysis code
from another deck of cards. If the user was on a large machine, some of
this processing might be done from a tape that had been previously written
by a different peripheral machine, a card-to-tape machine, off-line. The
statistical analysis would proceed and an answer deck or tape would be written which would then have to be taken to an off-line card-to-print or tape-to-
print machine for a "hard"printed copy to be produced by a slow printing
process.

The card input had to be in the form that the specific statistical routine
wanted. If the user wished to do some alternate statistical analysis, a re-
punching of the data might be called for since the alternate routine might
have been written by a different group using a different format. Very often
the routine did not even exist.

Another similar example can be constructed by substituting accounting
data for scientific data in the first example and substituting simple searching, sorting, and arithmetic for the statistical analyses. Search and retrieval were primitive and very time consuming.

On the real-time front, as in, for example, the monitoring of lab ani-
mals, much of the quick-response experimentation was controlled by analog
machines. Final output charts had to be hand punched into card decks for
digital analysis, as in the first example.

The time/effort investment decision problem had always existed from
the very first. In this early era, the user had to spend weeks learning the
machine language or assembler language of a machine and then spend weeks
or months of full-time effort coding the solution to a problem. The machine
language consisted of pure numbers (binary, octal, or decimal). The assembler was little more than a one-for-one translator of mnemonic words into
the machine language. The programmer still thought in terms of the actual
hardware registers of the machine and in terms of the internal representation of instructions and data in the machine.

Often, the mathematical subroutines one needed were not fully available.
Either the user had to write them or do without the desired analysis. In any
event, the user made the decision of whether to do this entire interfacing
process once only, or to continue doing it, in which case the user often became the local computer expert. From these ranks came the first generation of in-house computer competence and the first generation of computer
scientists. Very few were strong enough to both become expert with computers and be able to continue contributing to their original field directly.
Many contributed by their introduction and use of computers in their field.
Many of the statistical routines, scientific subroutines, and file routines
were, in fact, produced by these pioneers who were from fields other than
computing. As we will see, the advent of the inexpensive personal computer
is bringing a partial recapitulation of this history.

This era also saw major advances in the availability of utility processes.
An example is the multitape sort which became available as a standard product.

IV. MIDDLE HISTORY (LATE 1950s TO LATE 1960s)

In the late 1950s and early 1960s, operating systems (OS) were introduced.
Their purpose was to enable sharing of the scarce resource of main-frame
instruction cycles and to enable the more efficient use of new input/output
devices and of the new rotating disk storage media. At first, user jobs
were run in a "batch" (grouped together), but run one after the other with
little resource sharing allowed. Later in this era, user jobs were run in a
batch (grouped together) but with considerable overlap and with as much resource sharing as the OS could allow. The OS enabled simultaneous peripheral operation on-line (SPOOLing). "Peripherals" are the card-oriented
input/output devices that reside on the periphery of the main machine. This
SPOOLing included the card-to-tape, card-to-printer, tape-to-card, and
tape-to-printer processes.

The centralization of control enabled by the operating system placed the
user outside the machine room except in the smallest installations. These
latter, the very small machines, were still accessed in a one-person-on-
one-computer manner, especially for machines used in science and engineering. This mode of interfacing to scientific computation, one on one,
was less efficient than shared use, but the smaller scientific machines were
often not controlled by a nonuser administration, and the machines were not
intended to be a widely shared resource. The "inefficiency" was not real
because there was no actual contention for the limited computational resource in these installations.

The access by the user in this era was still mostly indistinguishable
from the access by a programmer. A job was usually submitted as a deck
of cards with some initial cards containing job-control commands. The data
were also in a deck of cards or sometimes on a tape created from a deck of

cards. More coherent systems of routines were appearing for given purposes. Examples included libraries of scientific subroutines that all took the same data formats for input and output, statistical packages that similarly were data-format compatible, and file systems that enabled the user to maintain large files of data with standard retrieval procedures.

The time/effort investment decision was not substantially different in this era from that of the earlier era. Major personal and departmental commitment had to be made to computing and the return on investment was only moderately better than before. Learning one machine helped in learning another, but the time and effort were large in the frame of a career.

A typical scenario is the following. A graduate student wants to take a sample of data from a biochemical kinetic system and then model the system by positing a system of differential equations, integrating them, and fitting the solution to the data as a measure of correctness of the model. The data, as in the early era, would most likely have to be hand punched into cards or paper tape. Cards would be better than paper tape because more printers were around to print out the cards for manual data verification.

Next, since there were only a very few nonlinear estimation routines and those were new, not widely known, and sometimes indefinitely reliable, the student might elect to fudge the model, integrate using a subroutine package, and fit using a different statistical package. The only difference between the early era and this middle era is that the integration and fitting routines would now exist in the library, be better documented, be better tested and so more reliable, and be more easily accessed. The computer installation now maintained both a reference shelf for the nonprogrammer user, and a formal consulting group of knowledgeable, experienced people.

There would be need of consultation which would be met by the availability of experienced users of the library routines. The output might be plotted off-line. The final report would be manually typed on a regular typewriter, and some printer output would be displayed in an appendix.

Where earlier this activity might have brought a Ph.D., it now would bring a Master's degree. The user would have learned to use a computer, but only as a user of library routines, or at most, an algebraic computer language such as FORTRAN (FORmula TRANslation). The student would have spent months of time, but mostly on the problem, not on programming as in the earlier era.

V. LATE HISTORY (LATE 1960s TO LATE 1970s)

This era, which precedes the present, is best summarized as proceeding in three directions (trichotomous). First, we find the maturing of the batch mode of operation which characterized the middle era. The central processing center provides service to users who are not necessarily programmers but who do know a large amount about their special set of processing tools. Of course there are always the inept naive users, but the total system does little for them other than providing a "user consultant" to guide them.

Second, we find the beginnings of "interactive" use. This is character-ized by on-line terminals, either hard-copy typewriters or display tubes. The interface to the user is often, but not always, "conversational," where the computation is arrived at after a verbal interchange which requires the user to properly respond to the text presented by the computer.

In its early phase, this mode of interfacing to computation is little more than a means of putting in the control cards from a terminal rather than sub-mitting them to a batch input station. In fact, there is an intermediate batch mode called "remote job entry," where the user submits the job via a card reader, which is remote from the central processor, rather than to a job window at the machine site.

In this second (interactive) mode, the user may set up a program or control statement sequence in a line-at-a-time manner using an editor. This editor is a processor that requires from the user specific commands that cause the input to be properly formatted for storage and further proc-essing. An example is the creation and running of a simple FORTRAN job. Instead of punching control cards, statement cards, and data cards, the card contents are typed in (one card becomes one line of text). The com-pleted text file is then sent by a typed command to the FORTRAN processor which treats the text file essentially as if it were a card deck being read in, as in the batch processing mode.

In the more mature ways of interactive computing, the computation sys-tem gives the user a truly interactive programming language such as APL. The typing of the program is interlaced with partial executions to test the logic of the code. Running code can be interrupted, saved, and recalled at a later time for restarting even in the very middle of an execution of the code! The programmer can write code that conversationally interacts with the program's user. In general both writing and running codes of this na-ture are interactive. The interface to the ultimate computation is thick with hidden user assists. In this era, the user is not always a programmer. The classic example of a nonprogrammer, purely interactive user is the reservation clerk at an airline terminal. This is not the usual mode of in-terfacing to computation for most technologists and scientists.

The third of the three modes of interfacing to computation is the anoma-lous return to the early era of one-on-one use. The advent of the much less expensive minicomputer (also called "mini") found the individual user sitting down at the bare console (now with a keyboard instead of just binary flip switches). The user had complete control of the entire machine. How-ever now the unit cost of a computation was low enough so that the pressure to share the raw resource was below the threshold that brings multiuser op-erating systems into use. The latest extension of this trend, beginning in the late 1970s, is the micro-based personal computer.

Without a complicated operating system as an intermediary be-tween the user and the machine, computing was easier than on a

large-operating-system-controlled batch machine. Of course, there had
been improvements in programming languages, which aided problem statement and debugging. Also, output devices such as plotters were easier to
use. This was due to the existence of libraries of plotting routines and more
easily controlled plotters (such as program-controlled on-line graphics devices with attached hardcopy printers).

As this era progressed, the minis became larger and more capable, the
operating system appeared on them, and in general, they became more like
the older computers, or "maxis," but were less costly. As the minis became maxis in capability, the micros replaced them at the low end. This
trend continues today. Instruction cycles and memory are becoming cheaper, but peripherals such as plotters and printers are decreasing in cost at
a slower rate. As televisionlike displays become reduced in cost, more intelligent units and color units replace them at the high end of the choice
spectrum.

The small single-user mini, on which only a simple control program
ran, began to look less desirable as a mode of interfacing to computation
since the owning department found themselves in the computer management
business. Maintenance is an administrative fact of life on all machines.
Staffing was sometimes a problem. Rental costs for space, power, auxiliary devices, and many other trivial but bothersome problems took the bloom
off of some of the earlier small installations as they matured.

There was often no viable alternative. The big central machines were
difficult to get to, costly per unit computation, and more of an intellectual
hurdle to master, with their layers of control programs that had to be pene-
trated before getting at the basic cycles of the machine.

The time/effort investment problem of this trichotomous era was similarly trichotomous. The time to learn to code or to use a code on a batch
system was often weeks. The time to become adept was many months of
nearly full time use. Of course, if the user only wanted to use someone
else's code, the investment was only the learning time for that code, the
time to learn how to access the machine, and the time to learn how to get
something out of the use of the code; this was often only days to weeks. But
there were not many fully coded applications. The user was often required
to create postanalysis routines or some other processing that required knowledge of the raw machine or at least some programming language.

The interactive mode of use enabled the user to more quickly access the
machine but did not reduce by much the amount the user had to learn to run
a code and get answers. The advent of single-user operating systems did
improve this situation, and it is rapidly getting better in the modern era.
There is a genuine emphasis on making access via interactive systems more
"user-friendly." Examples are found in the self-teaching features of
VM/CMS (IBM, 1982) and UNIX (Unix, 1978).

VM/CMS is an example of a single-user operating system providing interactive access to the full set of computational resources of a machine.
The user signs on to a physical machine but is presented a virtual machine

(VM), running the conversational monitor system (CMS) which is an operating system supporting one user on the VM. The user's conceptual model is that an entire machine is available. The user then interactively programs or uses an application code. The learning path through VM/CMS is shorter than most other operating systems since the virtual machine is a simple intellectual model around which various technical facts of operation can be associated.

The mini user was also faced with learning the hardware of a real machine plus whatever language was required for application development. This was a considerable task, of the same magnitude as learning a large machine. The saving grace was that the operating system or control program was often minimal.

Before proceeding to the present, notice the common feature of the time/effort investment problem across all eras. Most of the user's effort goes into learning how to control the machine to the point where its raw instruction cycles are brought to bear upon the problem which is the user's real concern. The eras differ in the amount of those raw cycles that are available to ease the user's access to the machine. As time passes, more facilities are built in to assist the user.

VI. THE PRESENT AND NEAR FUTURE (LATE 1970s TO EARLY 1980s)

The present is, of course, a continuation of the recent past. However there are four new trends that have already matured out of the three modes of interfacing, and that can be clearly defined as alternative submodes with significantly improved return on the time/effort investment. These are the improved interactive single-user operating system environment, the use of application packages (SRs) that run in that environment, the use of that environment for more than simple single-shot problem solving (e.g., text processing and electronic mail), and independently, the distribution of functions out of single processors to distributed compatible minis or intelligent terminals (work stations). The converse of this last trend is the separation of machines. That is, the distribution of processors out to meet function where it is needed. Both of these last trends off-load the central mainframe machine. A form of this latter mode is the independent personal computer that can be attached by communication lines to a large machine or to a central data base. The off-loaded functions are often those of editing text or creating data-base queries (formatted questions in a query language that cause data to be returned from the data base to the user; just the data asked for in the query is returned).

A good way to see some of the details of these trends is to concentrate attention on a single example which, while not exhaustive, is representative in covering the improvements in interfacing to computation that are occurring. The example is the same graduate student who wants to do a kinetic, say, a pharmacokinetic, model study. The example is real but idealized in that not all components are in place today to solve this one specific class of problems easily.

The student, once given the problem, discovers that there exists a modeling package for this class of problems. In fact, there are at least three packages. Two are out of the old batch environment. One is in the new mini environment, and it is mildly interactive. The student is familiar with a programming language (learned in 2 weeks of a FORTRAN IV class). The student can get access to an interactive system via a terminal in the department office. It takes a few days of part-time effort to learn the initial control procedures for accessing the interactive system. After receiving the newer pharmacokinetic modeling code (the application program) which the authors sent by mail, and which is not written in an interactive language yet the student attempts to have the central site load the data from the transmittal medium (tape or floppy disk). At this point, the incompatibility of storage media requires that the student get help from a consultant from the data processing center. Even today, it might be a major problem to transfer code from one manufacturer's machine to another's.

Assuming this problem is solved, the student next has the problem that the language in which the program is written is not quite compatible with the language running on the student's machine. The student now uses the interactive on-line editor to make changes in the source file of the program statements. Test runs are tried. The compiler's diagnostic routine highlights language errors The student iterates rapidly to a working version of the code. Quite likely the original code's documentation manual came on the same tape as the code itself. The student inserts comments in the manual about the changes made to the original code, and a new document is made available to local colleagues that represents the code running on their machine. Only incremental comments went into the documentation, making it reasonable for the student to spend that small, extra, professional effort for the community.

Data from lab or clinical trials need to be put in and formatted for the code to act upon. This is essentially a problem in cross-media communication or text editing. Either the data were generated in machine readable form from modern instrumentation, or they must be transcribed to machine readable form. In either case, after they get into the local machine, the student will have to examine them, possibly have to reformat them for the code, maybe plot them, possibly put them into other forms for premodel statistical summarization, and do other processes on them. These are essentially editing functions done easily with the interactive system editor—the same one used for writing changes to the original application code. At no time does the data leave the confines of the on-line rotating magnetic storage media. Even the manual for the code is on-line and perused via the display terminal. It is possible to print out this "soft" copy in "hard"-copy form (on paper).

Up to this point the student only had to have a rough idea of how the application code worked and what data input formats it required. In truth, before proceeding, the manual for the modeling system had better be read through thoroughly. This is a very important point in the scenario for our orientation purposes.

Any nontrivial process requires setup, control, monitoring, and review. The writers of nontrivial code document their code. They try to make it easy to use. Some do a good job of making their system user friendly. There is always the requirement that a user meet the code author halfway. There is a learning process that the user must go through to properly use someone else's code. This is a major component of the time/effort investment decision. Not only must our student learn the rudiments of the computer interface (the interactive system and its editor), but the student has to learn how to properly use the application code. The manuals on some application codes run into hundreds of pages and are comparable in complexity to the computer interface, or even exceed it in complexity.

The investment decision is further complicated by the usual truth that even today, most application systems are not complete. They often explicitly rely on pre- or postprocessing functions, such as plotting, that are themselves independent complex application systems to be learned. Therefore, the user must be able to foresee all allied processes in the complete function to be performed. This means insuring that all allied codes are available, data compatible, and easily accessed at a reasonable dollar cost.

The allied codes must be well documented with documentation easily accessed by the user. The codes must be reliable and supported. This means that if a critical failure crops up, the user can get to an expert who can troubleshoot the problem and fix it quickly. This support, documentation plus consultation, on really large application systems is absolutely essential, statistically. Most users get through without a problem. Once in a while, a user may be totally halted by a major problem. Progress is impossible without the complete documentation necessary for problem diagnosis and correction.

We return to our mythical student, who now has a running application code that certainly will do the job, data in the correct format that represent the problem under study, and a knowledge of this application code developed well enough to start some test runs.

The next steps are the setup and initial runs of the real data. The objective is to fit the data to a system of differential equations that might represent the metabolic system under study. The code does this rapidly. In some application codes, in the interactive environment, the user need not set a huge computational process, such as this, into unmonitored motion since it is an unwise use of a computer because of the potential for waste. These interactive systems allow the user to stop the computation arbitrarily to see how things are progressing. A well-designed application on an interactive system will make provision for viewing intermediate results, issuing noncatastrophic halts, and making corrective restarts which save as much of the computation as possible.

The runs proceed with minor interventions to correct missed data, errors, reruns due to the inherent learning curve, i.e., blunders caused by not reading the manual thoroughly, and then one day a fit is made that plots out acceptably. The underlying nonlinear parameter estimation technique was robust enough for this problem. The underlying integration schemes enabled a time simulation of the metabolic process that seems reasonable. The model works. Now to write it up.

The output data are in an on-line file, the student knows the system editor, and it would be cost effective if there were some kind of a document composition facility (a composer). Well, the student used something of this sort to update the code documentation. Let us now pause to look at the composer. The editor allows the student to type in text and manipulate it. What is required is a system that would also allow the student to imbed, in the text, commands of some sort that would force the text to be printed out in the highly articulated format of a research report. This means composing with the headings, indented text, tables of data, bibliographic citations, footnotes, and the other format requirements of the journal of the student's desire.

Most of these facilities are available today on interactive terminal-oriented computers and even on some of the very small home computers. Learning the composer is a task of a few days duration, with one large document being the minimum experience needed to develop some facility with the editor/composer. It is true of almost any of the systems that we have seen that they are usable in a few days, but true facility of use comes only with extensive experience running into weeks.

It is instructive for our orientation purposes to follow our student's career a little further. The pay out, or return on investment, has not yet begun in earnest. Our student moves on to a second institution as a researcher. After arranging for the proper access at both new and old locations and confirming that both machines can talk to each other over a data network, the researcher copies over the data, the code, and the report files from the first machine to the second. This takes a few minutes. A grant proposal is required in a hurry. The old report is used as a text base with some new prose added. The grant proposal is ready in a few days. The data and the code form the basis of a new but related study.

More importantly, the knowledge base developed by the researcher is not wasted. Because most computation systems are similarly designed, many of the concepts learned on one will carry over to another. It is true, today, that we are far from a universal system that needs to be learned once and for always. That is because we do not know the best way to do everything. We do not even know the list of the "everything" that we want to do.

The researcher, now a respected colleague and recognized computer expert, is called upon to help choose computation resources for the

department. What lessons have been learned that can be used in this decision process? What about distributed processing, personal computers, or
word processors? What about the large investment of time these things
seem to require? After all, the new departmental spectrographic analyzer
has a 223-page manual that most departmental members won't get around
to reading for months; so why learn a system editor? There are no formula
answers. There are unresolvable requirements for time/effort expenditures. Each individual has to make their own choice. The choices are less
one-sided today in that computer interfacing is becoming easier.

Our wizened junior staff member points out the need to include computer
training for the undergraduates. Fortunately there are introductory courses taught. All that the department has to do is not schedule work against
them which would cause student schedule conflicts.

The department establishes a computational lab for students. The lab
has a few micro-based personal computers for small computations. The
plan is to tie them into the local distributed processor and through it to the
central site. They will act as local intelligent terminals for problem preparation. The students will take problem sets off a central shared file, and
they will send answers back to the professor's machine for correction. Because the personal machines have attached floppy disks or tape cassettes,
students can be alloted half-hour sessions on the terminals. If they don't
finish their work during a session, they can save the current state of their
work on a floppy disk and put the removable disk in a notebook to carry with
them until their next session. A student can reload and begin where the last
session left off.

The student uses computing tools of the discipline such as the super-
routines, explained below. In this manner, a student can build experience
with computing in the discipline being studied. The computer is a tool,
much as a hand calculator is today, but the computer has much more to offer than simple numerical computation.

As for the senior staff, a small stand-alone distributed processor with
shared terminals in the immediate area would provide an easily accessed
introduction to computing at times convenient for them, and since it is compatible with other computational resources of the institution, it can also be
used as a work station for data transfer. For those more familiar with computing, a personal terminal with direct access to the local distributed processor makes sense. The same terminals connect through the local processor to significant computational resource at the central site. The secretar-
ial staff also can easily be trained to do standard correspondence on a word
processor. The only requirement is that the documents generated should
be available for access by the staff's terminals. This is so that all of the
department's resources can be brought to bear in whatever combination is
needed by various projects. There is, therefore, a requirement that the
word processor, the interactive terminals, the local distributed processor,
and the central site communicate compatibly.

Examples of cooperative departmental access are in the following scenarios. A multimember grant proposal requires combining sample reports from a number of people. Copies of previous correspondence appear as inserts. Data from research projects are inserted as tables and graphs for illustrative purposes. Another example is the cooperative study by a number of people wherein a division of labor requires interperson communication. A central file of memos and instructions is maintained for all to access. Data are kept in a central file where only the data owner can modify them, but all can copy them.

The cost of facilities supporting these scenarios is coming down each year. The relative payback on investment of time and effort is quite high for the individual. The payback for the institution is more intangible. Enabling the institution's members to access easily a significant computational resource with an easily learned interface which is coherent across many types of use is a goal which ought to be of the highest priority for administrative management. Unfortunately, in a dynamic technology there are obsolescence factors in every new modality. The plans for providing computation to an individual or an institution must factor-in the rapid change. Four-function hand-held calculators at $300 were only a novelty. Our $35 full-slide-rule-function, credit-card-sized pocket calculators are useful tools.

A few years ago, within a large organization, a 20-user timesharing computer was economically justifiable. Today a 100- to 200-user timesharing system is justifiable. Also today the possibility of off-loading the big system onto a compatible distributed processor is a real choice. The possibility of the distributed machine supporting all the user-friendly interactive features mentioned above is quite real. The demand for these services is growing, but it is not apparent to those planners who themselves may not have invested the personal time to learn the advantages of the new more easily learned interfaces to computation. Education plus first-hand experience will bring about the evolutionary switchover to these "intelligent nodes" or "intelligent terminals" and their interactive use. Personal intelligent machines will be used compatibly with them. Real world data inputs such as sensor inputs from instrumentation will be possible with the same machines. It is now time to plan how we will organize and invest our effort in mastering these useful tools.

VII. SUPER-ROUTINES (SRs)

Now that we have some orientation as to where the field of computation is coming from and going to, let us take a closer look at the contribution a food scientist or engineer might make to the FS&E discipline using computing.

Certainly, the personal decision to learn enough about modern interactive computing to use it effectively will enable the individual to progress faster and contribute more in a given time than otherwise possible. But there is an even more effective means of interfacing to computation and contributing to the field. It is the use and creation of "super-routines."

A super-routine is a set of independent routines that solve a specific problem. It is also much more. Given the highly interactive nature of modern computation, it is possible to create a two-way interface between a user and a computational process which allows the user to state rapidly the entire sequence of computational actions to be performed. This process of statement may be conversational or at least menu driven. The latter is a mode in which the application user is presented a menu of choices on a screen. The user inputs a few typed characters to indicate choices or, on some systems, merely presses a light pen against the screen, pointing at the appropriate choice. Having thus "commanded" the super-routine, the SR proceeds to execute the indicated process.

If the professional FS&E person had to program this kind of interface facility, there would be no time for development of the programs that actually execute discipline-dependent processes. Fortunately there are available, on many interactive systems, such menu facilities. They are called by many names such as "menu writer," "dialogue manager," or "full-screen support package." The point to remember is that in estimating the time/ effort investment for the individual, one must seriously consider the availability and use of such advanced facilities if more than passive use of computation is planned. A professional may only be a user of other's super-routines, but it is often possible with a modest amount of extra effort to create new usable ones.

So, the super-routine has a powerful command interface. What else distinguishes it from a subroutine package? It usually has its own means of handling data. Either it contains its own file handling subsystem, or it uses a standard file system in such a way as to unload, from the user, data creation and input/output problems. This often involves another major component of the SR, the report writer.

The report writer is a facility for generating a small set of standard formatted reports of the form usual to the application and often very specific to it. Again, if the FS&E professional had to create such a facility, no time would be left to do the real work. Fortunately, here too, often there are general report writers and file systems for the professional to use. The system editor and composer are also available to modify the standard reports or to create more complicated reports than those the report writer generates.

What is left to include in making an SR? We have the command facility, the data file manager, the report writer, the system editor, and the system composer. We need subroutine libraries and powerful higher-level programming languages that allow the succinct statement of algorithms and processes to a machine. We often need compatible graphic output codes.

To complete the SR component list, there is the programming code that solves a significant discipline-dependent problem. These codes are called "kernels" to indicate that they are the fertile heart of the overall code.

They, and they alone, are the reason for all the other codes. The kernels solve the user's problem. Everything else exists to ease the programmer's task and to ease the program user's task.

The nonkernel components should not be confused with ease-of-use or user-friendly features of modern interactive applications. The editor, composer, report writer, etc., would be easy to use and user friendly because they contain ease-of-use features.

The nonkernel components may be integral to an SR, or they may be part of the larger interactive environment on a given machine. The borders of the SR and its environment are not always clearly defined. When the user sees a machine with a single application development environment, i.e., one editor, one language, one report writer, and one way to do file handling, it can be called an application system or an application environment. We consider an SR to exist where there has been created a coherent master process that leads the process user, who is not a computing professional, completely through the statement of a problem, the manipulation of relevant data, the computation, and finally, report generation—all for a specific problem in a limited class of similar problems from some discipline. The user command language which controls the entire process of computation is discipline oriented; it may use discipline terminology. The SR may well be a discipline-oriented process using or running in an application environment.

A super-routine would exist even if many of the nonkernel components are shared with other SRs. To be considered an SR, it is necessary and sufficient that a software system solve more than one problem in a discipline by only changing the computational kernel.

What we need now is the specific knowledge of what problems are worth solving. We need the commitment of talented knowledgeable professionals to the job of creating those computational kernels that, when housed in the outer structure listed above, become powerful, easily used tools of the trade, i.e., SRs.

It is critical to know that the computing profession is continuing to provide the nonkernel components of the SRs but is not, and cannot be, relied upon to provide the discipline-oriented and discipline-dependent computational tools. This is best done by creating the kernels and imbedding them in a user-friendly super-routine. This, in turn, is most efficiently done by employing a standard architecture for the interfaces between the kernels and the support functions such as report writing or data management. Higher-level languages are required for transportability.

Each individual must recognize the personal responsibility to learn to interface to computation, learn to use the new super-routines as they become available, and support their creation, if not actually participate in the creative process. At the very least, one must know enough about the computational process to form educated judgments as to the usefulness of computational alternatives such as super-routines. This includes the ability to judge the amortization period of the time/effort investment in learning to use these SRs and the basic facilities of interactive computation.

VIII. SOME EXAMPLES RELATED TO SUPER-ROUTINES (SUPER-ROUTINES DO NOT YET COMPLETELY EXIST)

A. MPSX

Mathematical programming system extended (MPSX) (IBM, 1979) is a linear, mixed-integer, and separable programming system. It was defined during the middle era as a batch application and then updated to the interactive mode of use by running under VM/CMS. VM/CMS is an IBM interactive operating system that provides an application environment with easy-to-use text editors, report formatting, file management, and programming languages.

Under CMS, the user can transfer data to MPSX's internal file system. There is presently no menu interface. The user edits a command file. A problem can be set up using a powerful control language, ECL (or extended control language) (Slate and Spielberg, 1978). The problem, which can take a long running time, can be run in a background batch mode while foreground interactive work proceeds, or the problem can be run interactively, whereby a user can monitor the computation as it proceeds. There is a data preparation and report writing facility within MPSX that is tailored to its purposes.

Most important is that, from our point of view, the control language ECL allows the user to code algorithms using the components of MPSX. This is done by using ECL within the high-level language PL/I (Bates and Douglas, 1975). This implies that new more powerful super-routines can be produced using linear, mixed-integer, or separable programming as a component process. Since quadratic programming can be built up using MPSX, even some nonlinear models can be built.

It is possible for a computing center to customize the use of MPSX for a group of users. The input and reports can be prepared in a standard jargon. The total personal time investment of the individuals of the using group to learn to use MPSX for modeling is thereby reduced.

This would be a super-routine. By just changing the method kernel, different classes of linear, separable, or quadratic optimization problems could be stated, solved, and reported—all in a jargon and input/output format peculiar to a discipline. The users would need to spend relatively little time learning how to state their problems for solution. They would be communicating with a machine seemingly custom designed for their discipline.

B. EPLAN

EPLAN (IBM, 1975) is an econometric modeling system written in APL [cf. APL (A Programming Language) Sec. VIII.C]. One can also use APL under

VM/CMS. EPLAN is an APL library work space that, once loaded, allows
the user to quickly state linear econometric models calling upon time-series
files of historical data. These are used to find the coefficients of the model
by regression. The model can be used to project the economic trend under
study. A possible use is in the study of market trends and responses. Government time-series files are easily imported into an interactive system for
access by EPLAN.

There exist compatible SR components in APL such as a data manager,
a report writer, and a full-screen menu handler for menu-mediated dialogues
(see Sec. VIII. C), but EPLAN does not directly use them. EPLAN does not
come with a menu-mediated front-end user interface. It is well within the
current state of the art to use the menu manager and put such a front-end on
EPLAN, if the user wishes to.

Again, as in the MPSX example, the EPLAN program, along with the
other APL components, can be coordinated by a computing center to provide
a single SR for a group of users in a discipline. They can be provided with
a standard way of stating econometric modeling problems. The reports also
can be standardized. In this way, a user interface to computation is provided so that the user investment in learning time is minimized. Yet, the
user has a significant tool to use for a large set of problems. The tool communicates to the user in the jargon of his or her discipline, not in the jargon
of APL, econometrics, or computing.

C. APL (SR Generator and SR Environment)

Properly, APL (IBM, 1979) is not a super-routine but is a language/control
system for writing SRs. This means that the application kernel and the en-
veloping data manager, report formatter, report writer, and the menu manager are available, all in APL. We include APL because there are so many
desirable features for the casual computer user that it should be presented
in this discussion. APL, as a language, is one of the better interactive languages. It is interpretive in its implementations, which allows the nonprofessional programmer to easily chip away at a problem by trial and error.
It has a very extensive library of coherent routines and super-routine components. Its control structure allows the casual user to create quickly new
super-routines out of old components and new application kernels. APL imposes a design discipline by its very nature that allows one person's routines
to run easily with another's. There exists a proposed interchange standard
for exchanging APL work spaces (Cartwright, 1978) among machines of various manufacturers.

Examples from the library are the statistical packages (IBM, 1980);
the full-screen manager, which is part of the function editor (IBM, 1979)
and can be used to create menus; and the APL departmental reporting
system (ADRS) (IBM, 1979), which can be used for report generation.
There is also a data-base system ADI (IBM, 1979) and a graphic display
SR component (IBM, 1980).

IX. SOME SUGGESTED NEW SUPER-ROUTINES

A. Curriculum SRs

Rather than specify a specific SR, we refer here to any SR that can be used as a teaching tool. This concept involves the classroom use of small personal computers, such as the IBM Personal Computer (tm), which are language compatible with larger distributed processors or central site machines. The creation of problem sets that can be executed with little or no student programming is the first step. The student is taught to compute answers with much the same facilities that the researcher uses, but the student's path through problem statement and solution is guided by the instructor and the controlling code. This implies that the student can access a central data base of code, problems, problem data, and corrected previous problems. The student need not be continuously connected to the central facility. Small independent processors are useful for problem set-up and report preparation of the final output. Only student-chosen input data, student-chosen setup commands, and student-derived answers are sent over communication lines to the instructor for grading. If properly done, the SRs invoked in these learning experiences are the same ones used by the student at a later time when the student becomes an operational professional.

The use of SRs that can run on small student machines and on large processors enables a more economic use of computation in the classroom. This is not computer-aided instruction. This is the use of computation in the profession at the earliest level of professional education. This use of computation is simplified but completely parallel to the way computation is used by a researcher doing more difficult problems.

B. Pharmacokinetic Modeling SR

Let us look at a more specific example. The structure of pharmacokinetic modeling is well determined (Berman and Weiss, 1978; D'Argenio and Schumitzky, 1979; Jacquez, 1972; Notari, 1975). What we call for here is a new encasement of one of these systems in an interactive environment with a menu-driven front end using terminology closer to FS&E use. Being able to change nomenclature is a general requirement for SRs. A super-routine such as this one could then be used as a curriculum SR in nutrition courses, pharmacology courses, etc. It should be arbitrarily interruptible so that students can get on and off shared terminals at arbitrary points in the code's execution.

As an example, a class studying a specific metabolic pathway for which there exists a reasonable model could use the SR just as an evaluation tool. The instructor would first set up the code to request input in the terms of the specific textbook being used. Then the student could be expected to respond with the proper inputs without having to learn the SR in its full generality. In this way, beginners can use advanced computational tools of the laboratory without first being required to fully master them, thus avoiding

the attendant time/effort cost that full mastery entails. The purpose of the
curriculm at this early stage of the student's schooling is to learn the phenomena not the tools for investigating the phenomena.

The current state of research as reflected in a recent tutorial article
(Carson and Jones, 1979) seems to be such that even experienced researchers would find such an SR useful. Only one code is referred to in the article.
That one is a FORTRAN batch-oriented code. It is a very useful and important code. However it is now possible to take advantage of the new fully interactive interface for computation and to provide an even better tool for the
modeling of metabolism and similar problems.

C. Canning Plant Simulation SR

Another specific example is the problem of simulating a canning plant. For
purposes of teaching and research there is no reason why one of the standard
discrete simulators, such as general-purpose simulation system (GPSS)
(Schriber, 1974) can't be specialized to accept inputs describing a canning
operation. This underuse of the general simulator is a useful simplification.
It is a means of creating specific model subclasses that can easily be learned
by students and colleagues. The students need only learn the specific discipline-dependent input terms of the model class and only those control functions needed to get a specific output. They thereby get an introduction to
discrete simulation as a side benefit while studying a specific discipline-
related system, a canning plant's operation. Their investment in time and
effort is much less than if they first had to learn the general simulator and
then create a specific model themselves.

X. SUMMARY

The interface to computation has changed from one-to-one person-machine
dialogue via manual switches, through the many-to-one batch mode of operation, to the many-to-one interactive mode of operation. Today, the display device, used for interaction with a large system, may itself be a micro-
based computer allowing again one-to-one use in its off-line mode of operation. It may be connected to a distributed network via a local minicomputer.

More important than just hardware is the growing allocation of computational resource being dedicated to aiding and easing the communication
across the person-machine interface to computation. Components such as
dialogue managers, editors, composers, data-base management subsystems,
high-level languages, and general word processing systems are all accessed
through an interactive operating system layer that is becoming thinner.

The time/effort investment to get started in this new technology is relatively small. The amortization time is short, and the return on investment
can be huge, especially if the food scientist or engineer concentrates effort
by using super-routines where they exist.

The super-routine is a coherent set of those interfacing aids, just mentioned, that contains discipline-dependent computational kernels. Thereby is produced a coherent total system for the statement, solution, and reporting of a significant but limited class of specific problems.

Concomitant with learning to use SRs is the creative work that is required to produce new SRs for food science and engineering, for both academic and industrial environments. All the components for producing SRs may not yet be available on a given computer system. It is a new professional responsibility to find them and produce them or to at least recognize the need for these necessary interfacing aids and to communicate that need to those who can produce the aids.

This then is the opportunity and the challenge.

ACKNOWLEDGMENT

I thank Prof. Steve Tannenbaum of the MIT Nutrition and Food Science department for introducing me to the problem, in food science and engineering, of metabolic process modeling. This was done via his problem of modeling human nitrate/nitrite metabolism. He introduced me to other computational problems in FS&E and to some of the literature. I also thank Prof. Mark Karel of the same department, and Prof. Israel Saguy of the Volcani Center in Israel, for conversations on other computational problems in FS&E, especially control theoretic problems.

Dr. Yon Bard of the IBM Cambridge Scientific Center was helpful in developing a wider view of the nonlinear estimation literature and how it related, now and historically, to FS&E. He has written a book on nonlinear estimation which documents some of the early work on nonlinear estimation codes (see Bard, 1974). They were originally created for general modeling but quickly became employed in biokinetic modeling.

Love Seawright and Dick MacKinnon get credit for interpreting the charter of our Cambridge Scientific Center to include time in our project schedules for learning about the computational problems and needs of scientists and technologists in other fields and for including time to communicate these needs both into IBM and out to the users in documents such as this book.

Prof. Israel Saguy gets credit for cajoling me into writing my experiences and thoughts in a way that might be beneficial to professionals in FS&E.

Thanks to Dr. Judith M. Frank, my wife, for biotechnical and copy help.

XI. APPENDIX: THIS CHAPTER AS A 1980s
INTERFACING EXAMPLE

This chapter was written in a way illustrative of the modern mode of interfacing for computation. The initial draft of this document was entered into VM/CMS via a 3277 display terminal using the display editing system and the composer SCRIPT/VS. The text file was sent via RSCS, a data-networking system, to an IBM 6670, which is a laser printer and a communicating copier and which, although a node on the data network, is less than 15 ft from the author's office. The copy came out of the IBM 6670 within seconds in a camera-ready form.

Of course, the draft needed revision and some reformatting. Once, during the writing period, the author was called to California for a week-long meeting and so could not work on this document. The text file was sent over the network to our departmental word processing specialist, Barbara Whitehill, for rework. Her office is less than 50 ft away from mine. Both of our terminals are connected to the same physical machine. A red-penciled corrected copy of this document was given to her. Using this, she put the indicated text and format corrections into the text file using another compatible system editor, the VM system product editor, which she prefers to use. The printed copy was sent out on an IBM Office System/6 (also a node on the network) near her office for hard-copy review. The corrected source file was sent back over the network to my virtual machine (residing on the same physical machine).

Before leaving for California, I had sent messages over the network to friends at the lab I was to visit and arranged to get access to a terminal at their West Coast location during night hours. Since there is a 3-hr time difference, I could log onto my East Coast machine from the West Coast via the network, and add more to the document. Barbara Whitehill and I had a potential 3-hr worktime overlap after my West Coast meetings.

This is an example of cooperative shared access to files for word processing using an interactive system, VM/CMS. Not all the functions implied by this narrative are available as IBM products. The IBM Cambridge Scientific Center, where all of this occurred as an experiment, is an advanced technology group that develops this kind of support as part of our experimentation in computer science.

While working on this chapter, I used the editor to look at data files I had stored away a year previous to the editing but that I was considering using in an example of an SR. The data were the output of the dose/response computations that I did when I put the seminal Teorell (1937) computations on a modern machine. This was done for tutorial purposes. A continuous systems modeling program (CSMP) (IBM, 1975) was used to state and run the model. The model was stated in CSMP, in a grand total of 18 APL statements including input data. My computation was first done on an IBM 370/158 and then rerun on an independent IBM 5100 Personal (desk-top) Computer which I took home for this purpose. The output was plotted by an on-line graphic terminal to confirm the article's hand-computed graphics. CSMP, written in APL, runs on both machines in a compatible way.

After looking over the file and printing it out, I rejected it as inappropriate. If I had wanted to include the data file in this document, it would have been a matter of minutes to format it and imbed it into the composer source file for this document.

REFERENCES

Bard, Y. (1974). Nonlinear Parameter Estimation. Academic Press, New York.

Bates, F. and Douglas, M. L. (1975). Programming Language/One. Prentice-Hall, Englewood Cliffs, N.J.

Bell System (1978). Unix time sharing system, Bell System Technical Journal 57(6):Part 2, July-August.

Berman, M. and Weiss, M. F. (1978). SAAM27 (Simulation Analysis and Modeling) U.S. Dept. of Health Education and Welfare Publication NIH 78-180. Supt. of Documents Govt. Printing Office, Washington, D.C.

Carson, E. R. and Jones, E. A. (1979). Use of kinetic analysis and mathematical modeling in the study of metabolic pathways in vivo, N. Engl. J. Med. 300(18):1016-1027.

Cartwright, D. (1978). The Proposed Interchange Standard, prepared for the STAPL/SIGPLAN subcommittee chaired by Patrick E. Hagerty. APL QUOTE QUAD. 8(2), ACM.

D'Argenio, D. Z. and Schumitzky, A. (1979). A program package for simulation and parameter estimation in pharmacokinetic systems, Computer Programs in Biomedicine, Vol. 9, No. 2. North-Holland, Amsterdam, Netherlands, p. 115-134.

Eduards, P. and Broadwell, B. (1979). Data Processing (Computers in Action). Wadsworth, Belmont, Calif.

IBM DPD (1982). IBM Virtual Machine Facility/370: CMS User's Guide. IBM Manual GC20-1819. White Plains, N.Y.

IBM DPD (1979). IBM Mathematical Programming System Extended/370. Program Reference Manual. IBM Manual SH19-1095. White Plains, N.Y.

IBM DPD (1975). APL Econometric Planning Language. IBM Manual SH20-1620. White Plains, N.Y.

IBM DPD (1979). APL Language. IBM Manual GC26-3847. White Plains, N.Y.

IBM DPD (1980). APL Statistical Library; APL Multivariate Time Series Analysis; and APL Advanced Statistical Library. IBM Manuals SH20-1841, 2174, and 1948. White Plains, N.Y.

IBM DPD (1979). VSAPL Extended Editor and Full Screen Manager. IBM Manual SH20-2341. White Plains, N.Y.

IBM DPD (1979). A Departmental Reporting System, IBM Manual SH20-2165. White Plains, N.Y.

IBM DPD (1979). APL Data Interface. IBM Manual SH20-1954. White Plains, N.Y.

IBM DPD (1980). IBM 3277 Graphics Attachment Support. IBM Manual SH20-2138. White Plains, N.Y.

IBM DPD (1975). IBM System/370 APL: Continuous System Modeling Program. IBM Manual SH20-2115. White Plains, N.Y.

Jacquez, J. A. (1972). Kinetics of distribution of tracer-labeled materials, Compartmental Analysis in Biology and Medicine. Elsevier, New York.

Notari, R. E. (1975). Biopharmaceutics and Pharmacokinetics. Marcel Dekker, New York.

Schriber, T. J. (1974). Simulation Using GPSS. John Wiley & Sons, New York.

Slate, L. and Spielberg, K. (1978). The extended control language of MPSX/370, IBM Systems J. 17(1).

Teorell, T. (1937). Kinetics of Distribution of Substances Administered to the Body. (I. The Extravascular Modes of Administration). Archieves Internationales De Pharmacodynamie et de Therapie. G. Doin & Cie Paris. Vol. LVII 1937.

Glossary*

Absolute address
> The binary number that is assigned as the address of a physical memory storage location.

Absolute section
> The portion of a program in which the programmer has specified physical memory locations of data items.

Access time
> The interval between the instant at which data is required from or for a storage device and the instant at which the data actually begins moving to or from the device.

ADC (Analog to Digital converter)
> A circuit which converts analog signals to binary data.

Address
> A label, name or number that designates a location in memory where information is stored.

Algorithm
> A prescribed set of well-defined rules or processes for the solution of a problem in a finite number of steps.

Alphanumeric
> Referring to the subset of ASCII characters that includes the 26 alphabetic characters and the 10 numeric characters.

ANSI
> American National Standards Institute.

APL (A Programming Language)
> A condensed, high-level language capable of describing complex information processing in convenient notation. It uses arrays as basic data elements and manipulates them with a set of

---

*Reproduced by permission from the Introduction to RT-11, Digital Equipment Corporation, 1978, Maynard, Massachusetts. (Order No. DEC-11 ORITA-A-D, DN1.) The editor assumes responsibility for the accuracy of the material herein.

powerful operators. Statements are usually interpreted during execution and require no compilation whatsoever.

Application program (or package)

A program that performs a function specific to a particular end-user's (or class of end-user's) needs. An application program can be any program that is not part of the basic operating system.

Argument

A variable or constant value supplied with a command that controls its action, specifically its location, direction, or range.

Array

An ordered arrangement of subscripted variables.

ASCII

The American Standard Code for Information Interchange; a standard code using a coded character set consisting of 8-bit coded characters for upper and lower case letters, numbers, punctuation and special communication control characters.

Assembler

A program that translates symbolic source code into machine instructions by replacing symbolic operation codes with binary operation codes and symbolic addresses with absolute or relocatable addresses.

Assembly language

A symbolic programming language that normally can be translated directly into machine language instructions and is, therefore, specific to a given computing system.

Assembly listing

A listing, produced by an assembler, that shows the symbolic code written by a programmer next to a representation of the actual machine instructions generated.

Asynchronous

Pertaining to an event triggered by the occurrence of an un-related event rather than "synchronous" or related operations scheduled by time intervals.

Background program

A program operating automatically, at a low priority, when a higher priority (foreground) program is not using system resources.

Backup file

A copy of a file created for protection in case the primary file is unintentionally lost or destroyed.

Base address
> An address used as the basis for computing the value of some other relative address; the address of the first location of a program or data area.

BASIC (Beginner's All-purpose Symbolic Instruction Code)
> An interactive, "algebraic" type of computer language that combines English words and decimal numbers. It is a widely available, standardized, simple beginner's language capable of handling industry and business applications.

Batch processing
> A processing method in which programs are run consecutively without operator intervention.

Baud
> A unit of signaling speed (one bit per second).

Binary
> The number system with a base of two used by internal logic of all digital computers.

Binary code
> A code that uses two distinct characters, usually the numbers 0 and 1

Bit
> A binary digit. The smallest unit of information in a binary system of notation. It corresponds to a 1 or 0 and one digit position in a physical memory word.

Block
> A group of physically adjacent words or bytes of a specified size that is peculiar to a device. The smallest system-addressable segment on a mass-storage device in reference to I/O.

Bootstrap
> A technique or routine whose first instructions are sufficient to load the remainder of itself and start a complex system of programs.

BOT (Beginning of Tape)
> A reflective marker applied to the backside of magtape which identifies the beginning of the magtape's recordable surface.

Bottom address
> The lowest memory address into which a program is loaded.

Breakpoint
> A location at which program operation is suspended to allow operator investigation.

Buffer

> A storage area used to temporarily hold information being transferred between two devices or between a device and memory. A buffer is often a special register or a designated area of memory.

Bug

> A flaw in the design or implementation of a program which may cause erroneous results.

Bus

> A circuit used as a power supply or data exchange line between two or more devices.

Byte

> The smallest memory-addressable unit of information. In a PDP-11 computer system, a byte is equivalent to eight bits.

Call

> A transfer from one part of a program to another with the ability to return to the original program at the point of the call.

Calling sequence

> A specified arrangement of instructions and data necessary to pass parameters and control to a given subroutine.

Central processing unit (CPU)

> A unit of a computer that includes the circuits controlling the interpretation and execution of instructions.

Character

> A single letter, numeral, or symbol used to represent information.

Character pointer

> The place where the next character typed will be entered. (The character pointer is visible as a blinking cursor on VT-11 display hardware.) During editing, the character pointer indicates the place in an ASCII text file where the next character typed will be entered into the file.

Clear

> To erase the contents of a storage location by replacing the contents, normally with 0s or spaces.

Clock

> A device that generates regular periodic signals for synchronization.

Code
> A system of symbols and rules used for representing informa-
> tion — usually refers to instructions executed by computer.

Coding
> To write instructions for a computer using symbols meaningful
> to the computer itself or to an assembler, compiler, or other
> language processor.

Command
> A word, mnemonic, or character, which, by virtue of its syntax
> in a line of input, causes a computer system to perform a
> predefined operation.

Command language
> The vocabulary used by a program or set of programs that
> directs the computer system to perform predefined operations.

Command language interpreter
> The program that translates a predefined set of commands into
> instructions that a computer system can interpret.

Command string
> A line of input to a computer system that generally includes a
> command, one or more file specifications and optional
> qualifiers.

Compile
> To produce binary code from symbolic instructions written in a
> high-level source language.

Compiler
> A program that translates a high-level source language into a
> language suitable for a particular machine.

Computer
> A machine that can be programmed to execute a repertoire of
> instructions. Programs must be stored in the machine before
> they can be executed.

Computer program
> A plan or routine for solving a problem on a computer.

Computer system
> A data processing system that consists of hardware devices,
> software programs, and documentation that describes the
> operation of the system.

Concatenation
> The joining of two strings of characters to produce a longer
> string.

Conditional assembly
> The assembly of certain parts of a symbolic program only when certain conditions are met during the assembly process.

Configuration
> A particular selection of hardware devices or software routines or programs that function together.

Console terminal
> A keyboard terminal that acts as the primary interface between the computer operator and the computer system. It is used to initiate and direct overall system operation through software running on the computer.

Constant
> A value that remains the same throughout a distinct operation. (Compare with Variable.)

Context switching
> The saving of key registers and other memory areas prior to switching between jobs with different modes of execution, as in background/foreground programming.

Conversational
> See Interactive.

CPU
> See central processing unit.

Crash
> A hardware crash is the complete failure of a particular device, sometimes affecting the operation of an entire computer system. A software crash is the complete failure of an operating system usually characterized by some failure in the system's protection mechanisms or flaw in the executing software.

Create
> To open, write data to, and close a file for the first time.

Cross reference listing
> A printed listing that identifies all references in a program to each specific symbol in a program. It includes a list of all symbols used in a source program and the statements where they are defined or used.

Current location counter
> A counter kept by an assembler to determine the address assigned to an instruction or constant being assembled.

Data
> A term used to denote any or all facts, numbers, letters, and symbols. Basic elements of information that can be processed by a computer.

Data base
An organized collection of interrelated data items that allows one or more applications to process the items without regard to physical storage locations.

Data collection
The act of bringing data from one or more points to a central point for eventual processing.

Debug
To detect, locate, and correct coding or logic errors in a computer program.

Default
The value of an argument, operand, or field assumed by a program if not specifically supplied by the user.

Define
To assign a value to a variable or constant.

Delimiter
A character that separates, terminates, or organizes elements of a character string, statement, or program.

Device
A hardware unit such as an I/O peripheral, magnetic tape drive, card reader, etc. Often used erroneously to mean "volume."

Device control unit
A hardware unit that electronically supervises one or more of the same type of devices. It acts as the link between the computer and the I/O devices.

Device handler
A routine that drives or services an I/O device and controls the physical hardware activities on the device.

Device independence
The ability to program I/O operations independently of the device for which the I/O is intended.

Device name
A unique name that identifies each device unit on a system. It usually consists of a 2-character device mnemonic followed by an optional device unit number and a colon. For example, the common device name for RK05 disk drive unit 1 is "RK1:"

Device unit
One of a set of similar peripheral devices (e.g., disk unit 0, DECtape unit 1, etc.). May be used synonymously with volume.

Diagnostics
> Pertaining to a set of procedures for the detection and isolation of a malfunction or mistake.

Digit
> A character used to represent one of the non-negative integers smaller than the radix (e.g., in decimal notation, one of the characters 0 to 9; in octal notation, one of the characters 0 to 7; in binary notation, one of the characters 0 and 1).

Direct access
> See Random access.

Directive
> Assembler directives are mnemonics in an assembly language source program that are recognized by the assembler as commands to control a specific assembly process.

Directory
> A table that contains the names of and pointers to files on a mass-storage volume.

Directory-structured
> Refers to a storage volume with a true volume directory at its beginning that contains information (file name, file type, length, and date-of-creation) about all the files on the volume. Such volumes include all disks, diskettes, and DECtapes.

Disk device
> An auxiliary storage device on which information can be read or written.

Display
> A peripheral device used to portray data graphically (normally refers to some type of cathode-ray tube system).

Downtime
> The time interval during which a device or system is inoperative.

Echo
> The printing by an I/O device, such as terminal or CRT, of characters typed by the programmer.

Edit
> To arrange and/or modify the format of data (e.g.. to insert or delete characters).

Editor
> A program that interacts with the user to enter text into the computer and edit it. Editors are language independent and will edit anything in character representation.

Effective address
    The address actually used in the execution of a computer
    instruction.

Emulator
    A hardware device that permits a program written for a specific
    computer system to be run on a different type of computer
    system.

Entry point
    A location in a subroutine to which program control is
    transferred when the subroutine is called.

EOT (End Of Tape)
    A reflective marker applied to the backside of magtape which
    precedes the end of the reel.

Error
    Any discrepancy between a computed, observed, or measured
    quantity and the true, specified, or theoretically correct value
    or condition.

Execute
    To carry out an instruction or run a program on the computer.

Expression
    A combination of operands and operators that can be evaluated
    to a distinct result by a computing system.

Extension
    Historically-used synonym for file type.

External storage
    A storage medium other than main memory, e.g., a disk or tape.

Field
    A specified area of a record used for a particular category of
    data.

FIFO (first in/first out)
    A data manipulation method in which the first item stored is
    the first item processed.

File
    A logical collection of data treated as a unit, which occupies
    one or more blocks on a mass-storage volume such as disk or
    magtape, and has an associated file name (and file type).

File maintenance
    The activity of keeping a mass-storage volume and its directory
    up to date by adding, changing, or deleting files.

File name
> The alphanumeric character string assigned by a user to identify a file. It can be read by both an operating system and a user. A file name has a fixed maximum length that is system dependent. (The maximum in an RT-11 operating system is six characters, the first of which must be alphabetic. Spaces are not allowed.)

File type
> The alphanumeric character string assigned to a file either by an operating system or a user. It can be read by both the operating system and the user. System-recognizable file types are used to identify files having the same format or type. If present in a file specification, a file type follows the file name in a file specification, separated from the file name by a period. A file type has a fixed maximum length that is system dependent. (The maximum in an RT-11 operating system is three characters, excluding the preceding period and not including any spaces.)

File specification
> A name that uniquely identifies a file maintained in any operating system. A file specification generally consists of at least three components: a device name identifying the volume on which the file is stored, a file name, and a file type.

File-structured device
> A device on which data is organized into files. The device usually contains a directory of the files stored on the volume. (For example, a disk is a file-structured device, but a line printer is not.)

Flag
> A variable or register used to record the status of a program or device; the noting of errors by a translating program.

Floating point
> A number system in which the position of the radix point is indicated by the exponent part and another part represents the significant digits or fractional part (e.g., 5.39×10^8 – Decimal; 137.3×8^4 – Octal; 101.10×2^{13} – Binary).

Flowchart
> A graphical representation for the definition, analysis, or solution of a problem, in which symbols are used to represent operations, data, flow, and equipment.

FOCAL (FOrmula CALculator)
> An on-line interactive, service program designed to help scientists, engineers, and students solve numerical problems. The language consists of short imperative English statements which are easy to learn. FOCAL is used for simulating mathematical models, for curve plotting, for handling sets of simultaneous equations, and for many other kinds of problems.

Foreground
> The area in memory designated for use by a high-priority program. The program that gains the use of machine facilities immediately upon request.

FORTRAN (FORmula TRANslation)
> A problem-oriented language designed to permit scientists and engineers to express mathematical operations in a form with which they are familiar. It is also used in a variety of applications including process control, information retrieval, and commercial data processing.

Full duplex
> In communication, pertaining to a simultaneous, 2-way independent "asynchronous" transmission.

Function
> An algorithm accessible by name and contained in the system software which performs commonly-used operations. For example, the square root calculation function.

Garbage
> Meaningless signals or bit patterns in memory.

General register
> One of eight 16-bit internal registers in the PDP-11 computer. These are used for temporary storage of data.

Global
> A value defined in one program module and used in others. Globals are often referred to as entry points in the module in which they are defined and as externals in the other modules that use them.

Hack
> A seemingly inspired, but obscure, solution that is superior by some measure to a straightforward one.

Half duplex
> Pertaining to a communication system in which 2-way communication is possible, but only one way at a time.

Handler
> See device handler.

Hardware
> The physical equipment components of a computer system.

Hardware bootstrap
> A bootstrap that is inherent in the hardware and need only be activated by specifying the appropriate load and start address.

High-level language
> A programming language whose statements are typically translated into more than one machine language instruction. Examples are BASIC, FORTRAN and FOCAL.

High-order byte
> The most significant byte in a word. The high-order occupies bit positions 8 through 15 of a PDP-11 word and is always an odd address.

Image mode
> Refers to a mode of data transfer in which each byte of data is transferred without any interpretation or data changes.

Indirect address
> An address that specifies a storage location containing either a direct (effective) address or another indirect (pointer) address.

Indirect file
> A file containing commands that are processed sequentially, but that could have been entered interactively at a terminal.

Industry-standard
> A condition, format, or definition that is accepted as the norm by the majority of the (computer) industry.

Initialize
> To set counters, switches, or addresses to starting values at prescribed points in the execution of a program, particularly in preparation for re-execution of a sequence of code. To format a volume in a particular file-structured format in preparation for use by an operating system.

Input
> The data to be processed; the process of transferring data from external storage to internal storage.

Input/Output device
> A device attached to a computer that makes it possible to bring information into the computer or get information out.

Instruction
> A coded command that tells the computer what to do and where to find the values it is to work with. A symbolic instruction looks more like ordinary language and is easier for people to deal with. Symbolic instructions must, however, be changed into machine instructions (usually by another program) before they can be executed by the computer.

Interactive processing
> A technique of user/system communication in which the operating system immediately acknowledges and acts upon requests entered by the user at a terminal. Compare with batch processing.

Interface
> A shared boundary. An interface might be a hardware component to link two devices or it might be a portion of storage or registers accessed by two or more computer programs.

Internal Storage
> The storage facilities forming an integral physical part of the computer and directly controlled by the computer, e.g., the registers of the machine and main memory.

Interpreter
> A computer program that translates then executes a source language statement before translating (and executing) the next statement.

Interrupt
> A signal that, when activated, causes a transfer of control to a specific location in memory, thereby breaking the normal flow of control of the routine being executed.

Interrupt driven
> Pertaining to software that uses the interrupt facility of a computer to handle I/O and respond to user requests: RT-11 is such a system.

Interrupt Vector
> Two words containing the address of an interrupt service routine and the processor state at which that routine is to execute.

Iteration
> Repetition of a group of instructions.

Job
> A group of data and control statements which does a unit of work, e.g., a program and all of its related subroutines, data, and control statements; also, a batch control file.

Kluge
> A crude, makeshift solution to a problem.

Label
> One or more characters used to identify a source language statement or line.

Language
> A set of representations, conventions, and rules used to convey information.

Latency
> The time from initiation of a transfer operation to the beginning of actual transfer; i.e., verification plus search time. The delay while waiting for a rotating memory to reach a given location.

Library
>	A file containing one or more macro definitions or one or more relocatable object modules that are routines that can be incorporated into other programs.

LIFO (last in/first out)
>	A data manipulation method in which the last item stored is the first item processed; a push down stack.

Light pen
>	A device resembling a pencil or stylus which can detect a fluorescent CRT screen. Used to input information to a CRT display system.

Linkage
>	In programming, code that connects two separately-coded routines and passes values and/or control between them.

Linked file
>	A file whose blocks are joined together by references rather than consecutive locations.

Linker
>	A program that combines many relocatable object modules into an executable module. It satisfies global references and combines program sections.

Listing
>	The printed copy generated by a line printer or terminal.

Load
>	To store a program or data in memory. To place a volume on a device unit and put the unit on-line.

Load map
>	A table produced by a linker that provides information about a load module's characteristics (e.g., the transfer address, the global symbol values, and the low and high limits of the relocatable code).

Load module
>	A program in a format ready for loading and executing.

Location
>	An address in storage or memory where a unit of data or an instruction can be stored.

Locked
>	Pertaining to routines in memory that are not presently (and may never be) candidates for swapping or other shifting around.

Logical device name

> An alphanumeric name assigned by the user to represent a physical device. The name can then be used synonymously with the physical device name in all references to the device. Logical device names are used in device-independent systems to enable a program to refer to a logical device name which can be assigned to a physical device at run-time.

Loop

> A sequence of instructions that is executed repeatedly until a terminal condition prevails.

Low-order byte

> The least significant byte in a word. The low-order byte occupies bit positions 0 through 7 in a PDP-11 word and is always an even address.

Machine instruction

> An instruction that a machine can recognize and execute.

Machine language

> The actual language used by the computer when performing operations.

Macro

> An instruction in a source language that is equivalent to a specified sequence of assembler instructions, or a command in a command language that is equivalent to a specified sequence of commands.

Main program

> The module of a program that contains the instructions at which program execution begins. Normally, the main program exercises primary control over the operations performed and calls subroutines or subprograms to perform specific functions.

Manual input

> The entry of data by hand into a device at the time of processing.

Mask

> A combination of bits that is used to manipulate selected portions of any word, character, byte, or register while retaining other parts for use.

Mass storage

> Pertaining to a device that can store large amounts of data readily accessible to the computer.

Matrix

> A rectangular array of elements. Any matrix can be considered an array.

Memory
> Any form of data storage, including main memory and mass storage, in which data can be read and written. In the strict sense, memory refers to main memory.

Memory image
> A replication of the contents of a portion of memory, usually in a file.

Mnemonic
> An alphabetic representation of a function or machine instruction.

Monitor
> The master control program that observes, supervises, controls or verifies the operation of a computer system. The collection of routines that controls the operation of user and system programs, schedules operations, allocates resources, performs I/O, etc.

Monitor command
> An instruction or command issued directly to a monitor from a user.

Monitor command mode
> The state of the operating system (indicated by a period at the left margin) which allows monitor commands to be entered from the terminal.

Mount a volume
> To logically associate a physical mass storage medium with a physical device unit. To place a volume on a physical device unit (for example, place a magtape on a magtape drive and put the drive on-line).

Multiprocessing
> Simultaneous execution of two or more computer programs by a computer which contains more than one central processor.

Multiprogramming
> A processing method in which more than one task is in an executable state at any one time, even with one CPU.

Nondirectory-structured
> Refers to a storage volume that is sequential in structure and therefore has no volume directory at its beginning. File information (file name, file type, length, and date-of-creation) is provided with each file on the volume. Such volumes include magtape and cassette.

Nonfile-structured device
> A device, such as paper tape, line printer, or terminal, in which data cannot be organized as multiple files.

Object Code
> Relocatable machine language code.

Object module
> The primary output of an assembler or compiler, which can be linked with other object modules and loaded into memory as a runnable program. The object module is composed of the relocatable machine language code, relocation information, and the corresponding global symbol table defining the use of symbols within the module.

Object Time System
> The collection of modules that is called by compiled code in order to perform various utility or supervisory operations (e.g., FORTRAN object time system).

Octal
> Pertaining to the number system with a radix of eight; for example, octal 100 is decimal 64.

ODT
> On-line Debugging Technique: an interactive program for finding and correcting errors in programs. The user communicates in octal notation.

Off-line
> Pertaining to equipment or devices not currently under direct control of the computer.

Offset
> The difference between a base location and the location of an element related to the base location. The number of locations relative to the base of an array, string, or block.

One's complement
> A number formed by interchanging the bit polarities in a binary number: e.g., 1s become 0s; 0s become 1s.

On-line
> Pertaining to equipment or devices directly connected to and under control of the computer.

Op-code (operation code)
> The part of a machine language instruction that identifies the operation the instruction will ask the CPU to perform.

Operand
> That which is operated upon. An operand is usually identified by an address part of an instruction.

Operating system
> The collection of programs, including a monitor or executive and system programs, that organizes a central processor and peripheral devices into a working unit for the development and execution of application programs.

Operation

> The act specified by a single computer instruction. A program step undertaken or executed by a computer, e.g., addition, multiplication, comparison. The operation is usually specified by the operator part of an instruction.

Operation code

> See op-code.

Operator's console

> The set of switches and display lights used by an operator or a programmer to determine the status of and to start the operation of the computer system.

Option

> An element of a command or command string that enables the user to select from among several alternatives associated with the command. In the RT-11 computer system, an option consists of a slash character (/) followed by the option name and, optionally, a colon and an option value.

Output

> The result of a process; the transferring of data from internal storage to external storage.

Overflow

> A condition that occurs when a mathematical operation yields a result whose magnitude is larger than the program is capable of handling.

Overlay segment

> A section of code treated as a unit that can overlay code already in memory and be overlaid by other overlay segments when called from the root segment or another resident overlay segment.

Overlay structure

> A program overlay system consisting of a root segment and optionally one or more overlay segments.

Page

> That portion of a text file delimited by form feed characters and generally 50-60 lines long. Corresponds approximately to a physical page of a program listing.

Parameter

> A variable that is given a constant value for a specific purpose or process.

Parity

> A binary digit appended to an array of binary digits to make the sum of all bits always odd or always even.

Patch
> To modify a routine in a rough or expedient way, usually by modifying the binary code rather than re-assembling it.

PC
> See Program counter.

PDP
> Programmed data processor.

Peripheral device
> Any device distinct from the computer that can provide input and/or accept output from the computer.

Physical device
> An I/O or peripheral storage device connected to or associated with a computer.

Priority
> A number associated with a task that determines the preference its requests for service receive from the monitor, relative to other tasks requesting service.

Process
> A set of related procedures and data undergoing execution and manipulation by a computer.

Processor
> In hardware, a data processor. In software, a computer program that includes the compiling, assembling, translating, and related functions for a specific programming language (e.g., FORTRAN processor).

Processor status word
> A register in the PDP-11 that indicates the current priority of the processor, the condition of the previous operation, and other basic control items.

Program
> A set of machine instructions or symbolic statements combined to perform some task.

Program counter (PC)
> A register used by the central processor unit to record the locations in memory (addresses) of the instructions to be executed. The PC (register 7 of the 8 general registers) always contains the address of the next instruction to be executed, or the second or third word of the current instruction.

Program development
> The process of writing, entering, translating, and debugging source programs.

Program section
   A named, contiguous unit of code (instructions or data) that is
   considered an entity and that can be relocated separately
   without destroying the logic of the program.

Programmed request
   A set of instructions (available only to programs) that is used to
   invoke a monitor service.

Protocol
   A formal set of conventions governing the format and relative
   timing of information exchange between two communicating
   processes.

PSW
   See Processor status word.

Queue
   Any dynamic list of items; for example, items waiting to be
   scheduled or processed according to system or user assigned
   priorities.

Radix
   The base of a number system; the number of digit symbols
   required by a number system.

RAM (random access memory)
   See Random access.

Random access
   Access to data in which the next location from which data is to
   be obtained is not dependent on the location of the previously
   obtained data. Contrast Sequential access.

Read-only memory (ROM)
   Memory whose contents are not alterable by computer
   instructions.

Real-time processing
   Computation performed while a related or controlled physical
   activity is occurring so that the results of the computation can
   be used in guiding the process.

Record
   A collection of related items of data treated as a unit; for
   example, a line of source code or a person's name, rank, and
   serial number.

Recursive
   A repetitive process in which the result of each process is
   dependent upon the result of the previous one.

Re-entrant

Pertaining to a program composed of a shareable segment of pure code and a non-shareable segment which is the data area.

Register

See General register.

Relative address

The number that specifies the difference between the actual address and a base address.

Relocate

In programming, to move a routine from one portion of storage to another and to adjust the necessary address references so that the routine, in its new location, can be executed.

Resident

Pertaining to data or instructions that are normally permanently located in main memory.

Resource

Any means available to users, such as computational power, programs, data files, storage capacity, or a combination of these.

Restart

To resume execution of a program.

ROM

See Read-only memory.

Root segment

The segment of an overlay structure that, when loaded, remains resident in memory during the execution of a program.

Routine

A set of instructions arranged in proper sequence to cause a computer to perform a desired operation.

Run

A single, continuous execution of a program.

Sector

A physical portion of a mass storage device.

Segment

See Overlay segment.

Sequential access

Access to data in which the next location from which data is to be obtained sequentially follows the location of the previously obtained data. Contrast Random access.

Software

The collection of programs and routines associated with a computer (e.g., compilers, library routines).

Software bootstrap

A bootstrap that is activated by manually loading the instructions of the bootstrap and specifying the appropriate load and start address.

Source code

Text, usually in the form of an ASCII format file, that represents a program. Such a file can be processed by an appropriate system program.

Source language

The system of symbols and syntax easily understood by people that is used to describe a procedure that a computer can execute.

Spooling

The technique by which I/O with slow devices is placed on mass storage devices to await processing.

Storage

Pertaining to a device into which data can be entered, in which it can be held, and from which it can be retrieved at a later time.

String

A connected sequence of entities such as a line of characters.

Subprogram

A program or a sequence of instructions that can be called to perform the same task (though perhaps on different data) at different points in a program, or even in different programs.

Subroutine

See Subprogram.

Subscript

A numeric valued expression or expression element that is appended to a variable name to uniquely identify specific elements of an array. Subscripts are enclosed in parentheses. There is a subscript for each dimension of an array. Multiple subscripts must be separated by commas. For example, a two-dimensional subscript might be (2,5).

Supervisory programs

Computer programs that have the primary function of scheduling, allocating, and controlling system resources rather than processing data to produce results.

Swapping

> The process of moving data from memory to a mass storage device, temporarily using the evacuated memory area for another purpose, and then restoring the original data to memory.

Synchronous

> Pertaining to related events where all changes occur simultaneously or in definite timed intervals.

Syntax

> The structure of expressions in a language and the rules governing the structure of a language.

System program

> A program that performs system-level functions. Any program that is part of or supplied with the basic operating system (e.g., a system utility program).

System volume

> The volume on which the operating system is stored.

Table

> A collection of data into a well-defined list.

Terminal

> An I/O device, such as an LA36 terminal, that includes a keyboard and a display mechanism. In PDP-11 systems, a terminal is used as the primary communication device between a computer system and a person.

Timesharing

> A method of allocating resources to multiple users so that the computer, in effect, processes a number of programs concurrently.

Toggle

> To use switches on the computer operator's console to enter data into the computer memory.

Translate

> To convert from one language to another.

Trap

> A conditional jump to a known memory location performed automatically by hardware as a side effect of executing a processor instruction. The address location from which the jump occurs is recorded. It is distinguished from an interrupt which is caused by an external event.

Truncation

> The reduction of precision by ignoring one or more of the least significant digits; e.g., 3.141597 truncated to four decimal digits is 3.141.

Turnkey
> Pertaining to a computer system sold in a ready-to-use state.

Two's complement
> A number used to represent the negative of a given value in many computers. This number is formed from the given binary value by changing all 1s to 0s and all 0s to 1s and then adding 1.

Underflow
> A condition that occurs when a mathematical operation yields a result whose magnitude is smaller than the smallest amount the program can handle.

User program
> An application program.

Utility program
> Any general-purpose program included in an operating system to perform common functions.

Variable
> The symbolic representation of a logical storage location that can contain a value that changes during a processing operation.

Vector
> A consecutive list of associated data.

Volume
> A mass storage medium that can be treated as file-structured data storage.

Wildcard operation
> A shorthand method of referring to all files with a specific characteristic in their name.

Word
> Sixteen binary digits treated as a unit in PDP-11 computer memory.

Write-enable
> The condition of a volume that allows transfers that would write information on it.

Write-protect
> The condition of a volume that is protected against transfers that would write information on it.

Author Index

Underscored numbers give the page on which the complete reference is listed.

E

F

G

S

Saaty, T. L., 230, 239
Saguy, I., 119, 122, 124, 129, 130, 135, 259, 261, 264, 308, 319, 320, 322, 323, 330, 338, 340, 341, 342, 357, 358, 359
Saltmarch, M., 74, 112, 114
Sargent, R. W. H., 303, 320
Sasieni, M., 239
Savitzky, A., 391, 404
Scandora, A. E., 403, 404
Schechter, R. S., 276, 277, 278, 282, 299, 303, 305, 316
Scheffe, H., 56, 70
Schlueter, D. L., 114, 134
Schneider, H. W., 375, 384
Schoeber, W. J. A. H., 138, 148
Schroeder, E., 124, 134
Schubert, H., 102, 108, 112, 115
Schwimmer, S., 105, 106, 115
Seber, G. A. F., 56, 57, 70
Segars, R. A., 246, 261
Segerlind, L. J., 44, 46, 250, 253, 260, 261
Shannon, D. F., 303, 320
Shapiro, M. B., 399, 403
Sheridan, T. B., 375, 385
Sherif, S. M., 253, 261
Sherman, P., 242, 261
Shindo, A., 330, 338, 359
Shinodo, A., 308, 320
Shinskey, F. G., 368, 379, 383, 384, 385
Shipman, J. S., 333, 358
Siemens, N., 239
Silivius, J. R., 88, 115
Smith, H., 52, 54, 58, 59, 70, 294, 295, 320
Smith, M. L., 201, 238
Smoot, J. M., 124, 135
Snee, R. D., 59, 70
Sokhansanj, S., 255, 256, 257, 260

Sontag, D. A., 72, 115
Sood, V. K., 294, 320
Spencer, J. L., 324, 340, 357
Spendley, W., 295, 296, 320
Stainforth, P. T., 258, 261
Stanley, D. W., 242, 260, 295, 319
Starr, M. K., 239
Stassen, H. G., 375, 384
Stegun, I. A., 145, 147
Steinberg, D., 274, 317
Steinbrenner, K., 16, 21
Steir, H. L., 239
Stern, N. B., 13, 22
Stern, R. A., 13, 22
Steward, G., 124, 135
Stewart, G. W., 15, 21
Stoner, D. L., 250, 261
Storey, C., 336, 359
Stults, B., 264, 285
Svendsen, P. J., 402, 403, 403
Swann, W. H., 278, 282, 285, 289, 292, 294, 299, 302, 305, 306, 316, 320
Swartz, P. A., 297, 319
Sweat, V. E., 250, 261
Szekeley, J., 276, 319

T

Talburt, W. F., 122, 135
Tanford, C., 399, 403
Tannenbaum, S. R., 113
Taylor, J., 403, 404
Teixeira, A. A., 130, 135, 138, 145, 148
Terborgh, G., 208, 239
Thomas, E. L., 294, 319
Thome, R. J., 383, 385
Thompson, D. R., 77, 79, 115, 255, 256, 257, 260
Timbers, G. E., 145, 148
Tiru, M., 102, 113
Torgersen, 195, 239

Subject Index

A

A Programming Language, see APL

Access time, 7

Accuracy, numerical techniques and, 139–141

Ada, 14

Adams' formulas, 34–35

ALGOL, 13

Algorithms, 8, 19, 60
  computerized linear programming, 177–180
  simplex, using to solve linear programming problems, 175–177

Alternating variable search, 290

American Standard Code for Information Interchange (ASCII), 3

Analog computers, 2
  food rheology research and, 248–249

Analog to digital (A/D) converter, 8

ANOVA, 56

Anthropometrics, 374

APL, 13, 413–415, 434, 444–445

Arithmetic and logic unit, 7

Arrhenius equation, 87–90, 119–120, 142

Arrhenius form for temperature function, 102

Arrhenius relation, goodness of fit of, 89–90

ASCII, see American Standard Code for Information Exchange

"Assembly language," 11

Assignment problems, 184–192
  labor, 184–188
  material, 188–192

Auxiliary routines, 18

B

Backward-difference discretization, 140

Backward differences, 24–25

Bar chart, 193–194

BASIC, 13, 208–210, 237, 399

BCD, see Binary-coded decimal

Beginner's All-Purpose Symbolic Instruction Code, see BASIC

BFGS method, 303